advanced analysis of

gene expression

microarray data

SCIENCE, ENGINEERING, AND BIOLOGY INFORMATICS

Series Editor: Jason T. L. Wang
(New Jersey Institute of Technology, USA)

Published:

Vol. 1: Advanced Analysis of Gene Expression Microarray Data
(Aidong Zhang)

Vol. 2: Life Science Data Mining
(Stephen T. C. Wong & Chung-Sheng Li)

Vol. 3: Analysis of Biological Data: A Soft Computing Approach
(Sanghamitra Bandyopadhyay, Ujjwal Maulik & Jason T. L. Wang)

Vol. 4: Machine Learning Approaches to Bioinformatics
(Zheng Rong Yang)

Vol. 5: Biodata Mining and Visualization: Novel Approaches
(Ilkka Havukkala)

Vol. 6: Database Technology for Life Sciences and Medicine
(Claudia Plant & Christian Böhm)

Vol. 7: Advances in Genomic Sequence Analysis and Pattern Discovery
(Laura Elnitski, Helen Piontkivska & Lonnie R. Welch)

Vol. 8: Biological Data Mining and Its Applications in Healthcare
(Xiaoli Li, See-Kiong Ng & Jason T. L. Wang)

advanced analysis of

gene expression

microarray data

aidong zhang

(State University of New York at Buffalo, USA)

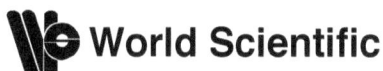

World Scientific

NEW JERSEY · LONDON · SINGAPORE · BEIJING · SHANGHAI · HONG KONG · TAIPEI · CHENNAI

Published by

World Scientific Publishing Co. Pte. Ltd.

5 Toh Tuck Link, Singapore 596224

USA office: 27 Warren Street, Suite 401-402, Hackensack, NJ 07601

UK office: 57 Shelton Street, Covent Garden, London WC2H 9HE

British Library Cataloguing-in-Publication Data
A catalogue record for this book is available from the British Library.

Science, Engineering, and Biology Informatics — Vol. 1
ADVANCED ANALYSIS OF GENE EXPRESSION MICROARRAY DATA

ISBN-13 978-981-256-645-4
ISBN-10 981-256-645-7

For my parents, Ming-cai Zhang and Shu-xiang Li

Preface

My interest in studying gene expression microarray data originated in 1998, when a faculty member from the School of Pharmacy at the State University of New York at Buffalo approached me for assistance with his microarray data. I was immediately attracted to the interesting problems presented by these data sets. Since that time, I have been closely following this area of research and have been fascinated by the enormous quantity of data being generated and the need for effective approaches to its analysis.

The total volume of microarray data has increased rapidly over the past several years, a trend which is likely to continue. In addition to traditional time-series and gene-sample data sets, microarray data have also appeared in new and challenging formats. For example, some recent microarray data operate in three dimensions, simultaneously addressing the time, gene, and sample components. Many existing analytical approaches focus on various aspects of data preprocessing, including the processing of scanned images, missing-data estimation, normalization, and summarization. Some apply conventional statistical and machine-learning techniques to pattern analysis and class prediction, but these methods often do not transfer well to use with microarray data. More recently, advances in data-mining techniques have been fruitfully applied to the analysis of complex patterns in microarray data.

This book is intended to provide an understanding of the most current methods available for the analysis of gene expression microarray data, with a particular focus on data-mining techniques. Data mining is well-established area of research which develops scientific approaches to the extraction of knowledge from large data sets. These approaches are more conventionally applied in industrial settings, especially in retail, financial, and telecommunications contexts, but have also recently gained acceptance for biomedical applications such as the analysis of genomic data. The development of techniques for the effective analysis of genomic datasets is a crucial step in the medical application of bioinformatics. This unique merging of computer-science and biomedical expertise is expected to provide the synergy needed to advance biomedical research to the next level. This book is intended to provide a useful in-depth survey for bioinformatics researchers which I hope will guide and stimulate further investigation.

The book assumes some knowledge on the part of the reader of the fields of molecular biology, data mining, and statistics, as well as a basic understanding of microarray technology. To bridge the gap between these disparate fields, brief overviews are provided of each. Chapter 2 borrows from several standard texts on molecular biology to set forth the fundamental concepts of that field. Chapter 3 summarizes the nature of microarray experiments and data-gathering, and Chapter 4 is a brief tutorial on the techniques of statistical analysis typically applied to microarray data.

Some of the materials in this book have been amassed over several years of teaching a bioinformatics course at an advanced graduate level at the State University of New York at Buffalo. Many of the approaches and research papers referenced in this book can be found at http://www.cse.buffalo.edu/DBGROUP/bioinformatics/research.html.

Acknowledgments. I would like to express my deepest thanks to Dr. Daxin Jiang, whose dedicated and indefatigable efforts were essential to the preparation of this book. I would also like to thank my other former doctoral students, Dr. Chun Tang, Xian Xu, and Dr. Li Zhang, for their excellent technical contributions. I am also highly appreciative of the editorial work of Rachel Ramadhyani in recasting my text into proper and idiomatic English.

The inspiration for this book was an invitation from Dr. Jason T. L. Wang to prepare a contribution to his Science, Engineering, and Biology Informatics (SEBI) series. I would like to express my special thanks to Dr. Wang. I also would like to thank the commissioning editor, Ms. Yubing Zhai, for guiding the development of this book and to Mr. Ian Seldrup of World Scientific Publishing Co., Inc. for his overall editorial supervision.

Aidong Zhang
Buffalo, New York

Contents

Chapter 1

Introduction

1.1 The Microarray: Key to Functional Genomics and Systems Biology

The new field of bioinformatics employs computer databases and data mining algorithms to analyze proteins, genes, and complete collections of deoxyribonucleic acid (DNA) on a genome-wide level [222]. As defined by National Institutes of Health (NIH), bioinformatics consists of "research, development, or application of computational tools and approaches for expanding the use of biological, medical, behavioral or health data, including those to acquire, store, organize, analyze, or visualize such data."

The rapid development of biological technologies over the last decade has resulted in the complete sequencing of many important model organisms. These include *Haemophilus influenzae* (the first sequenced genome of a free-living organism, 1,830,137 base pairs [100]), *Saccharomyces cerevisiae* (the first eukaryotic genome sequence with 12,068 kilo bases [112]), *Caenorhabditis elegans* (the first complete sequence of a multicellular organism [279]), *Drosophila melanogaster* (120 mega bases [1]), *Arabidopsis thaliana* (an important plant model [195]), and more than 30 microbial genomes. By 2001, the first draft version of the sequence of base pairs in human DNA had been released (human chromosome 22 in 1999 [89] and chromosome 21 in 2000 [127]). We are now moving from the *pre-genomic era* characterized by the effort to sequence the human genome, to a *post-genomic era* that concentrates on harvesting the fruits hidden in the genomic text.

An overarching challenge in this post-genomic era is the management and analysis of enormous quantities of sequence data. The aim of any organizational scheme would be to provide biologists with an inventory of all genes used to assemble a living creature, analogous to the Periodic Table of chemistry [172]. Understanding the biological systems with tens or hundreds of thousands of genes will similarly require organizing the constituents by properties and will reflect similarities at diverse levels such as:

- Time and place of RNA expression during development;
- Subcellular localization and intermolecular interaction of protein products;
- Physiological response and disease;
- Primary DNA sequence in coding and regulatory regions; and
- Polymorphistic variation within a species or subgroup [173].

Recently, the advent of microarray technology has made it possible to monitor the expression levels of thousands of genes in parallel. Arrays offer the first promising tool for addressing the challenges of the post-genomic era, by providing a systematic way to survey variation in DNA and RNA. There have been widespread applications of this technology during the last several years, and it seems likely to become a standard tool both of research in molecular biology and of clinical diagnostics.

1.2   Applications of Microarray

Microarrays have already been extensively used in biological research to address a wide variety of questions. As noted by Collins [61], when applied to expression analysis, this approach facilitates the measurement of RNA levels for the complete set of transcripts of an organism. When applied to genotyping, microarrays usher in the possibility of determining alleles at hundreds of thousands of loci from hundreds of DNA samples, allowing the contemplation of whole genome-association studies to determine the genetic contribution to complex polygenic disorders. Moreover, the application of microarrays to mutation screening of disease genes with pronounced allelic heterogeneity is likely to move the possibility of genetic testing for disease susceptibility of individuals, or even entire populations, into the realm of practical reality. To motivate our subsequent discussion, we shall begin by presenting a few examples of some related research. We emphasize that this discussion is by no means exhaustive and in fact represents only a fraction of the universe of potential applications.

1.2.1 Gene Expression Profiles in Different Tissues

Cells from different tissues perform different functions. Although they can be easily distinguished by their phenotypes, a detailed understanding of the mechanisms of these different behaviors remains elusive [10]. Since cell function is determined by individual proteins and protein synthesis is dependent on which genes are expressed by the cell, the expression pattern of a gene provides indirect information about cell function. For example, a gene expressed only in the kidney is unlikely to be directly involved in the pathology of schizophrenia [67]. Microarray experiments can be used to identify those genes which are preferentially expressed in various tissues. This would enable scientists to gain valuable insights into the mechanisms that govern the functioning of genes and cells [10]. Moreover, the highly selective tissue expression of a drug target is attractive as a means to reduce the potential for unwanted side effects [10].

1.2.2 Developmental Genetics

Amaratunga and Cabrera [10] described the application of microarrays to developmental genetics. The genes in an organism's genome express differentially at different stages of its developmental process. Interestingly, it has been found that there is a subset of genes involved in early development that is used and reused at different stages in the development of the organism, generally in different order in different tissues, with each tissue having its own combination. Crucial to these processes are growth factors, which can also, later in an organism's development, be involved in causing or promoting cancer; these genes are known as *proto-oncogenes*. Microarrays can, in principle, be used to track the changes in the organism's gene expression profile, tissue by tissue, over the series of stages of the developmental process, beginning with the embryo and up to the adult. Supplementary applications of this line of research include deducing evolutionary relationships among species and assessing the impact of environmental changes on the developmental process of an organism.

1.2.3 Gene Expression Patterns in Model Systems

Detailed profiling of gene expression in model systems (such as *Saccharomyces cerevisiae*, *Caenorhabditis elegans*, *Drosophila melanogaster*, *Arabidopsis thaliana*) will yield valuable insights into the functions of genes and the mechanisms of important cellular processes. This type of analysis has already been described using the yeast *Saccharomyces cerevisiae*. For example, Spellman and his colleagues [263] used DNA microarrays to create a comprehensive catalog of yeast genes whose transcript levels vary periodically within the cell cycle. They reported 800 genes that meet an objective minimum criterion for cell cycle regulation.

In another example of such analysis, Gasch et al. [107] explored genomic expression patterns in the yeast *Saccharomyces cerevisiae* in response to diverse environmental transitions. DNA microarrays were used to measure changes in transcript levels over time for almost every yeast gene as cells responded to a variety of environmental impacts. These included temperature shocks, hydrogen peroxide, the superoxide-generating drug menadione, the sulfhydryl-oxidizing agent diamide, the disulfide-reducing agent dithiothreitol, hyper- and hypo-osmotic shock, amino acid starvation, nitrogen source depletion, and progression into a stationary phase. They found a large set of genes ($\sim$ 900) which showed a similar drastic response to almost all of these environmental changes, while additional features of the genomic responses were specialized for specific conditions.

In another study, Gasch et al. [106] used DNA microarrays to observe genomic expression in yeast *Saccharomyces cerevisiae* responding to two different DNA-damaging agents. The genome-wide expression patterns of wild-type cells and defective mutants were compared in Mec1 signaling under normal growth conditions and in response to the methylatingagent methylmethane sulfonate (MMS) and ionizing radiation. The comparative study identified specific features of these gene expression responses that are dependent on the Mec1 pathway. Among the hundreds of genes whose expression was affected by Mec1p, one set of genes appeared to represent an MEC1-dependent expression signature of DNA damage, cell cycle, mutations, and stimulus.

1.2.4 *Differential Gene Expression Patterns in Diseases*

One of the most attractive applications of microarray technology is the study of differential gene expression in disease. There are many *genetic diseases* that are the result of mutations in a gene or a set of genes. The mutations may cause genes to express inappropriately or to fail to express. For example, cancer could occur when certain regulatory genes, such as the p53 tumor suppressor gene, become always transcribed regardless of any regulatory factors [10].

Microarray experiments can be used to identify which genes are differentially expressed in diseased cells versus normal cells. The up- or down-regulation of gene activity can either be the cause of the pathophysiology or the result of the disease. Although targeting disease-causing genes is desirable to achieve disease modification, interfering with genes that are expressed as a consequence of disease can lead to the alleviation of symptoms. The opportunity to compare the expression of thousands of genes between "diseased" and "normal" cells will allow the identification of multiple potential targets [67]. This would enable the development of drugs aimed directly at the difference between diseased and normal cells. Such drugs can be designed to specifically target a particular gene, protein, or signaling cascade, and they are therefore less likely to cause undesirable side effects [10].

There are abundant examples of such microarray applications in the literature. Rheumatoid tissue was analyzed using a microarray of about 100 genes known to have a role in inflammation [130]. Among others, genes encoding interleukin-6 and several matrix metalloproteinases, including matrix metallo-elastase (HME), were markedly up-regulated. The latter result was unexpected, as the distribution of HME was previously thought to be limited to alveolar macrophages and placental cells [67]. Golub et al. [114] studied the expression profiles of patients with two subtypes of leukemia, ALL (acute lymphoblastic leukemia) and AML (acute myeloid leukemia). The clinical diagnosis of these two subtypes of cancer is extremely difficult due to their clinical similarity. Microarray experiments can be used to identify which genes are differentially expressed in the two different types of cancer patients, thereby creating specific disease profiles on the basis of their gene expression patterns [10].

1.2.5 Gene Expression Patterns in Pathogens

As noted by Debouck [67], activity in the sequencing of bacterial genomes is intense, with a new bacterial genome seemingly sequenced in its entirety every month. The small size of these genomes will allow the easy construction of individual microarrays in which every gene from a given microbe is represented. For microbiologists, confined for years to studying bacteria one gene at a time in a test tube under artificial growth conditions, the horizons appear unlimited. Microarray technology will identify genes that are turned on in vitro but not at the site of infection in vivo, and vice versa, and those genes that are only turned on during infection in vivo. Such genes encode virulence determinants that are regulated by environmental signals such as the transition from ambient temperature to body temperature [196]. Since traditional genetic techniques used to identify virulence genes are time consuming, they will be quickly supplanted by microarray methods. A similar approach will be used to study viral gene expression during the time course of acute infection or during latency. Microarrays can also be used to study the response of the host to challenges from the pathogen.

1.2.6 *Gene Expression in Response to Drug Treatments*

Microarrays are potentially powerful tools for investigating the mechanism of drug action. For example, Interferon-β (IFN-β) is the most widely prescribed immunomodulatory therapy for multiple sclerosis (an autoimmune disease of the brain and spinal cord). The therapy is known to exert all its biological effects via gene transcription but there are no validated markers for its long-term efficacy in multiple sclerosis. Although double blind, randomized, placebo-controlled clinical trials have established that IFN-β treatment reduces the progression of disability in multiple sclerosis, only 30-40% of patients respond well to the therapy. To define the mechanism of IFN-β and investigate the partial responsiveness of various patients, the expression levels of large numbers of genes were monitored for thirteen multiple sclerosis patients during a ten-point time-series [300].

In [67], Debouck gave two other applications of high-density microarrays to examine the effects of drugs on gene expression in yeast as a model system. In one, the effect of potent kinase inhibitors was analyzed on a yeast-genome-wide scale by measuring changes in mRNA levels before and after treatment with inhibitors [115]. The second study reported a gene expression pattern (or "signature") characteristic of the immunosuppressive drug FK506. This same signature was also observed in yeast cells carrying a null mutation in the FK506 target, establishing that genetic and pharmacological ablation of a gene function results in similar changes in gene expression. Treatment of the null mutants with FK506 also revealed additional pathways distinct from the drug's primary target [191]. It is possible that yeast will provide a high-throughput platform for studying cellular responses to drugs. However, a similar method applied to human cells and tissues would have even more direct utility in the identification and validation of novel therapeutics.

1.2.7 Genotypic Analysis

Variation in DNA sequence underlies most of the differences we observe within and between species. Locating, identifying and cataloging these genotypic differences represent the first steps in relating genetic variation to phenotypic variation in both normal and diseased states [184]. Lipshutz et al. [184] described a specific type of array that can be designed for this purpose. Given a reference sequence for a region of DNA, four probes are designed to interrogate a single position. One is designed to be perfectly complementary to a short stretch of the reference sequence; the other three are identical to the first, except at the interrogation position, where one of the other three bases is substituted. Upon incubation with the reference sequence, the probe complementary to the reference sequence will obtain the highest fluorescence intensity. In the presence of a sample with a different base at the interrogation position (a substitution variant), the probe containing the complementary variant base will obtain the highest fluorescence intensity.

Single-nucleotide polymorphisms (SNPs) are the most frequent type of variation in the human genome; this, and the ease with which they can be identified, recommend them for this type of analysis [184]. For example, the study by David Wang and colleagues [295] identified 3,241 candidate SNPs contained in STSs collected at the Whitehead Institute and Sanger Center and mapped 2,227 of them. Using conventional gel-based sequencing and high-density oligonucleotide arrays, a total of 2.3 million bases of sequence were screened for variations among eight individuals to identify candidate SNPs and create a third-generation genetic map for the human genome. More than two thousands of these SNPs were selected, and sets of appropriate probes were synthesized on a high-density array. The "SNP chips" are intended for commercial distribution and can be used for linkage, linkage disequilibrium and loss of heterozygosity studies. Arrays can also be used to scan the genome for new SNPs. Shrinking the area occupied by each of the interrogating oligonucleotide probes (or "synthesis features") to approximately $15 \times 20 \mu m$, a total of 50,000 nucleotides (on both strands) can be screened for the presence of polymorphisms on a single array [184].

1.2.8 Mutation Screening of Disease Genes

In [10], Amaratunga and Cabrera described using microarrays to study *complex diseases*. Complex diseases are not caused by a few errors in genetic information but by a combination of small genetic variations (polymorphisms) which predisposes an individual to a serious problem. The risk of such an individual contracting a complex disease tends to be amplified by non-genetic factors, such as environmental influences, diet and lifestyle. Coronary artery disease, multiple sclerosis, diabetes, and schizophrenia are complex diseases in which the genetic makeup of the individual plays a major role in predisposing the individual to the disease. The genetic component of these diseases is responsible for their increased prevalence within certain groups such as families, ethnic groups, geographic regions, and genders. Microarray experiments can be used to identify the genetic markers, usually a combination of SNPs, that may predispose an individual to a complex disease.

1.3   Framework of Microarray Data Analysis

The enormous quantity of gene expression microarray data and its significant applications in biomedicine require effective approaches to its analysis. Figure 1.1 presents a flowchart illustrating the typical components of microarray data processing and analysis. The framework consists of three major steps, determination of the biological problem and sample preparation, array generation, and data analysis. In the first step, the RNA sources are collected from the tissues of model systems or diseased/normal patients or from a cultivated homogeneous cell population as appropriate to the particular problem being investigated. RNAs are then extracted from these cells.

In the second step, a microarray experiment is carried out. Although there are different types of microarrays, all follow these common basic procedures [278]:

- *Chip manufacture:* A microarray is a small chip (made of chemically-coated glass, nylon membrane or silicon) onto which tens of thousands of DNA molecules (*probes*) are attached in fixed grids. Each grid cell relates to a DNA sequence.
- *mRNA preparation, labeling and hybridization:* Typically, two mRNA samples (a test sample and a control sample) are reverse-transcribed into cDNAs (*targets*), labeled using either fluorescent dyes or radioactive isotopics, and then hybridized with the cloned sequences on the surface of the chip.

- *Chip scanning:* Chips are scanned to read the signal intensity that is emitted from the labeled and hybridized targets.

Once the raw microarray data are obtained, several pre-processing steps may need to be performed prior to any further data analysis. These pre-processing steps include data transformation, estimation of missing values, and data normalization. After data pre-processing, the microarray data can usually be represented by a two-dimensional matrix $\{x_{ij}\}$, where each row r_i in the data matrix corresponds to one gene, each column c_j corresponds to each experimental condition, and each cell x_{ij} is a real value recording the expression level of gene i under condition j. Finally data analysis and visualization algorithms can be applied to the pre-processed data sets.

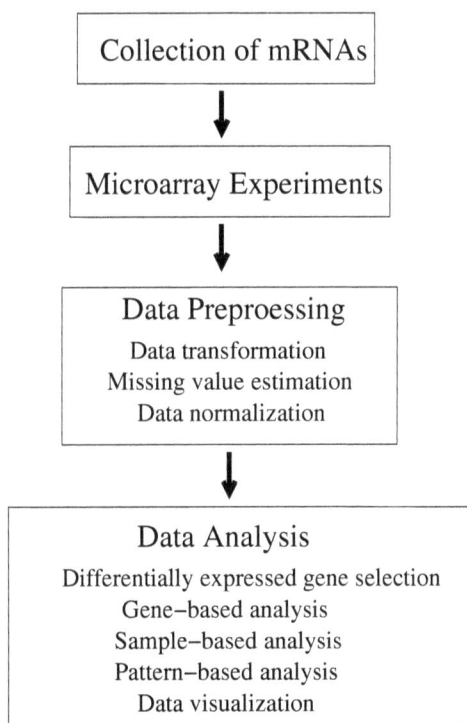

Collection of mRNAs

↓

Microarray Experiments

↓

Data Preproessing
Data transformation
Missing value estimation
Data normalization

↓

Data Analysis
Differentially expressed gene selection
Gene–based analysis
Sample–based analysis
Pattern–based analysis
Data visualization

Fig. 1.1   Framework of microarray data analysis.

In this book, we will not discuss in detail the manufacture of microarray chips or the procedures involved in the microarray experimentation. Neither will we elaborate on data pre-processing or statistical approaches to differential analysis. Instead, we refer readers to the work of Schena [241] and Speed [262] for details on these topics. This book will primarily focus on the development and application of advanced data mining, machine learning, and visualization techniques for the identification of interesting, significant, and novel patterns in microarray data.

The remainder of this book is organized as follows:

- Chapter 2 briefly introduces some concepts central to molecular biology, providing readers with an understanding of the basic mechanisms of microarray technology and the related analysis.
- Chapter 3 discusses various approaches to microarray manufacture, the basic procedures of microarray experimentation, and the pre-processing of the raw data.
- Chapter 4 explores several statistical methods for the identification of differentially expressed genes.
- Chapter 5 focuses on gene-based analysis, introducing various clustering algorithms to identify co-expressed genes and coherent patterns. In particular, an interactive approach which integrates users' domain knowledge into the clustering process is presented. Different cluster validation approaches are also reviewed.
- Chapter 6 presents approaches to sample-based analysis. Since a microarray experiment typically involves a much larger number of genes than that of samples, meaningful sample-based analysis must effectively reduce the extremely high dimensionality of the data set. We will discuss both supervised and unsupervised methods for the reduction of the dimensionality. A series of approaches to disease classification and discovery will then be surveyed.
- Chapter 7 explores methods for ascertaining the relationship between (subsets of) genes and (subsets of) samples. Several classical approaches to mining frequent itemsets, as well as a post-mining method, are introduced to identify interesting association rules among genes and samples. We will also discuss pattern-based clustering algorithms to find the coherent patterns embedded in the sub-attribute spaces. A novel approach will be presented to uncover the inherent correlation among genes, samples, and time-series.
- Chapter 8 looks at various methods for data visualization. The visualization process is intended to transform the data set from high-dimensional space into a more easily understood 2- or 3-

dimensional space.
- Chapter 9 discusses some new trends in mining gene expression microarray data. We focus on combining various data sources and integrating domain knowledge into the data mining process.
- Chapter 10 concludes the book.

1.4 Summary

As discussed in this chapter, effective approaches are demanded to analyze gene expression microarray data. Recently, a variety of data-mining techniques have gained acceptance for analyzing gene expression microarray data. This book is intended to provide researchers with a working knowledge of many of the advanced approaches currently available for this purpose. These approaches can be classified into five distinct categories: gene-based analysis, sample-based analysis, pattern-based analysis, visualization, and integration of domain knowledge. Chapters 5 through 9 will treat each of these approaches individually, each starting with an overview of current methods and moving to a more detailed discussion of a selected technique of particular interest. Applications of these approaches to specific gene expression microarray data sets will be discussed and experimental results will be presented. It is hoped that the discussion of new trends offered in Chapter 9 may stimulate further explorations in the field on the part of interested readers.

Chapter 2

Basic Concepts of Molecular Biology

2.1 Introduction

The reader's understanding of the methods and concepts discussed in later chapters assumes some familiarity with the fundamentals of molecular biology. This chapter is intended to provide a basic primer for those without this background; it does not purport to be a comprehensive exegesis of the subject. Since this will be a very general presentation, many exceptions to and deviations from general principles will not be explored here. For a more in-depth examination of molecular biology, readers are referred to Life: the Science of Biology by William K. Purves et al. [224], Genes VIII by Benjamin Lewin [180], and Molecular Biology by Robert F. Weaver [299].

2.2 Cells

The literal meaning of "biology" is "the study of all living things" [264]. In a biological context, "living things" are defined by the presence of certain characteristics. Some of these qualities include the ability to grow and develop, maintain internal homeostasis, reproduce themselves, detect and respond to stimuli, acquire and release energy, and interact with their environment as well as with each other [96]. In short, living things have an active participation in their environment, as opposed to the inactive status of nonliving things [249].

Current paleontological thinking places the origins of life on Earth about 3.5 billion years ago, shortly (in geological terms) after the Earth itself was formed (more than 4 billion years ago) [264]. From the first simple life-forms, the action of *evolutionary processes* over billions of years resulted in change and diversification, so that today we find both highly complex and very simple organisms.

Both complex and simple organisms are made up of cells, which can, in turn, be decomposed into organelles, organelles into molecules, and thus downward through a diminishing hierarchy of size. The *Cell Theory* was first formally articulated in 1839 by Theodor Schwann and Matthias Schleiden and was subsequently elaborated by Rudolph Virchow and other biologists. This theory has three parts: 1) all living things are composed of one or more cells; 2) cells are basic units of structure and function in an organism; 3) cells come only from the reproduction of existing cells [264].

The two basic classes of cells–*prokaryotic* and *eukaryotic* cells–are distinguished by their size and the types of internal structures, or *organelles*, which they contain. The structurally simpler prokaryotic cells are represented by bacteria and blue algae. All other organism types–protists, fungi, plants, and animals–consist of structurally more complex eukaryotic cells [159].

The internal structure and functions of eukaryotic cells are much more complex than those of prokaryotic cells. Figure 2.1(a) and (b) illustrates the internal structure of an animal cell and a plant cell, respectively. Both eukaryotic and prokaryotic cells contain a nuclear region which houses the cell's genetic material. The genetic material of a prokaryotic cell is present in a *nucleoid*, which is a poorly-demarcated region of the cell lacking a boundary membrane to separate it from the surrounding cytoplasm. In contrast, eukaryotic cells possess a *nucleus*, a region bounded by a complex membranous structure called the *nuclear envelope*. Figure 2.2 shows the structure of a nucleus. This difference in nuclear structure is the basis for the terms prokaryotic (pro = before, karyon = nucleus) and eukaryotic (eu = true, karyon = nucleus) [159].

Both prokaryotic and eukaryotic cells share a similar molecular chemistry. The most important molecules in the chemistry of life are proteins and nucleic acids. Roughly speaking, proteins determine what a living being is and does in a physical sense, while nucleic acids are responsible for encoding the genetic information and passing it along to subsequent generations [249]. In the following sections, we will provide a succinct description of the current state of knowledge regarding these two molecules.

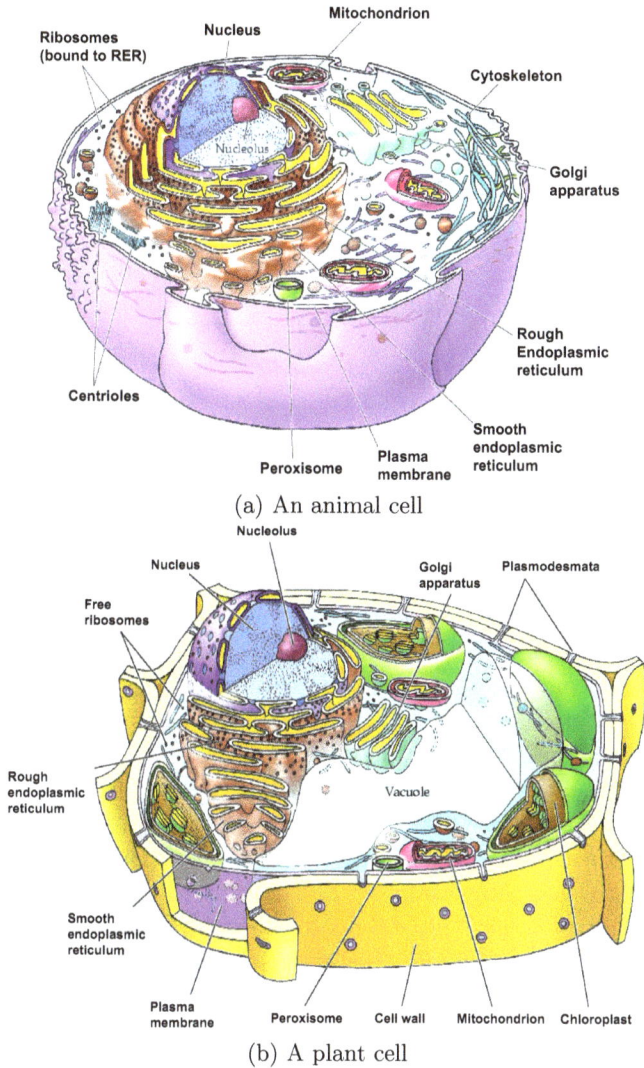

Fig. 2.1   Structure of a eukaryotic cell (Figure is from [224] with permission from Sinaucr Associates).

2.3   Proteins

All living organisms are composed largely of proteins, and their importance was well stated by the distinguished scientist Russell Doolittle, who wrote "we are our proteins." There are several types of proteins. *Structural*

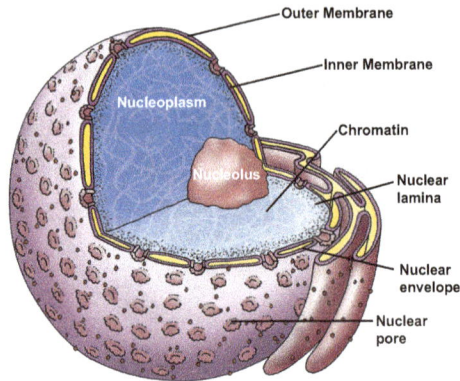

Fig. 2.2   Structure of a nucleus (Figure is from [224] with permission from Sinaucr Associates).

proteins form part of a cellular structure, *enzymes* catalyze almost all the biochemical reactions occurring within a cell, *regulatory proteins* control the expression of genes or the activity of other proteins, and *transport proteins* carry other molecules across the cell membrane or around the body [10].

A protein is composed of a chain of *amino acids*. Figure 2.3 illustrates the structure of an amino acid. Every amino acid is organized around one central carbon atom, known as the alpha carbon, or C_α. Other components of an amino acid include a hydrogen atom, an amino group (NH_2), a carboxyl group (COOH), and a side chain (R-group) [224]. It is the side chain that distinguishes amino acids from each other. Twenty amino acids occur in nature, and these are listed in Table 2.1. All living things (and even viruses, which do not fully meet the criteria for "life") are made up of various combinations of the same twenty amino acids.

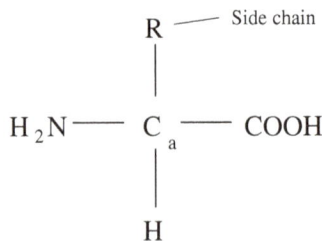

Fig. 2.3   General structure of an amino acid (Figure adapted from [224] with permission from Sinaucr Associates).

Table 2.1   The twenty amino acids commonly found in proteins (Table from [224] with permission from Sinaucr Associates).

	Name	One-letter abbreviation	Three-letter abbreviation
1	Alanine	A	Ala
2	Cysteine	C	Cys
3	Aspartic Acid	D	Asp
4	Glutamic Acid	E	Glu
5	Phenylalanine	F	Phe
6	Glycine	G	Gly
7	Histidine	H	His
8	Isoleucine	I	Ile
9	Lysine	K	Lys
10	Leucine	L	Leu
11	Methionine	M	Met
12	Asparagine	N	Asn
13	Proline	P	Pro
14	Glutamine	Q	Gln
15	Arginine	R	Arg
16	Serine	S	Ser
17	Threonine	T	Thr
18	Valine	V	Val
19	Tryptophan	W	Trp
20	Tyrosine	Y	Tyr

Structurally, proteins are polypeptidic chains, in that their amino acids are linked together by *peptide bonds*. A peptide bond is formed by the attachment of the carboxyl end of one amino acid to the amino end of the adjacent amino acid, with a water molecule liberated in the process. Figure 2.4 illustrates the dehydration process and the structure of the peptide bond. As a result of the loss of water, the polypeptidic chain is actually made up of the dehydrated residues of the original amino acids. Therefore, we generally speak of a protein as being made up of a certain number of residues rather than of amino acids. The peptide bond forms a *backbone* for every protein, which is made up of repetitions of the basic block - N - C_α - (CO) - [249].

The sequence of residues in a polypeptide is called the *primary structure* of a protein. Furthermore, proteins actually fold in three dimensions, resulting in secondary, tertiary, and quaternary structures. The *secondary structure* of a protein is formed through the tendency of the polypeptide to coil or pleat due to H-bonding between R-groups. This results in "local" structures such as alpha helices and beta sheets. *Tertiary structures* are the

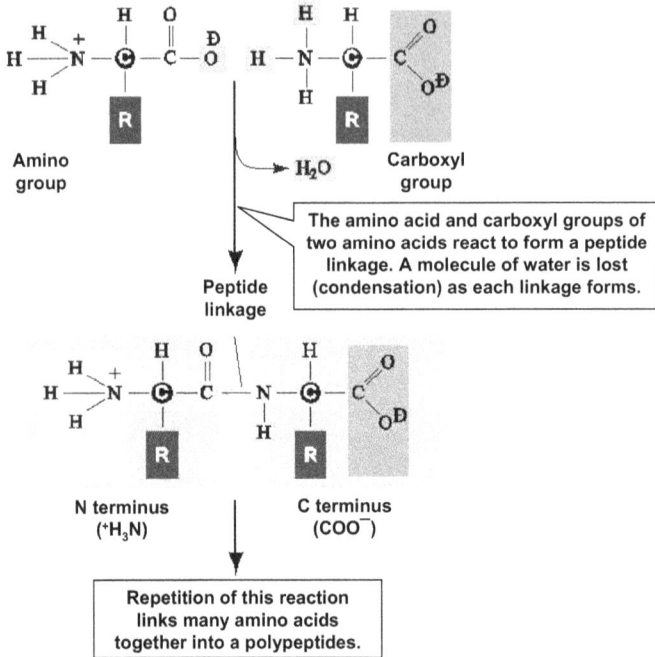

Fig. 2.4   Formation of a peptide bond between two amino acids by the dehydration of the amino end of one amino acid and the acid end of the other amino acid (Figure is from [224] with permission from Sinaucr Associates).

result of secondary-structure packing on a more global level, such as bonding or repulsion between R-groups. Many proteins are formed from more than one polypeptide chain; hemoglobin is an example of such a protein. These polypeptides pack together to form a higher level of packing called the *quaternary structure*.

The linear sequence of residues (the primary structure) of a protein determines its three-dimensional structure, and the three-dimensional shape of a protein determines its function. The rationale is that a folded protein has an irregular shape with nooks and bulges which enable closer contact with other molecules [249]. For example, structural proteins, such as collagens, have regular, repeated primary structures. They perform a variety of functions in living things; for example, they form the tendons, hide, and corneas of a cow. Microtubules, important in cell division and in the structures of flagella and cilia, are composed of globular structural proteins.

2.4 Nucleic Acids

Nucleic acids encode the information necessary to produce proteins and are responsible for passing along this "recipe" to subsequent generations [249]. There are two basic types of nucleic acid: ribonucleic acid (RNA) and deoxyribonucleic acid, or DNA. The polypeptidic sequence which forms the primary structure of a protein is directly related to the sequence of information in the RNA molecule, which, in turn, is a copy of the information in the DNA molecule (please refer to Section 2.5 for a detailed description of this process).

2.4.1 *DNA*

A DNA molecule consists of two *strands* of simpler molecules. Each strand has a *backbone* consisting of repetitions of the same basic unit. Figure 2.5 (right panel) illustrates the structure of the DNA backbone. The basic unit of DNA is formed by a sugar molecule, 2'-deoxyribose, attached to a phosphate residue. The sugar molecule contains five carbon atoms, labeled 1' through 5' (see Figure 2.7). The backbone is created by a series of bonds between the 3' carbon of one unit's sugar molecule, the phosphate residue, and the 5' carbon of the next unit. DNA strands have an orientation determined by the numbering of the carbon atoms, which, by convention, starts at the 5' end and finishes at the 3' end. A single-stranded DNA sequence is therefore always written in this canonical 5'→3' direction, unless otherwise stated [249].

The molecules attached to each 1' carbon in the backbone are called *bases*. There are four kinds of bases: adenine (A), guanine (G), cytosine (C), and thymine (T) (Figure 2.6). Bases that have two rings of carbon and nitrogen atoms, such as adenine and guanine, are called *purines*. *Pyrimidines* are bases that have one ring of carbon and nitorgen atoms, such as cytosine and thymine [264]. Since nucleotides are differentiated only by their bases, a DNA molecule can be referred to interchangeably as having a particular number of either bases or nucleotides, although bases and nucleotides are not otherwise synonymous. DNA molecules in nature are very long. In a human cell, DNA molecules have hundreds of millions of nucleotides [249]. A short fragment of a DNA molecule, typically between five and 50 base pairs long, is called an oligonucleotide (or oligo).

As mentioned before, DNA molecules consist of double strands. The double strands are tied together in a helical structure (see Figure 2.5,

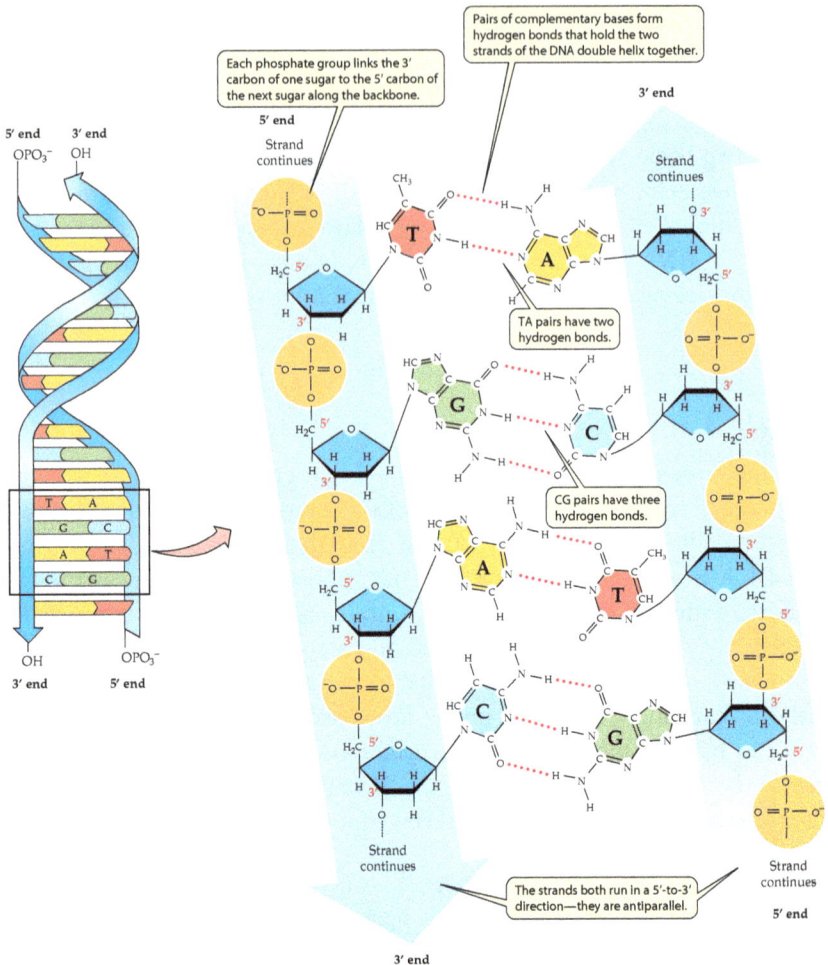

Fig. 2.5   The backbone and double-helix structure of DNA (Figure is from [224] with permission from Sinaucr Associates).

left panel), with bases to the center (like rungs on a ladder) and sugar-phosphate units along the sides of the helix (like the side rails of a twisted ladder). This double helix structure was discovered by James Watson and Francis Crick in 1953. The bases on the two strands are paired according to the *complementary base pairing rules* (also called the *Watson-Crick base pairing rules*): adenine (base A) only pairs with thymine (base T), and guanine (base G) only pairs with cytosine (base C). The pairs so formed are

Pyrimidines

Cytosine (C)   Thymine (T)   Uracil (U)

Purines

Adenine (A)   Guanine (G)

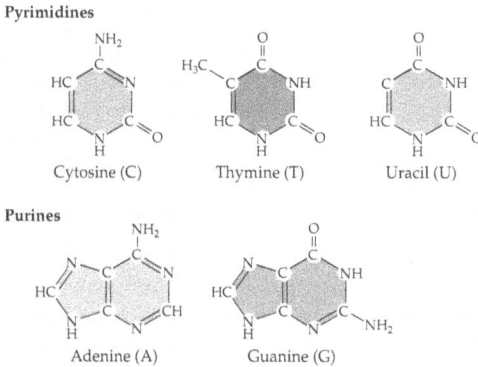

Fig. 2.6   DNA and RNA bases (Figure is from [224] with permission from Sinaucr Associates).

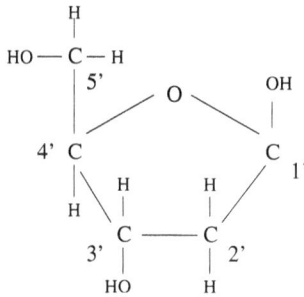

Fig. 2.7   Structure of a deoxyribose sugar.

called *base pairs*; they provide the unit of length most used when referring to DNA molecules, abbreviated to *bp*. Thus, we can state that a certain piece of DNA is 100,000 bp long, or 100kbp [249]. The force that holds a base pair together is a weak hydrogen bond. Although each individual bond is weak, the cumulative effect of many such bonds is sufficiently strong to bind the two strands tightly together. As a result, DNA is chemically inert and is a generally stable carrier of genetic information [10].

As mentioned above, each DNA strand has a 5' to 3' orientation, indicated by the sequence of its carbon atoms. In the double-stranded structure of DNA, each strand maintains its own orientation, with the 5' end of one strand aligning with the 3' end of the other strand. In other words, the two strands are *antiparallel*. As a consequence, it is possible to infer the sequence of one strand if we know the sequence of the other through an operation called *reverse complementation*. For example, given strand $s = AGCTAAC$ in the 5'→3' direction, we first reverse s, obtaining s' = CAATCGA, and then replace each base by its complement, obtaining $\bar{s} = GTTAGCT$, which is the reverse compliment of s [249].

This reverse complementation is precisely the mechanism that allows DNA in a cell to *replicate*. Even prior to the discovery of DNA, it had been recognized that any carrier of hereditary material must be able to replicate itself so that information could be passed on from generation to generation. However, the actual mechanism of self-replication was unknown. When the structure of DNA was deduced, it was understood that the complementary structure of the DNA molecule would permit exact self-replication, thus fulfilling this requirement [10].

2.4.2 *RNA*

RNA molecules are similar to DNA molecules, with the following basic compositional and structural differences [249]:

- The sugar component of RNA is ribose rather than deoxyribose;
- In RNA, thymine (T) is replaced by uracil (U), which also binds with adenine (see Figure 2.6);
- RNA does not form a double helix. RNA-DNA hybrid helices sometimes occur, or parts of an RNA molecule may bind to other parts of the same molecule by complementarity. The three-dimensional structure of RNA is far more varied than that of DNA.

DNA and RNA also differ in that while DNA performs essentially one function (that of encoding information), cells contain a variety of RNA types, each performing different functions [249]. This will be discussed in more detail below.

2.5   Central Dogma of Molecular Biology

DNA molecules are responsible for encoding the information necessary to build each protein or RNA molecule found in an organism. In this sense, DNA is sometimes referred to as "the blueprint of life." The information flow from DNA via RNA and thus to the protein is described by the so-called *central dogma of molecular biology*, which includes the following four major stages (see Figure 2.8) [65]:

(1) The information contained in DNA is duplicated via the *replication process*.
(2) DNA directs the production of encoded messenger RNA (mRNA) through a process called *transcription*.
(3) In eukaryotic cells, the mRNA is then *processed* and migrates from the nucleus to the cytoplasm of the cell.
(4) In the final stage of the information-transfer process, messenger RNA carries the encoded information to protein-synthesizing structures called *ribosomes*. Through a process called *translation*, the ribosomes use this coded information to direct protein synthesis.

In this section, we will describe DNA encoding mechanism and how a protein is built out of DNA.

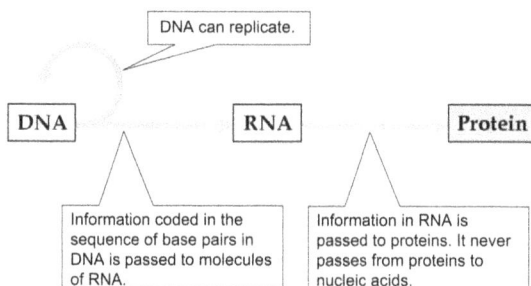

Fig. 2.8   The central dogma of molecular biology (Figure is from [224] with permission from Sinaucr Associates).

2.5.1   Genes and the Genetic Code

Each cell of an organism has one or more DNA molecules. Each DNA molecule forms a *chromosome*. The complete set of chromosomes inside a cell is called a *genome*. The number of chromosomes in a genome is characteristic of a particular species. For example, every cell in *Homo sapiens* (modern humans) has 46 chromosomes, whereas this number is 8 in *Drosophila melanogaster* (the fruit fly) and 32 in *Saccharomyces cerevisiae* (yeast).

A DNA molecule contains certain contiguous stretches which encode information for building proteins. However, some portions of the DNA molecule do not contain encoded information but rather are termed "junk DNA." This may actually be a misnomer, as it has been suggested that junk DNA may indeed perform unrecognized and valuable functions. A *gene* is a contiguous stretch of DNA that contains the information necessary to build a protein or an RNA molecule. Gene lengths vary, but human genes normally have about 10,000 bp. The starting and ending points of genes can be recognized by certain cell mechanisms.

As described in Section 2.3, a protein is composed of a chain of amino acids. The mechanism by which genes specify the sequence of amino acids in a protein is called the *genetic code*. To be specific, a triplet of nucleotides is used to specify each amino acid. Such a triplet is called a *codon*. Figure 2.9 illustrates the correspondence between each possible triplet and each amino acid. In this figure, nucleotide triplets are denoted using RNA rather than DNA bases, since, it is RNA that provides the link between DNA and actual protein synthesis. This will be discussed in more detail in Section 2.5.2.

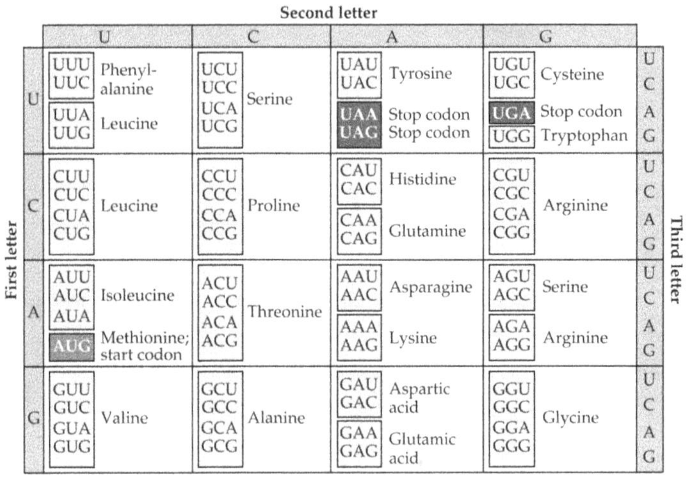

		Second letter			
	U	**C**	**A**	**G**	
U	UUU UUC Phenyl-alanine / UUA UUG Leucine	UCU UCC UCA UCG Serine	UAU UAC Tyrosine / UAA Stop codon UAG Stop codon	UGU UGC Cysteine / UGA Stop codon UGG Tryptophan	U C A G
C	CUU CUC CUA CUG Leucine	CCU CCC CCA CCG Proline	CAU CAC Histidine / CAA CAG Glutamine	CGU CGC CGA CGG Arginine	U C A G
A	AUU AUC Isoleucine AUA / AUG Methionine; start codon	ACU ACC ACA ACG Threonine	AAU AAC Asparagine / AAA AAG Lysine	AGU AGC Serine / AGA AGG Arginine	U C A G
G	GUU GUC GUA GUG Valine	GCU GCC GCA GCG Alanine	GAU GAC Aspartic acid / GAA GAG Glutamic acid	GGU GGC GGA GGG Glycine	U C A G

First letter (left axis) / Third letter (right axis)

Fig. 2.9   The genetic code mapping codons to amino acids (Figure is from [224] with permission from Sinaucr Associates).

Given the four base types, the total number of possible base combinations within nucleotide triplets is 64. However, these 64 combinations can only refer to the twenty amino acids which actually occur. There is therefore redundancy in coding, and several different triplets will correspond to the same amino acid. For example, both AAG and AAA code for lysine (see Table 2.1). Moreover, three of the possible codons (UGA, UAG, and UAA) do not code for any amino acid and are used instead to signal the end of a gene. Such redundancy is actually a valuable feature of the genetic code, rendering it more robust in the event of small errors in the transcription process. This will be discussed in the next subsection.

2.5.2 *Transcription and Gene Expression*

Transcription is the process of synthesizing RNA using genes as templates. A gene is *expressed* when, through the transcription process, its coding is transferred to an RNA molecule. To initiate a transcription process, the DNA double helix is "unzipped," starting at the promoter site of a gene. The *promoter site* is a region on the 5' side of the DNA strand which indicates that a gene is forthcoming. The codon AUG, which codes for methionine, also serves as the signal for the start of a gene. Once the DNA double helix has been opened at this starting point, one DNA strand serves as a *template strand*. An RNA molecule is constituted by binding together ribonucleotides complementary to the template strand until the STOP codon is met. The composing process always builds mRNA molecules from the 5' end to the 3' end, whereas the template strand is read from 3' to 5'. This resulting RNA is called the *messenger RNA*, or, briefly, mRNA. Since the two strands of the original DNA helix were also complementary, the new mRNA molecule will have the same ribonucleotide sequence as the unused DNA strand, with the base U substituted for T. The selection of the template from the two strands available in original DNA pair varies from gene to gene, as signaled by the location of the promoter site for each gene. After the transcription process, the mRNA will be transported to cellular structures called *ribosomes* to guide the manufacture of proteins.

Transcription as described above is valid for prokaryotes. For eukaryotes, many genes are composed of alternating parts called *introns* and *exons*. After transcription, the introns are *spliced* out from the mRNA. This means that only the exons will participate in protein synthesis. *Alternative splicing* occurs when the same genomic DNA can give rise to two or more different mRNA molecules on the basis of alternative selection of introns and exons, generally resulting in the production of different proteins (please refer to [180] for details of introns, exons, and alternative splicing). Because of the changes which result through the splicing of introns and exons, the entire gene as found in the chromosome is usually called the *genomic DNA*, and the spliced sequence consisting of exons only is called the *complementary DNA* or *cDNA* [249]. The cDNA can be obtained by a *reverse transcription process* which transforms mRNA back into DNA.

Within a DNA or RNA sequence, the bases may be parsed in alternative ways to generate codon groupings. For example, in the sequence TAATCGAATGGGC, adjacent bases could be grouped as codons TAA, TCG, AAT, GGG, omitting the final C. It would also be possible to ignore the initial T, producing codons AAT, CGA, ATG, GGC. Yet another reading frame would yield codons ATC, GAA, TGG, through the omission of the two initial bases (TA) and the two final bases (GC). An *open reading frame* (ORF) selects one of these approaches to reading a DNA sequence and parses the bases into a sequence which begins at the start codon, contains an integral number of codons, and does not include any STOP codons within the sequence [249].

Scientists involved in gene expression research usually find it easier to work with *expressed sequence tags* (ESTs) instead of the entire gene. An EST is a unique short subsequence (only a few hundred base pairs in length), generated from the DNA sequence of a gene, that acts as a "tag" or "marker" for the gene. An advantage of ESTs is that they can be back-translated into genetic code that is coded for or expressed as exons rather than including introns or other non-coding DNA [10].

2.5.3   Translation and Protein Synthesis

Once the transcription process has generated properly-encoded mRNA, the *translation process* which synthesizes proteins is initiated. As mentioned before, the synthesis of proteins takes place inside cellular structures called *ribosomes* (see Figure 2.10). Another type of RNA, *transfer RNA* or tRNA, makes the connection between a codon and the corresponding amino acid. As illustrated in Figure 2.11, each tRNA molecule has, on one side, an anticodon that has high affinity for a specific codon and, on the other side, an amino acid attachment site that binds easily to the corresponding amino acid. As the messenger RNA passes through the interior of the ribosome (see Figure 2.12), a tRNA which matches the current codon will bind to it and bring along the corresponding amino acid. The attached amino acid falls in place just next to the previous amino acid in the protein chain being formed. A suitable enzyme then catalyzes the addition of this current amino acid to the protein chain, releasing it from the tRNA (see Figure 2.13). In this way, a protein is constructed residue-by-residue. When a STOP codon appears, no tRNA associates with it, and the synthesis ends. The messenger RNA is released and degraded by cell mechanisms into ribonucleotides,

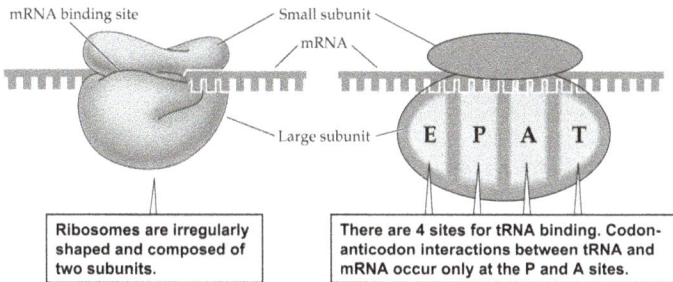

Fig. 2.10 The structure of a ribosom (Figure is from [224] with permission from Sinaucr Associates).

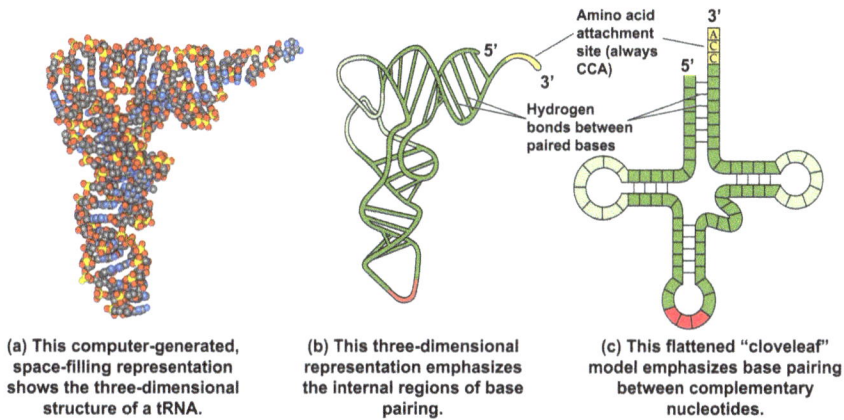

Fig. 2.11 The structure of tRNA (Figure is from [224] with permission from Sinaucr Associates).

which will then be recycled to make other RNA [249].

2.6 Genotype and Phenotype

Genomes belonging to the same species vary slightly from organism to organism in a phenomenon known as *genome variation* (or *genetic variation*). It is this subtle variability in genomes that is responsible for the evolution and diversity of organisms. Some genome variations are unique to an organism, while others are passed on through generations via reproductive cells [10].

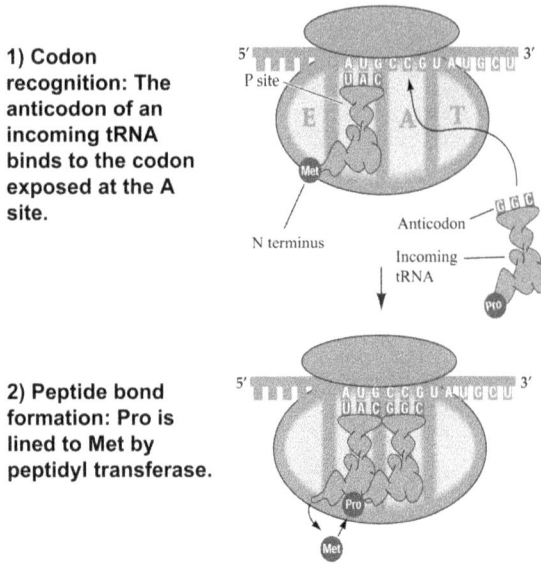

1) Codon recognition: The anticodon of an incoming tRNA binds to the codon exposed at the A site.

2) Peptide bond formation: Pro is lined to Met by peptidyl transferase.

Fig. 2.12   The translation process (Figures are from [224] with permission from Sinaucr Associates).

Most genome variations involve only a few bases. Some common varia-tions include the replacement of one base by another (*substitution*), the exci-sion of a base (*deletion*), the addition of a base (*insertion*), and the removal of a small subsequence of bases and their reinsertion in the opposite order (*inversion*) or in another location (*translocation*). Such genome variations are due to mutations and polymorphisms. A *polymorphism* is a genome variation in which every possible sequence is present in at least one percent of a population, whereas a *mutation* refers to a genome variation that is present in less than one percent of a population. In practice, the terms "mutation" and "polymorphism" are often used interchangeably [180].

As mentioned earlier, DNA resides in chromosomes. A cell may contain a single set of chromosomes (the *hoploid state*) or two chromosome sets (the *diploid state*); in the latter case, each chromosome is represented by two copies. An exception is the pair of sex chromosomes, which draws one copy from the father and one from the mother. The two members of a pair of chromosomes are called *homologous chromosomes*. The existence of genome variation means that some genes may differ slightly from individual to individual. When this happens, each alternate version of a gene is called

3) Elongation:
Free tRNA is
released from the
P site, and the
ribosome shifts
by one codon, so
the growing
polypeptide
moves to the P
site. The free
tRNA is released
via the E site.

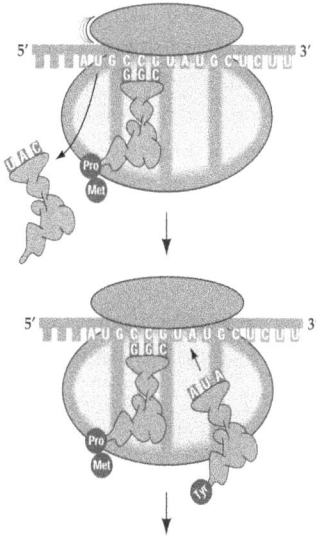

4) The process
repeats: Codon
recognition,
peptide bond
formation, and
elongation.

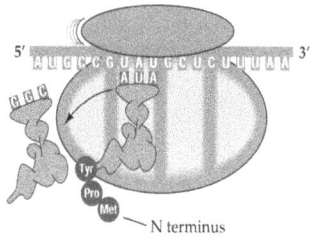

Fig. 2.13   The translation process continued (Figures are from [224] with permission from Sinaucr Associates).

an *allele*. In fact, every diploid cell carries two alleles of each gene, one in each of a pair of homologous chromosomes. When both alleles are the same, the organism is said to be *homozygous* for that gene; otherwise, the organism is said to be *heterozygous* for that gene. In the latter case, only one of the alleles, the *dominant allele*, may be expressed, while the other, *recessive allele* is not expressed [10].

The common and properly-functioning version of a gene is referred to as its *wild-type allele*; a version with a mutation is called a *mutant allele* [180]. As with the redundancy present in genetic coding (see Section 2.5.1), the presence of two versions of each gene is another protective mechanism provided by nature. If one copy should happen to be defective, the other copy is available to compensate. Due to this protective mechanism, many genome variations do not produce any noticeable effects. However, the effects of a small percentage of genome variations are noticeable, with both and beneficial deleterious results. Much current research is focused on elucidating genotype-phenotype relationships, as the relationships between diseases and their genomic basis are termed. *Genotype* refers to the genetic makeup of an individual, while the outward characteristics of the individual are its *phenotype*. They are, naturally, connected, as the phenotype is shaped, develops, and functions on the basis of the information provided by and encoded in the genotype [10]. The recent development of microarray technology, which is the focus of this book, has proved to be a useful tool to help delineate the genotype-phenotype relationship [67].

2.7   Summary

This chapter has provided a brief introduction to the salient characteristics of cells, proteins, DNA, RNA, and other biological concepts which will underlie the discussions in subsequent chapters. This presentation has largely been excerpted from a variety of standard textbooks and websites [224, 249, 299].

Chapter 3

Overview of Microarray Experiments

3.1 Introduction

The recent development of high-density DNA microarray technology enables researchers to capture the "snap-shots" of cells on a genome-wide scale at the transcriptional level. The expression levels of thousands, or even tens of thousands, of genes can be monitored using a single microarray chip. Prior to the advent of high-throughput technology, the examination of gene expression levels was limited to a much smaller scale per experiment [193].

Modern microarray technology was originated from the *Southern blot* (named after E. M. Southern, a British biologist), which was the first array of genetic material. This technique employs (radioactively) labeled DNA (or RNA) probes to hybridize to identify very similar DNA sequences placed on a nitrocellulose filter called a *blot*. The number of DNA bands that hybridize to a short probe gives an estimate of the number of closely related genes in an organism [299]. Similar techniques are the *Northern blot* and the *Western blot*, which, respectively, employ RNA strands and proteins in lieu of DNA sequences. The basic principle behind these techniques is that DNA and RNA strands can be labeled for detection and then used to probe other nucleic acid molecules that have been attached to a solid surface.

McLachlan [193] briefly reviewed the history of the microarray technology. In the 1980s, a group led by R. P. Ekins in the Department of Molecular Endocrinology at the University College, London was the first to use simple *microspotting* techniques to manufacture arrays for high-sensitivity immunoassay studies [93]. Numerous groups of researchers have furthered the technology introduced by Ekins and his colleagues. In the United States, notable research has been accomplished by Stephen P.A. Fodor and his colleagues at Affymetrix, Inc. (Santa Clara, California) [101], as well as groups at Stanford University, particularly Patrick O. Brown, in the Department of Biochemistry and Biophysics [242]. Brown and his colleagues at Stanford are credited with engineering the first DNA microarray chip, while Stephen Fodor and colleagues at Affymetrix, Inc., created the first patented DNA microarray wafer chip, the GeneChip. Numerous commercial entities and academic groups have since contributed to advancements in DNA microarray technology.

In the most general form, a DNA array is a chip made of nylon membrane, glass or plastic. Usually, the chip is arranged in a regular grid-like pattern and segments of DNA strands are either deposited or synthesized within individual grids. Once the array is prepared, a microarray experiment involves three basic steps: sample preparation and labeling, sample hybridization and washing, and microarray image scanning and processing.

In this chapter, we will first introduce and compare the various technologies available for microarray manufacture in Section 3.2. In Sections 3.3 and 3.4, we will then describe the steps involved in a microarray experiment. The raw data resulting from a microarray experiment may contain noise, missing values, and experimental bias. Therefore pre-processing and normalization of the raw data is indispensable before any further analysis can be performed. We will discuss these two issues in Sections 3.5 and 3.6, respectively.

3.2   Microarray Chip Manufacture

There are two main approaches to manufacture of microarray chips: deposition of DNA fragments by robotic spotting and *in situ* synthesis [173]. Manufacture by robotic deposition may proceed through the deposition of PCR-amplified cDNA clones or the printing of already-synthesized oligonucleotides. *In situ* fabrication can be divided into photolithography, ink jet printing, and electrochemical synthesis [83]. In this section, the two currently most widely-used chips, the cDNA microarray and the Affymetrix GeneChip, will be described as examples for the deposition-based and in situ microarray manufacture.

3.2.1 *Deposition-Based Manufacture*

The manufacture of deposition-based arrays involves the consideration of three issues: the selection of DNA "probes"[1], preparation of the probes, and the printing process. Let us first consider which "probes" are to be printed on the array. In many cases, these are chosen directly from databases including GenBank [24], dbEST [33] (see Section 2.5.2 for the concept of *EST*), and UniGene [245], the resource backbones of the array technologies [38, 88]. Additionally, full-length cDNAs, collections of partially sequenced cDNAs (or ESTs), or randomly chosen cDNAs from any library of interest can be used [88].

In the process of deposition-based manufacture, the DNA probes are prepared away from the chip. Probes can either be polymerase chain reaction (PCR) products or oligonucleotides. The PCR technique was developed in 1983 through the work of Kary B. Mullis and his colleagues at Cetus Corporation in Emeryville, California (Mullis, 1990) [202]. This technique creates billions of copies of specific fragments of DNA from a single DNA molecule. After the amplification, the PCR products are partially purified by precipitation and/or gel-filtration to remove unwanted salts, detergents, PCR primers and proteins present in the PCR cocktail [88]. Alternatively, DNA probes can be prepared by pre-synthesizing DNA oligonucleotides for use on the array [266].

Once the DNA probes are determined and prepared, a typical printing process follows five steps:

(1) Robots dip thin pins into the wells of solutions to collect the first batch of DNA.
(2) The pins touch the surface of the arrays to spot the DNA. Usually, the DNA is spotted onto a number of different arrays, depending on the number of arrays to be made and the amount of liquid the pins can hold.
(3) The pins are washed to remove any residual solution and ensure no contamination of the next sample.
(4) The pins are dipped into the next set of wells.
(5) Return to Step (2) and repeat until the array is complete [266].

Figure 3.1 illustrates some of the instrumentation used to fabricate a microarray.

---

[1]Throughout this book, we will follow the nomenclature proposed by Duggan et al. [88] and refer to the DNA on the array as *probes* and the labeled DNA in solution as the *target*.

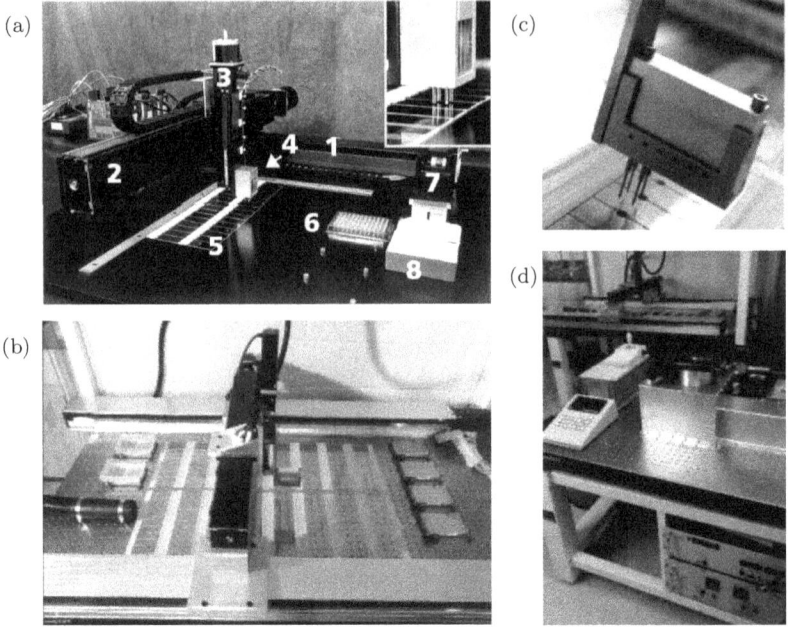

Fig. 3.1   Microarray instruments. (a) The microarray robot at the University of Pennsylvania. (b) The microarray robot at the Albert Einstein College of Medicine (AECOM). (c) Four of the possible twelve pen tips in use. (d) The laser scanner of AECOM. Images are from [53].

3.2.2   *In Situ Manufacture*

Arrays synthesized *in situ* are fundamentally different from spotted arrays in the following aspects [83]:

- **Selection of probes.** Probe selection is performed based on sequence information alone. Therefore, every probe synthesized on the array is known. In contrast, with cDNA arrays, which deal with expressed sequence tags, the function of the sequence corresponding to a spot is often unknown. Additionally, since this selection method avoids duplicating identical sequences among gene family members, this approach can distinguish and quantitatively monitor closely-related genes.

- **Preparation of probes.** The probes are photochemically synthesized base-by-base on the surface of the array. There is no cloning and no PCR process involved.
- **Printing process.** Since the probes are synthesized on the surface of the array, no printing process is needed. The elimination of cloning, amplification, and printing of DNA reduces many sources of potential noise in the cDNA system and thus constitutes an advantage of this approach to array fabrication.

In situ synthesis takes place via a covalent reaction between the 5' hydroxyl group of the sugar of the last nucleotide to be attached to the array and the phosphate group of the next nucleotide. Each nucleotide added to the oligonucleotide probe anchored to the glass chip has a protective group at its 5' position to prevent the addition of more than one base during each round of synthesis. This protective group is then converted to a hydroxyl group, using either acid or light, prior to the next round of synthesis [266].

3.2.2.1　*The Affymetrix GeneChip*

At Affymetrix, the construction of high-density DNA probe arrays relies on light-directed synthesis using two techniques: the *photolithography* (akin to the technology for building very large scale integrated (VLSI) circuits) and *solid-phase DNA synthesis* [184]. Described simply, the Affymetrix technique uses light to convert the protective group on the terminal nucleotide into a hydroxyl group to which further bases can be added. The light is directed through a photolithographic mask which allows light to pass to specific areas of the array but not to others. In this way, a specific base can be added to a probe at a specific location. A series of such masks allows the base-by-base construction of sequences [83, 266] (see Figure 3.2).

The gene expression arrays employ a match/mismatch probe strategy [184] (see Figure 3.3). The probes that exactly match the target sequence are called *reference probes*. For each reference probe, there is a corresponding *mismatch probe* which contains an altered nucleotide at the central base position. These two probes, reference and mismatch, are always synthesized adjacent to each other to control spatial differences in hybridization. To enhance the confidence level of signal detection, each gene has several reference/mismatch pairs, where each pair corresponds to various parts of the gene [83]. The use of the PM (perfect match) minus MM (mismatch) differences averaged across a set of probes for each gene greatly reduces the contribution of background and cross-hybridization

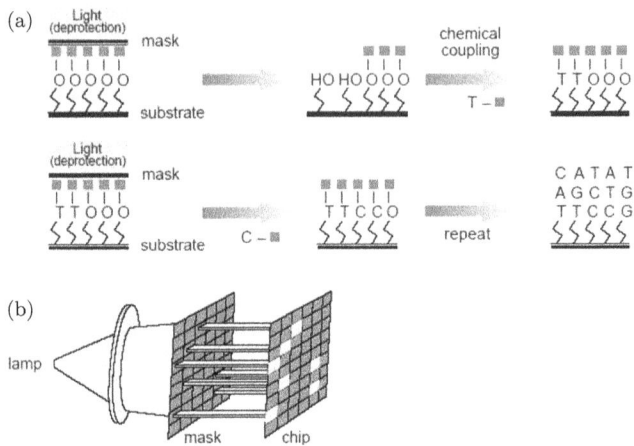

Fig. 3.2   (a) Light directed oligonucleotide synthesis. A solid support is prepared with a covalent linker molecule terminated with a photolabile protecting group. Light is directed through a mask to deprotect and activate selected sites, and protected nucleotides couple to these activated sites. The process is repeated, activating different sets of sites and coupling different bases allowing arbitrary DNA probes to be constructed at each site. (b) Schematic representation of the lamp, mask and array. Figures are from [184].

and increases the quantitative accuracy and reproducibility of the measurements [184].

3.3   Steps of Microarray Experiments

Irrespective of the specific technology employed, a microarray experiment consists of three basic steps: sample preparation and labeling, sample hybridization and washing, and microarray image scanning and processing. In the following section, the cDNA microarray will serve as a basis for a general discussion of these steps. In general, experiments based on other platforms, such as the Affymetrix GeneChip, follow similar principles unless specifically stated. The general schema of a cDNA microarray is illustrated in Figure 3.4.

3.3.1   *Sample Preparation and Labeling*

Sample preparation involves extracting and purifying the mRNAs from the tissue of interest. Due to a number of challenges, this procedure can be quite variable [10, 266]. First, the target mRNA typically accounts for only

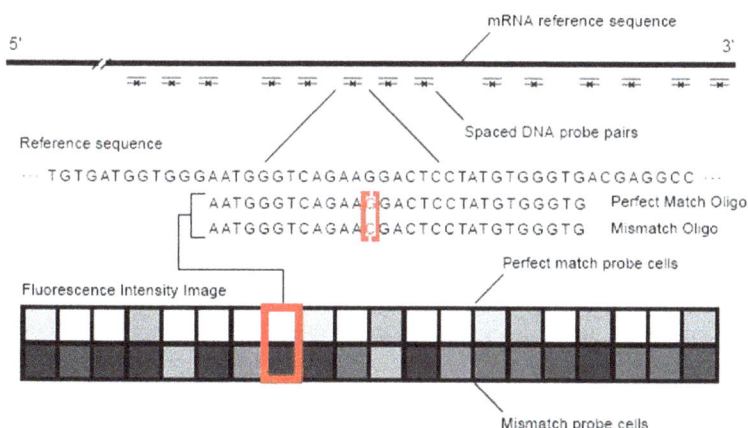

Fig. 3.3 Expression probe and array design of Affymetrix. Oligonucleotide probes are chosen based on uniqueness criteria and composition design rules. For eukaryotic organisms, probes are chosen typically from the 3' end of the gene or transcript (nearer to the poly(A) tail) to reduce problems that may arise from the use of partially degraded mRNA. The use of the PM minus MM differences averaged across a set of probes greatly reduces the contribution of background and cross-hybridization and increases the quantitative accuracy and reproducibility of the measurements. Figure is from [184].

a small fraction (less than 3%) of all mRNA in a cell. Second, it could be very difficult to isolate mRNA specific to the study from a heterogeneous range of cells (for example, diseased tissue contains a mixture of normal tissue, inflammatory cells, necrotic tissue, and, in cancer samples, areas of different grades). Finally, captured mRNA degrades very quickly. To address this rapid degradation, the mRNA is usually reverse-transcribed into more stable cDNA (for cDNA microarrays) immediately after extraction [10].

To allow detection of which cDNAs are bound to the microarray, the sample goes through a platform-dependent labeling process. For Affymetrix platform, a biotin-labeled complementary RNA is constructed for hybridizing to the GeneChip. The protocols are very carefully defined by Affymetrix to ensure that every Affymetrix laboratory follows identical steps. Experimental results obtained in different Affymetrix laboratories should therefore be reliably comparable [266].

Detection of DNA on cDNA microarrays had previously been performed using radioactively-labeled DNA, but it is now more common to use dyes which fluoresce when exposed to a specific wavelength of light. In the most common experiments, two samples are hybridized to arrays, each labeled

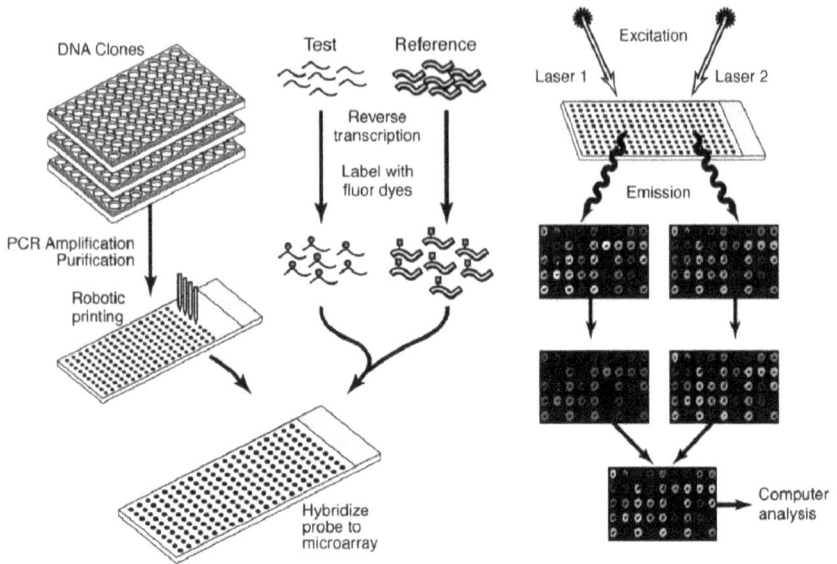

Fig. 3.4   cDNA microarray schema. Templates for genes of interest are obtained and amplified by PCR (Polymerase Chain Reaction). Following purification and quality control, aliquots (~5 nl) are printed on coated glass microscope slides using a computer-controlled, highspeed robot. Total RNA from both the test and reference sample is fluorescently labeled with either Cye3- or Cye5-dUTP (Deoxyuridine Triphosphate) using a single round of reverse transcription. The fluorescent targets are pooled and allowed to hybridize under stringent conditions to the clones on the array. Laser excitation of the incorporated targets yields an emission with a characteristic spectra, which is measured using a scanning confocal laser microscope. Monochrome images from the scanner are imported into software in which the images are pseudo-colored and merged. Information about the clones, including gene name, clone identifier, intensity values, intensity ratios, normalization constant and confidence intervals, is attached to each target. Data from a single hybridization experiment is viewed as a normalized ratio (that is, Cye3/Cye5) in which significant deviations from 1 (no change) are indicative of increased (> 1) or decreased (< 1) levels of gene expression relative to the reference sample. In addition, data from multiple experiments can be examined using any number of data mining tools. Figure is from [88].

with Cy3 or Cy5 dyes, which are excited by green and red lasers, respectively [10]. This results in a *two-channel cDNA microarray* and allows the simultaneous measurement of both samples. In the future, it is possible that more than two labeled samples may be used, producing a *multichannel cDNA microarray*.

3.3.2   *Hybridization*

Hybridization is the step in which the DNA probes on the microarrays and the labeled DNA (or RNA) target forms heteroduplexes according to the Watson-Crick base-pairing rule (see Section 2.4.1) [266]. The essential principle here is that a single-stranded DNA molecule will bind to another single-stranded DNA molecule with a precisely matching sequence with much higher affinity than that to an imperfectly matching sequence [10]. In reality, however, hybridization is a complex process and a DNA segment may also bind well to a sequence similar but not identical to its complementary target, a phenomenon called *cross-hybridization*. This is influenced by many conditions, including temperature, humidity, salt concentration, formamide concentration, target solution volume, and hybridization operator [266].

Hybridization may be performed either manually or by a robot. Robotic hybridization provides much better control over the temperatures of the target and slide. The consistent use of a single hybridization station also reduces the variability which arises from multiple hybridizations and various operators [266]. After hybridization, the microarray is removed from the chamber or station and is then washed to eliminate any excess labeled sample so that only the DNA complementary to the probes remains bound on the array. Finally, the microarray is dried using a centrifuge or by blowing clean compressed air [10].

3.3.3   *Image Scanning*

After the completion of hybridization, the surface of the hybridized array is scanned to produce a microarray image. As previously mentioned, samples are labeled with biotin or fluorescent dyes that emit detectable light when stimulated by a laser. The emitted light is captured by the photo-multiplier tube (PMT) in a scanner, and the intensity is recorded. Most scanners contain one or more lasers that are focused onto the array (for two-channel microarrays, the scanner uses two lasers) [266].

Although the scanner is only intended to detect light emitted by the target DNA strands which are bound to their complementary probes, it also will capture incidental light from various other sources. These other sources may include labeled DNA sample which has hybridized non-specifically to the glass slide, residual (unwashed) labeled sample which has adhered to the slide, various chemicals used in processing the slide, and even the slide itself. This incidentally-captured light is called background [10].

The scanned output of an Affymetrix chip is usually a monochrome image (Figure 3.5). With two-channel microarrays, the output is a pair of monochrome images. Each image is from one of the lasers in the scanner. The two monochrome images are combined to create the *false color images* of microarrays (Figure 3.6). Both monochrome and two-color images are usually stored in the tagged image file format (TIFF).

Fig. 3.5   An example of an Affymetrix image (Figure is from [184]).

3.4   Image Processing

The microarray image generated by the scanner forms the raw data of a microarray experiment. Prior to data analysis, the image must be converted from the TIFF format into the numerical information that quantifies gene expression. The manner in which this is accomplished will have a major impact on the quality of the resulting data and the success of further analysis. For *in situ* synthesized microarrays, both Affymetrix and Agilent have integrated tailored image-processing algorithms into their software packages, allowing end-users to directly generate quantified microarray data.

(a)

(b)

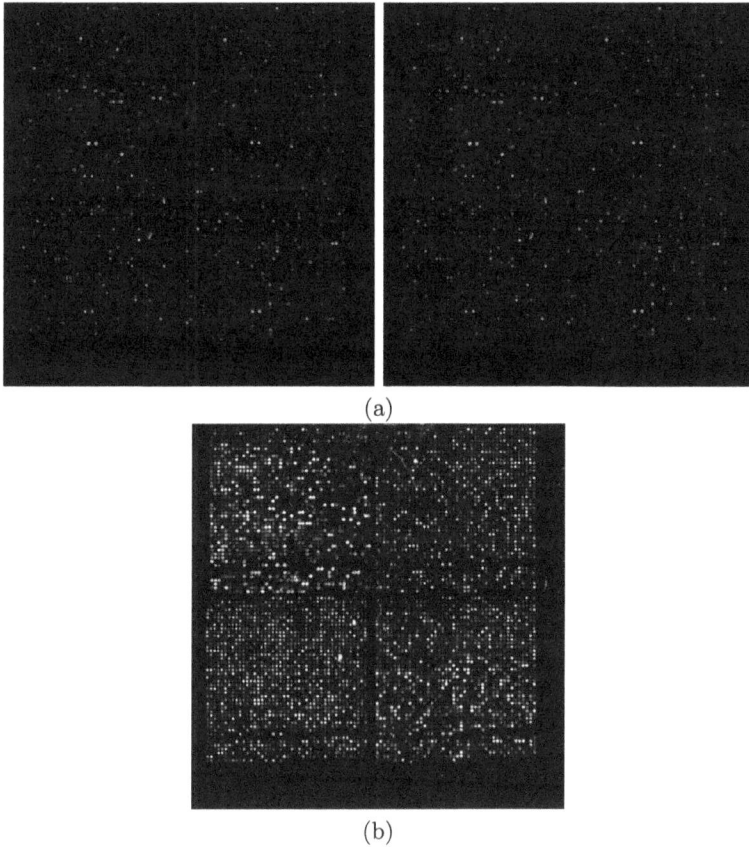

Fig. 3.6   Output of scanners for cDNA microarray. (a) The raw monochrome images for individual channels. (b) The combined *false color image*. Green spots in the combined image correspond to spots that are expressed more in channel one. Red spots correspond to those expressed more in channel two. Yellow spots have a similar level of expression in both channels. Dark spots are low expressed in both channels. Images are from http://genome-www.stanford.edu/cellcycle/.

For cDNA microarrays, the images consist of spots arranged in regular grid-like patterns. The processing of these images consists of four basic steps:

(1) **Spot identification**. The process of spot identification involves locating the position of individual signal spots in an image and estimating their size.

(2) **Image segmentation**. This process involves decomposing an image into a set of non-overlapping regions. Specifically, this step involves differentiation of those pixels which form the spot and should be included in the calculation of the signal from those pixels which are background or noise and should be eliminated.

(3) **Spot quantification**. After the pixels belonging to the signal and the background have been distinguished, this process involves calculating the intensity for each spot. Here, pixel intensity values are combined into a unique value representing the expression level of a gene deposited in a given spot.

(4) **Spot quality assessment**. This process involves calculating some quality-control measures which evaluate the quality of both the entire array and the individual spots on the array. These measures can assist human inspectors in determination of the data reliability and identification of those spots with questionable quality values.

3.5   Microarray Data Cleaning and Preprocessing

3.5.1   *Data Transformation*

It is common practice to transform DNA microarray data from the raw intensities into log intensities before proceeding with analysis. There are several objectives of this transformation [266]:

- There should be a reasonable even spread of features across the intensity range.
- Variability should be constant at all intensity levels.
- The distribution of experimental errors should be approximately zero.
- The distribution of intensities should be approximately bell-shaped.

Figure 3.7 shows the histogram of intensities of a typical microarray data set before and after log transformation. We can see that the raw data is very heavily clumped together at low intensities and sparsely distributed at high levels. By contrast, the data is more evenly spread over the intensity range after the log transformation. The transformation greatly reduces the skewness of the distribution and simplifies visual examination.

Microarray data analysis typically uses logarithms to base 2 [266]. In processing, the ratio of the raw Cy5 and Cy3 intensities is transformed into the difference between the logs of the intensities of the Cy5 and Cy3 channels. Therefore, 2-fold up-regulated genes correspond to a log ratio

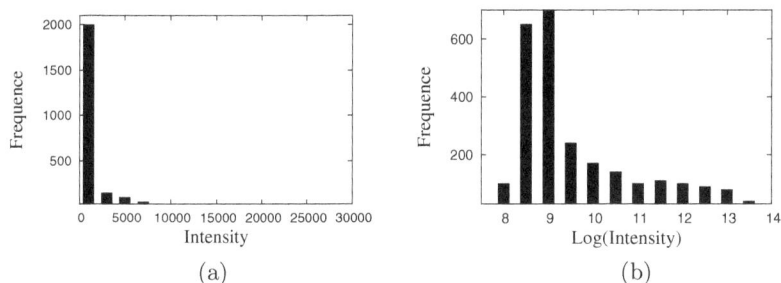

Fig. 3.7 Histogram of the intensities (a) before and (b) after the log transformation of an example data set.

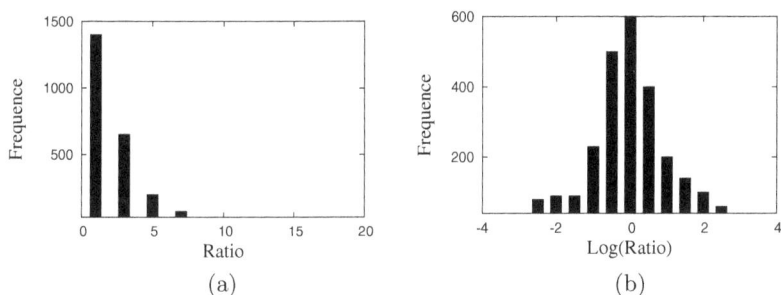

Fig. 3.8 Histogram of the ratios of intensity (a) before and (b) after the log transformation of an example data set.

of +1, and 2-fold down-regulated genes correspond to a log ratio of -1. Genes that are not differentially expressed have a log ratio of 0. These log ratios have a natural symmetry which reflects the biological structure and is not present in the raw fold differences. Figure 3.8 illustrates histograms of ratios of the intensity data set before and after log transformation.

3.5.2 *Missing Value Estimation*

DNA microarray experiments often generate data sets with multiple missing expression values. Missing values occur for diverse reasons, including insufficient resolution, image corruption, or slide contamination by dust or scratches. Missing data may also occur systematically as a result of the robotic methods employed in generating the microarrays [193]. Unfortunately, many algorithms for gene expression analysis require a complete data set as input. Therefore, methods for estimating missing data are needed before these algorithms can be applied.

Suppose a micraorray data set is represented by a matrix where each row corresponds to one gene and each column represents an experimental condition. A simple approach to imputing missing values is to replace a missing entry with the average expression over the rows (*the Row Average Method*). This method is not optimal since it does not take into account the correlation structure of the entire data set. In [284], Troyanskaya et al. propose two more complex algorithms based on K-nearest neighbors (KNNimpute) and singular value decomposition (SVDimpute). They also evaluated the performance of these two algorithms and the Row Average Method.

Imputation based on K-nearest neighbors (KNN). In simple terms, the KNN imputation algorithm estimates missing values by selecting K genes with expression profiles most similar to the gene of interest. Suppose that, for gene i, the expression value x_{ij} is missing in the jth experiment. The algorithm selects the K genes with non-missing values for experiment j which have closest expression profiles to gene i in the remaining experiments. A weighted average of values in experiment j from the K genes is then used as an estimate for x_{ij}. Metrics such as Euclidean distance or Pearson's correlation coefficient can be applied to measure the similarity or distance between gene profiles. In [284], the authors found that the Euclidean distance was a sufficiently accurate measure for the log-transformed data. In the weighted average, the contribution of each gene is weighted by distance or the similarity of its expression to that of gene i.

Imputation based on Singular Value Decomposition (SVD). This method first imputes all missing values in matrix A using the row average method in a preliminary step. The Singular Value Decomposition (SVD) is then applied to produce a set of mutually orthogonal expression patterns called *eigengenes*. These eigengenes can be combined linearly to approximate the gene expressions in a $N \times M$ microarray data matrix A, where N and M are the numbers of genes and experiments, respectively.

To be specific, the singular value decomposition of A is

$$A = U\Sigma V^T.$$

The columns of V^T form the eigengenes of $A^T A$, whose contribution to the expression in the eigenspace is quantified by corresponding eigenvalues on the diagonal of matrix Σ. The k most significant eigengenes are selected to form the basis for the imputation process. The value of k is usually determined empirically. The imputation process involves a regression of the missing value x_{ij} against the selected k eigengenes (while ignoring all expression values corresponding to experiment j). That is, the missing x_{ij} is obtained from a linear combination of the k eigengenes weighted by the regression coefficients. This process is iterated until the total change in the matrix A converges to a sufficiently small arbitrary value.

Troyanskaya et al. [284] compared the performance of KNN imputation, SVD imputation, and the Row Average Method in terms of both computational complexity and estimation accuracy. They concluded that although row averaging is the fastest method, it does not perform well in terms of accuracy. They recommend the KNN imputation method as the most robust against the increasing fraction of missing data. However, they also suggested exercising caution when drawing critical biological conclusions from data that is partially imputed; estimated data should be flagged where possible and its significance to the formulation of biological conclusions should be assessed in order to avoid unwarranted assumptions.

3.6   Data Normalization

The complexity of the microarray experimentation process often introduces systematic bias into intensity measurements. Among other sources of variability, systematic bias can be caused by the concentration and amount of DNA placed on the microarrays, wear to arraying equipment such as spotting pins, the quantities of mRNA extracted from samples, reverse transcription bias, lack of spatial homogeneity of the slides, scanner settings, saturation effects, background fluorescence, linearity of detection response, and ambient conditions [10].

In addition, dye bias is present in almost all multichannel experiments. Generally, Cy5 (red) intensities tend to be higher than Cy3 (green) intensities, but the magnitude of the difference generally depends on the overall intensity [10]. The reasons for the imbalance between the channels are as follows [266]:

- The Cy3 and Cy5 labels may be differentially incorporated into DNA samples with varying frequencies of occurrence.
- The Cy3 and Cy5 dyes may have different emission responses to the excitation laser at different frequencies of occurrence.
- The Cy3 and Cy5 emissions may be differentially measured by the photomultiplier tube at different intensities.
- The Cy3 and Cy5 intensities measured at various areas on the array

may differ due to a tilt in the array which results in variations in focus.

The purpose of normalization is to remove the effects of any systematic source of variation to the extent possible. For Affymetrix microarrays, normalization allows direct comparison of individual gene expression levels from one array. For a multi-channel microarray (such as the cDNA microarray), normalization can be applied to adjust the bias among multiple channels.

In general, normalization methods can be divided into *global normalization schemes* and *intensity-dependent normalization approaches*. The global normalization schemes assume that the spot intensities on each pair of arrays being normalized are linearly related. Therefore, the lack of comparability can be corrected by adjusting every single spot intensity on the same array or channel by an identical amount, called the *normalization factor*. By contrast, the intensity-dependent normalization methods determine the normalization factor for different spots according to their individual intensities. Normalization therefore relies on a nonlinear, intensity-dependent *normalization function* $X \to f(X)$ [10].

3.6.1   *Global Normalization Approaches*

3.6.1.1   *Standardization*

Data sets are standardized to ensure that the mean and the standard deviation of each data set are equal. The method is simple; from each measurement on the array, subtract the mean measurement of the array and divide by the standard deviation. After this transformation, the mean of the measurements on each array will be zero, and the standard deviation will be one. An alternative to using the mean and standard deviation is to use the median and median absolute deviation from the median (MAD). This has the advantage of being more robust to outliers than simply using the mean and standard deviation.

3.6.1.2   *Iterative linear regression*

In essence, this method iteratively performs a linear regression on the given pair of data sets $\{x_{1i}\}$ and $\{x_{2i}\}$. The approach assumes that most genes in two data sets are unchanged. The variation in the data sets is caused by systematic bias and can be described by a linear correspondence. Processing involves the following steps [83]:

(1) Perform a simple linear regression. This involves fitting a straight line of the form: $x_{1i} = x_{2i} \cdot m + b$ through the data in such a way that the errors (residuals) are minimal, where m and b are regression parameters.

(2) Find the residuals $e = x_{1i} - (x_{2i})_c$, where $(x_{2i})_c$ is the estimated value calculated as the linear function of x_{1i}.

(3) Remove all genes that have residuals greater than 2σ from 0, where σ is the standard deviation of the residuals.

(4) Repeat the steps above until the changes between consecutive steps are below a given threshold.

(5) Normalize x_{2i} using the regression line found above: $(x_{2i})_n = x_{2i} \cdot m + b$.

3.6.2 Intensity-Dependent Normalization

3.6.2.1 LOWESS: Locally weighted linear regression

Several reports have indicated that the $log_2(ratio)$ values can have a systematic dependence on the intensity [315, 317]. This most commonly appears as a deviation from zero for low-intensity spots. *Locally weighted linear regression* (LOWESS) [58] analysis has been proposed [315, 317] as a normalization method that can remove such intensity-dependent effects in the $log_2(ratio)$ values. In essence, LOWESS divides the data into a number of overlapping intervals and fits a polynomial function of the form:

$$y = a_0 + a_1 x + a_2 x^2 + a_3 x^3 + \cdots . \tag{3.1}$$

In more detail, the LOWESS procedure divides the data domain into a series of continuous short intervals using a sliding-window approach. This compensates for the fact that the polynomial approximation performs well only within a local neighborhood around the chosen point. A sliding window of a given width w is deployed starting from the left extremity of the data. The data points falling within this window are used to fit the polynomial function (Equation 3.1). Since the polynomial approximation is very prone to over-fitting when higher-degree polynomials are used, LOWESS usually limits the degree of Equation 3.1 to 1 or 2. The procedure continues by sliding the window to the right until the entire data range has been processed. For each local window, a new polynomial is fitted. This results in a smooth curve that provides a model for the data. The smoothness of the curve is directly proportional to the width of the sliding window; the smaller the width, the smoother the curve.

The effects of the LOWESS normalization are illustrated in Figure 3.9.

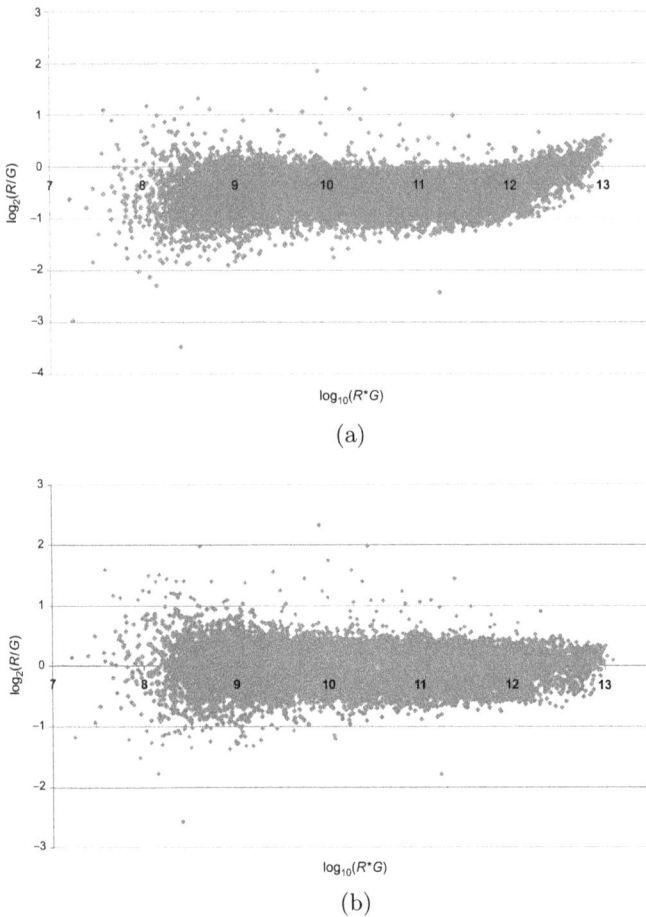

Fig. 3.9 The effect of LOWESS normalization. (a) The R-I plot before LOWESS normalization. (b) The R-I plot after LOWESS normalization. Please refer to Section 8.2.3 for a detailed discussion of R-I plot. Figures are from [225].

In this plot (called a *ratio-intensity plot* or *R-I plot* for short), the horizontal axis represents the sum of the log intensities $log_{10}(Cy3 \cdot Cy5) = log_{10}(Cy3) + log_{10}(Cy5)$ which is a quantity directly proportional to the overall intensity of a given spot. The vertical axis represents $log_2(Cy3/Cy5) = log_2(Cy3) - log_2(Cy5)$ which is the usual log-ratio of the two samples. Note the strong non-linear dye distortion in Figure 3.9(a) and how this is corrected by LOWESS in Figure 3.9(b).

3.6.2.2 *Distribution normalization*

While the purpose of LOWESS is to correct the mean of the data sets, the objective of *distribution normalization* is to make the distributions of the transformed spot intensities as similar as possible across the arrays.

In [34], Bolstad et al. propose a method for distribution normalization. Processing involves the following steps [83, 266]:

(1) Standardize the data.
(2) For each array D_i, order the standardized measurements from lowest to highest. Let D_{i1} be the smallest measurement in D_i, and D_{in} be the greatest measurement, where n is the number of measurements in D_i.
(3) Compute a new distribution D' whose lowest value D'_1 is the average of the lowest values of all the arrays being normalized, i.e., $D'_1 = \text{avg}\{D_{11}, \ldots D_{m1}\}$, where m is the number of arrays; whose second-lowest value is the average of the second-lowest values from each of the arrays, i.e., $D'_2 = \text{avg}\{D_{12}, \ldots D_{m2}\}$; and so on until the highest value is the average value of the highest values of each of the arrays, i.e., $D'_n = \text{avg}\{D_{1n}, \ldots D_{mn}\}$.
(4) Replace each measurement on each array with the corresponding average in the new distribution according to its rank. For example, if a particular measurement of array D_i is the 100th smallest value in the array, replace it with the 100th smallest value D'_{100} in the new distribution.

Distribution normalization is an alternative to LOWESS normalization. It is useful where the different arrays have different distributions of values. The assumption behind this method is that given a series of arrays, a small number of genes may be differentially expressed, however, the overall distribution of spot intensities should not vary too much.

3.7 Summary

In this chapter, we have provided an overview of the generation and processing of gene expression microarray data. The materials in this chapter have largely been excerpted from many publications and websites which address these topics [10, 83, 266]. As we have seen, microarray technology offers an efficient means to measure the expression levels of thousands of genes in a single experiment, across different conditions and over time [44, 49, 73, 94, 131, 147, 243, 301]. Experimental foci may include types of cancers, diseased organisms, or normal tissues. Arrays are now common in basic biomedical research for mRNA expression profiling and are increasingly used to explore patterns of gene expression in clinical research [39, 147, 243, 250, 301]. The application of this technology to the investigation of gene-level responses to drug treatments has the potential to provide deep insights into the nature of many diseases and guidance toward the development of new drugs.

Chapter 4

Analysis of Differentially-Expressed Genes

4.1 Introduction

One basic purpose of a microarray experiment is to identify those genes which demonstrate a significant change in expression level under the impact of certain experimental conditions, such as the presence of cancerous tumors. The experiment explores those genes which are *differentially expressed* in one set of samples relative to another, establishing potentially meaningful correlations between genes and specific conditions. Moreover, filtering out non-differentially expressed genes can reduce the dimensionality of the data set and facilitate further analysis such as sample classification (discussed in Section 6.3). Although simple in principle, the identification of differentially-expressed genes can be complex in practice, as there may be multiple experimental conditions or a lack of biological replicates. In this chapter, we will use four sample data sets to illustrate the problems of identifying differentially-expressed genes under various experimental design conditions.

Example data set A. Samples were taken from human T (Jurkat) cells grown at $37°C$ (for control samples) and $43°C$ (to explore the influence of heat shock) [243]. The expression levels of 1,046 genes were monitored by cDNA microarrays to identify heat-shock-regulated genes in human T cells. This data set is an example of *paired data*. There are two related measurements for a given sample, one being a control and the other exposed to heat shock. We are interested in the difference between the two measurements, as expressed by the log ratio, to determine whether a gene has been up-regulated or down-regulated by exposure to heat shock.

Example data set B. Samples were taken from 14 multiple sclerosis (MS) patients. The expression levels of 4,132 genes were measured by cDNA

51

microarrays for each patient prior to and 24 hours after interferon-β (IFN-β) treatment [206]. This data set is also an example of paired data. However, unlike example data set A, this data set contains 14 *biological replicates*. Each replicate presents two related measurements corresponding to pre- and post-treatment conditions. Here, we wish to identify genes that were differentially expressed in multiple sclerosis following treatment.

Example data set C. Samples were taken from 15 MS patients and 15 age- and sex-matched controls [206]. The expression profiles of 4,132 human genes were monitored by cDNA microarrays. This data set is an example of *unpaired data*. It contains two groups of individuals (MS and Controls), and our goal is to observe whether a gene is differentially expressed between the two groups. Unlike example data set B, there is no inherent relationship between the individuals in the two groups.

Example data set D. This data set is the union of Example data sets B and C and contains three groups: MS, Controls, and IFN-β treatment individuals. This data set is an example of *multi-group data*. Here, we intent to identify genes that are differentially expressed in one or more of these three groups.

Typically, early attempts to analyze differentially-expressed genes simply established a fixed cut-off k and selected those genes whose expression underwent a k-fold change [56, 74, 306]. However, the specification of k was often arbitrary and it did not take into account the overall distribution of the measurements. Several variants of these simple fold change methods have been proposed to fine-tune this approach [178, 242, 243, 276].

When replicates of the samples are available (such as in data sets B, C, and D), researchers can turn to some common statistical tests. For instance, t-test is a standard statistical test for detecting significant change of a variable between repeated measurements in two groups (such as data sets B and C); this can be generalized to multiple groups (such as data set D) via the ANOVA F statistic [257]. Many variants of the t-statistic for microarray analysis have been developed (e.g., [114, 198, 289]). In addition, non-parametric-based statistics are also commonly applied (e.g., [21, 86, 213, 324]).

Regardless of the specific approach used, the significance of the statistical measure must be determined. A microarray data set typically consists of thousands of genes, and the significance test will be carried out for each gene. A drawback of this *multiple testing* is the increased probability of observing a false positive, which rises with the number of statistical tests performed [35]. Therefore, when multiple tests are involved, methods should be applied to correct the significance level of individual tests.

Since this chapter assumes an understanding of statistical approaches, Section 4.2 will start by providing an overview of basic concepts in statistics. Fold change methods, parametric tests, and non-parametric tests will be discussed in Sections 4.3, 4.4, and 4.5, respectively. The problems associated with multiple testing and the available correction methods will be discussed in Section 4.6. Finally, we will briefly introduce ANOVA in Section 4.7.

4.2 Basic Concepts in Statistics

4.2.1 *Statistical Inference*

In the statistical context, the term *population* denotes the entire collection of individuals or objects about which information is desired [76]. Consider the question exemplified by data set B; here, we are interested in identifying genes that are differentially expressed in MS patients following IFN-β treatment. In this case, the population would be the set of gene expression data of all MS patients in the world. It is clearly impossible in practice to treat all MS patients with IFN-β and then measure their gene expression levels after treatment. Rather, we can take a subset, called a *sample*, of the total population. In the case of example data set B, this sample contains 14 patients, which we hope will be representative of the entire MS patient population.

The readout of a microarray experiment can be represented by *random variables*. For example, the expression level of a specific gene i in MS patients before and after IFN-β treatment can be represented by two random variables x_i and y_i, respectively. A random variable does not describe the actual outcome of a particular experiment. Instead, it associates the possible but as-yet-undetermined outcomes with a *probability distribution*. The probability distribution of a random variable can often be characterized by some *parameters*. For example, the mean μ_i of x_i is a parameter of the probability distribution of x_i.

Unfortunately, in most cases, the entire population is not available for analysis, so the actual value of the parameters remains unknown to the experimenter. However, we may gain some insights into a parameter of interest by applying a numerical descriptive measure, called a *statistic*, to the sample. For example, we can calculate the average intensity value $\bar{x}_i$ of gene i of patients before treatment. We intend to estimate the parameter value μ_i through the statistic value $\bar{x}_i$. This generalizing procedure is called *statistical inference*. That is, we hope to generalize our result from the small sample set of 14 patients to the entire population of MS patients.

4.2.2 Hypothesis Test

The problem of determining whether a gene is differentially expressed can be approached by a classical statistical procedure called the *hypothesis test*. To carry out a hypothesis test, we first need to clearly define the problem of interest. For example, we may be particularly interested in gene i from example data set B, and we may expect this gene to be up- or down-regulated after the patient has been treated. We would therefore like to determine whether the observed data support this hypothesis.

After we define the problem, we generate two hypotheses. These two hypotheses should be *mutually exclusive* and *all inclusive* [83]. Hypotheses which are mutually exclusive cannot be both true or both false at the same time. The all-inclusive stipulation means that the union of the hypotheses must encompass all possibilities. One of the postulated hypothesis will be the *null hypothesis*, denoted by H_0, which is a claim about a population characteristic that is initially assumed to be true. The other hypothesis will be the *alternative hypothesis*, denoted by H_a, which is the competing claim. We then consider the evidence (the observed sample data), and we only *reject* the null hypothesis in favor of the competing hypothesis if there is convincing evidence against the null hypothesis [76].

Taking the example of testing whether gene i is differentially expressed, the null hypothesis is $H_0 : \bar{x}_i = \bar{y}_i$ and the alternative hypothesis is $H_a : \bar{x}_i \neq \bar{y}_i$, where $\bar{x}_i$ and $\bar{y}_i$ refer to mean expression levels of two groups of samples, respectively. The null hypothesis H_0 will be rejected in favor of H_a if the observed measurements strongly suggest that H_0 is false. In other words, this outcome will indicate that this gene appears to be differentially expressed. Conversely, if the observed data do not contain such evidence, H_0 will not be rejected. With that outcome, it will be unclear whether or not this gene is differentially expressed.

In practice, DNA microarray experiments often involve great deal of variation and noise in the data. Moreover, the number of available experiments is usually limited. Hypothesis-testing under these conditions is therefore error-prone, with errors falling into two broad categories. A *Type I error* involves the rejection a null hypothesis when it is in fact true; its

Table 4.1   The possible outcomes of hypothesis testing.

Decision	Truth	
	H_0 is true	H_0 is false
H_0 was rejected	false positive (Type I error) α	true positive (correct decision) $1 - \beta$
H_0 was not rejected	true negative (correct decision) $1 - \alpha$	false negative (Type II error) β

counterpart, the *Type II error*, refers to not rejecting H_0 when is in fact false. The probability of a Type I error is usually denoted by α, while the probability of a type II error is denoted by β [76]. Clearly, $1 - \alpha$ corresponds to the probability of "true negatives," while $1 - \beta$ corresponds to the probability of "true positives." In our example, the latter corresponds to the portion of truly up- or down-regulated genes among those reported as differentially expressed. The possible outcomes of hypothesis testing are summarized in Table 4.1 [83].

To control the Type I error, it is common to consider two probability values. The *significance level* is the probability of a Type I error [76]; simply stated, it is the quantity of errors we are prepared to accept in our studies. For instance, in setting $\alpha = 0.1$, we acknowledge that we may be incorrect in a maximum of one out of every ten cases. The *p-value* (sometimes called the *observed significance level*) is the probability, assuming that H_0 is true, of obtaining a test statistic value at least as contradictory to H_0 as what actually resulted [76]. In other words, it is the observed probability of wrongly rejecting a null hypothesis when it is actually true. Small *p*-values suggest that the null hypothesis is unlikely to be true. To be specific, if the *p*-value is smaller than the significance level, the null hypothesis will be rejected. In fact, the *p*-value quantitatively indicates the strength of evidence for rejecting the null hypothesis H_0, providing nuanced input to a decision to reject or not to reject H_0.

To sum up, the procedure of a hypothesis test involves the following steps:

(1) Define the problem and specify the significance level.
(2) Generate the hypotheses.
(3) Choose an appropriate statistic.
(4) Calculate the statistic value based on the observed data.
(5) Calculate the corresponding *p*-value.

(6) Reject or not reject the null hypothesis based on the calculated p-value and the pre-specified significance level.

4.3 Fold Change Methods

4.3.1 *k-fold Change*

In general, the fold change for a gene is calculated as the average expression over all samples in a condition divided by the average expression over all samples in another condition. Using the fold change method, finding genes that are differentially expressed can be done by simply considering those genes which demonstrate a significant change between the experiment samples of particular interest (such as tumor samples) and controls. This approach is most suitable for data sets without biological duplicates (as exemplified by data set A). Typically, an arbitrary threshold such as a two- or three-fold change is chosen, and the difference (in log form) is considered to be significant if it is larger than the threshold (e.g., [56, 74, 306]). To facilitate the selection process, the ratio between the two expression levels for each gene is first calculated. Since most genes in a typical microarray experiment do not change, the ratios between experiment samples of particular interest and controls (will be referred to as *experiment/control ratios* in the following) of most genes will be around one, and their logs will be around zero.

The experiment/control ratios can be plotted into a histogram (this pro-

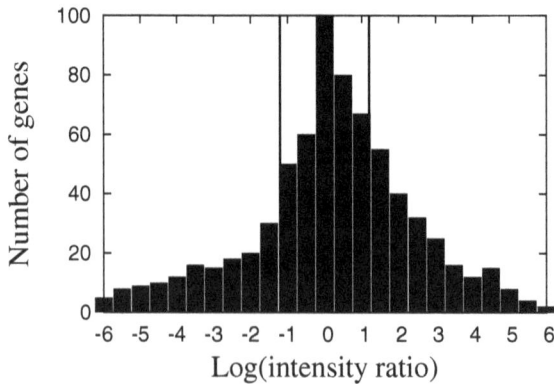

Fig. 4.1   Histogram of log ratios and selection of genes with 2-fold change ($\log_2 2 = 1$).

cedure will be described in Section 8.2.2). The horizontal axis of such a plot represents the log ratio values. Using this histogram, selecting differentially-regulated genes based on fold change corresponds to setting thresholds (vertical bars) at the desired minimum fold change and selecting the genes in the tails of the histogram [82]. For instance, to select genes that have a fold change of two, we place the threshold bars at $+/-1$ ($\log_2 2 = 1$) and select the genes outside the vertical bars (Figure 4.1).

Alternatively, the log expression levels of the experiment *vs.* the control can be plotted in a *scatter plot* (Figure 4.2(a)); further details are provided in Section 8.2.3. In a scatter plot, genes with a four-fold change will be at a distance of at least two units from the diagonal $y = x$. Therefore, given a threshold τ, the fold change method reduces to drawing lines parallel to the diagonal at a distance $\pm\tau$ and selecting the genes outside the lines. In a *ratio-intensity* plot (see Section 8.2.3 for detailed description), the horizontal line $y = 0$ corresponds to unchanged genes (Figure 4.2(b)), and the genes outside the boundaries $y = \pm\tau$ will be selected [83].

4.3.2 Unusual Ratios

The fold change approach simply selects those genes that exhibit the greatest change between the control and experiment samples. Although uncomplicated and intuitive, the naïve k-fold change method described above is frequently ineffective, since the value of k is usually chosen arbitrarily and may often be inappropriate [82]. An alternative method, called *unusual ratios*, considers the distribution of measurements within the data. Instead of blindly specifying the value of k, this method involves selecting those genes with experiment-to-control ratios at a specified distance from the mean experiment-to-control ratio (e.g., [242, 243, 276]). For example, this distance can be taken to be $\pm 1.5\sigma$ where σ is the standard deviation of the ratio distribution.

In practice, selecting genes $\pm 1.5\sigma$ away from the mean can be accomplished by standardizing the ratios and plotting them in a histogram (see Section 3.6.1 for a discussion of data standardization). Since the standardized data will have a mean of zero and a standard deviation of one, a histogram of the standardized values will be centered around zero, and the units on the horizontal axis will represent the standard deviation (Figure 4.3). Therefore, setting thresholds at $\pm 1.5\sigma$ will correspond to selecting those genes outside the vertical bars in Figure 4.3 [82].

Compared with the k-fold change method, this method has the advan-

(a)

(b)

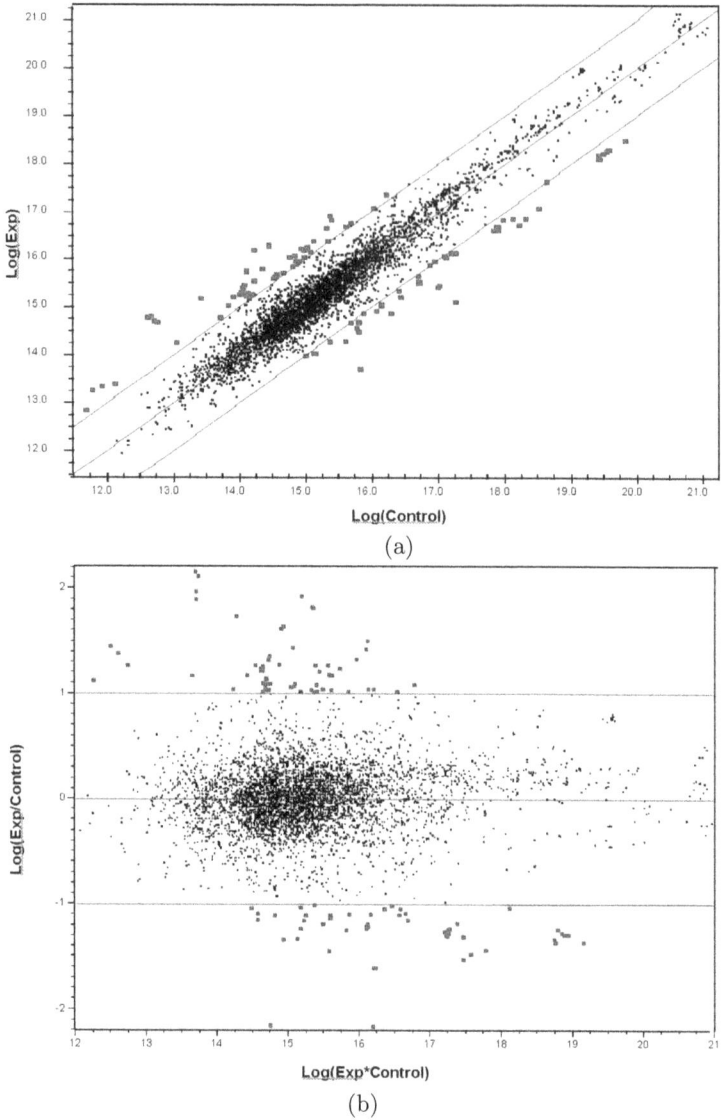

Fig. 4.2   (a) An example of the scatter plot. (b) An example of the R-I plot. Figures are from Kerry Bemis, Indiana Centers for Applied Protein Sciences.

tage of automatically adjusting the cut-off threshold. That is, the thresholds determined by this method are dependent on the distribution of all ratios in the given data set, allowing a more tailored selection than the

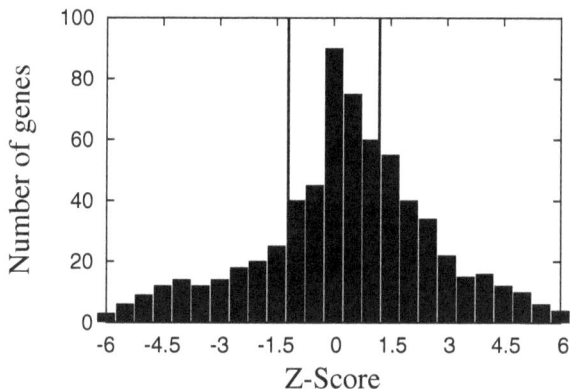

Fig. 4.3   Histogram of standardized log ratios and selection of genes with unusual ratios ($\pm 1.5\sigma$).

uninformed choice of a fixed threshold. However, this method also has an intrinsic drawback, in that the top k-percent of most-affected genes will always be selected, regardless of the number of genes regulated or the extent of regulation [82, 83]. The value of k is related to the standard deviation threshold specified by the user. For example, if the ratio is close to a normal distribution and the thresholds are set at $\pm 2\sigma$, the method will select 4.6% of the data set as the most regulated genes. This particular percentage arises because, under the assumption of normal distribution, the probability is $P(Z < -2) + P(Z > 2) = 0.0228 + 0.0228 = 0.0456$.

Given these assumptions, the unusual ratios method will always report 4.6% of the set of genes as differentially expressed even if the set actually contains a greater or lesser proportion of truly-regulated genes. For example, a data set containing no differentially-regulated genes will still return a 4.6% result. All microarray data sets exhibit a certain amount of variability due to noise, and this variability will be interpreted by the method as differential expression, rather than the product of chance. The method may also under-report the incidence of differential expression. Given the conditions mentioned above, this method will select only 4.6% of the set as differentially regulated, even if many more genes should have been included. In summary, the unusual ratio method uses a fixed-proportion threshold that will always report the same proportion of the genes as being differentially regulated [82, 83], regardless of the real underlying distribution.

4.3.3   *Model-Based Methods*

In this subsection, we will describe a model-based approach [178] for the selection of differentially-expressed genes which has the potential to overcome
some of the issues discussed above. To simplify explanation, we will first
apply the approach to the basic task of distinguishing expressed from non-
expressed genes. We will later generalize this approach to the identification
of up-regulated, down-regulated, or unchanged genes. In the model-based
approach, two events are considered: E_g represents the event that gene g
is expressed while $\overline{E}_g$ represents event that g is unexpressed. Let p denote
the prior probability of E_g, then $1 - p$ is the prior probability of $\overline{E}_g$.

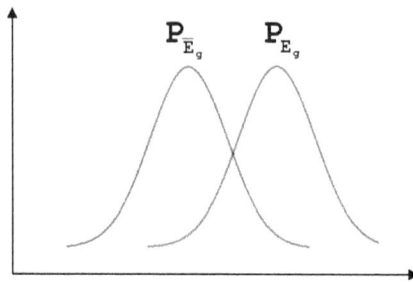

Fig. 4.4   Probability distributions of p_{E_g} and $p_{\overline{E}_g}$. Both distributions are assumed to
be normal with equal variance.

The model-based method assumes that the expressed genes and the
unexpressed genes follow two *probability distributions*, respectively: the expressed genes are associated with p_{E_g}, while the unexpressed genes are
associated with $p_{\overline{E}_g}$. An observed expression y may rise from either of
the two distributions. If y belongs to an expressed gene, it follows p_{E_g},
otherwise, it follows $p_{\overline{E}_g}$. This is illustrated in Figure 4.4, where the horizontal axis represents the intensity values and the vertical axis represents
the probabilities. The curve to the left represents $p_{\overline{E}_g}$, and the curve to the
right represents p_{E_g}.

Our purpose is to determine whether the observed ratio y arises from
an expressed gene or an unexpressed gene. To make this determination,
we estimate the likelihood that gene g is expressed given that the observed
ratio of g is y; i.e., we calculate the conditional probability $P(E_g|Y_g = y)$.
Clearly, the probability that g is unexpressed given y is $P(\overline{E}_g|Y_g = y) =$
$1 - P(E_g|Y_g = y)$. From Bayes theorem, the conditional probability can be

expressed as:

$$Pr(E_g|Y_g = y) = \frac{p \cdot p_{E_g}(y)}{p \cdot p_{E_g}(y) + (1-p) \cdot p_{\overline{E}_g}(y)}. \qquad (4.1)$$

We must now estimate the prior probability p and the distributions p_{E_g} and $p_{\overline{E}_g}$. To simplify the problem, we assume that p_{E_g} and $p_{\overline{E}_g}$ are normal distributions with equal variance. In this case, the mixture model can be completely characterized by four parameters, p, σ, μ_{E_g} and $\mu_{\overline{E}_g}$. These parameters can be estimated by a maximum likelihood approach, called the *EM algorithm* [69]. The EM algorithm (see 5.3.4) searches various combinations of the parameters and converges to a local maximum-likelihood parameter setting.

A more complex and realistic situation involves a data set with genes that are up-regulated, down-regulated, and unchanged. The mixture model can be expressed as:

$$p_g(y) = p_1 \cdot p_{Up_g}(y) + p_2 \cdot p_{Down_g}(y) + p_3 \cdot p_{Unchanged_g}(y).$$

Similarly, the probability distributions of up-regulated genes, down-regulated genes, and unchanged genes can be modeled by the normal distribution, and the EM algorithm can be applied to estimate the model parameters.

Draghici [83] noted that the model-based method offers a number of advantages over the fold change and unusual ratio approaches discussed previously. Here, the maximum-likelihood estimators (MLE) become unbiased minimum-variance estimators as the sample size increases; in addition, the likelihood functions can be used to test hypotheses about models and parameters. However, the disadvantage of the maximum-likelihood estimate approach is that the results quickly become unreliable as the sample size decreases. Indeed, for small samples, the estimates may be grossly different from the real underlying distribution. Moreover, MLE estimates can become unreliable when the data deviate considerably from normality [83].

It should be noted that the model-based, fold change, and unusual ratio approaches are all best suited to data sets without replication (such as example data set A). For data sets with replication, the tests discussed in the following sections are usually more appropriate.

4.4 Parametric Tests

The classical method for performing an hypothesis test on data of the type exemplified by data sets B and C is the t-test. This test was developed by W. S. Gossett [1876-1937] and was originally termed the "student's t-test." Two versions of this test, the *paired t-test* and *unpaired t-test*, are applicable to data sets containing two groups of observations. In addition, several variants on the classical t-statistic have been proposed.

4.4.1 Paired t-Test

The *paired t-test* is applicable to paired data; e.g., data sets in which each data point has a pair of observations. For instance, example data set B contains a pair of measurements for each patient, one prior to and one subsequent to treatment. The pair of measurements are often combined to generate a single log ratio. If the measured expression value before treatment is x_1 and the value after treatment is x_2, the combined ratio is $\log_2(\frac{x_1}{x_2})$. In this situation, the observed measurements for each gene form one vector of log ratios, and the paired t-test is reduced to a *one-sample t-test*. Here, the null hypothesis is that the gene is not differentially expressed, or the mean of the log ratios μ equals to 0, denoted by $H_0 : \mu = 0$. From the observed log ratios, we can use the following formula to calculate the *t-statistic*:

$$t = \frac{\bar{x}}{s/\sqrt{n}}, \tag{4.2}$$

where $\bar{x}$ is the average of the log ratios, s is the standard deviation, and n is the number of the patients in the experiment. A p-value can then be obtained by looking up a t-distribution with $n - 1$ *degrees of freedom*. Finally, the null hypothesis is rejected or not rejected based on the p-value and a pre-specified significance level.

The t-test is more sophisticated than the fold change methods. The significance of differentially-expressed genes depends not only on the average log ratio but also on both the population variability and the number of individuals in the study [266]. In general, the accuracy of the determination of differentially-expressed genes increases with the number of individuals in the experiment.

The t-test should be clearly understood as different from the unusual ratio method. In the latter, the entire set of genes in a *single* microarray is regarded as the sample set, and the most-changed genes in this sample set are considered to be differentially expressed. In contrast, the t-test takes the group of all patients as its sample set. A single t-test observation of expression level from one array may not yield reliable results for a specific gene, since perceived variability in that gene may actually arise from experimental noise. However, a conclusion based on the application of a t-test to multiple patients (biological replicates) may often be more reliable than the results of the unusual ratio method.

4.4.2 *Unpaired t-Test*

The *unpaired t-test* can be applied to unpaired data such as example data set C, where there are two unrelated groups of patients. Here, the null hypothesis states that the means of the expression levels of a given gene in the two samples will be equal; i.e., $H_0 : \mu_1 = \mu_2$ or $H_0 : \mu_1 - \mu_2 = 0$. Unpaired t-tests may be *equal-variance* and *unequal-variance*. As suggested by the names, the equal-variance t-test assumes that the two samples are taken from distributions with equal variances, while the unequal-variance test assumes that the two distributions have different variances.

Both the equal-variance and unequal-variance unpaired t-test use the following formula:

$$t = \frac{\overline{x}_1 - \overline{x}_2}{\sqrt{\frac{s_1^2}{n_1} + \frac{s_2^2}{n_2}}}, \tag{4.3}$$

where $\overline{x}_1$ and $\overline{x}_2$ are the means, s_1^2 and s_2^2 are the variances, and n_1 and n_2 are the sizes of the two groups, respectively.

Since the number of samples in a microarray experiment is often limited, the following *pooled sample variance* is often used to estimate of the sample variability in an equal-variance test,

$$s_1^2 = s_2^2 = s_p^2 = \frac{(n_1 - 1) \cdot s_1^2 + (n_2 - 1) \cdot s_2^2}{n_1 + n_2 - 2}.$$

Then the t-statistic for equal variance can be rewritten as:

$$t = \frac{\overline{x}_1 - \overline{x}_2}{\sqrt{s_p^2(\frac{1}{n_1} + \frac{1}{n_2})}}. \tag{4.4}$$

The degrees of freedom are given by the number of measurements minus the number of intermediate values we need to calculate. In Formula 4.4, there are $n_1 + n_2$ measurements and two intermediate values s_1^2 and s_2^2, so the number of degrees of freedom is $n_1 + n_2 - 2$.

In the case of unequal variance, the degrees of freedom need to be adjusted as:

$$\nu = \frac{(\frac{s_1^2}{n_1} + \frac{s_2^2}{n_2})^2}{\frac{(\frac{s_1^2}{n_1})^2}{n_1-1} + \frac{(\frac{s_2^2}{n_2})^2}{n_2-1}}. \tag{4.5}$$

The value of Formula 4.5 is usually not an integer and thus need to be truncated (rounded down) [76].

Now we must determine whether the variances in the two distributions are equal. This can be established through the use of another hypothesis test, where the hypotheses are $H_0' : \sigma_1^2 = \sigma_2^2$ and $H_a' : \sigma_1^2 \neq \sigma_2^2$. To test the null hypothesis, the F statistic can be used, as follows:

$$F = \frac{s_1^2}{s_2^2}. \tag{4.6}$$

A p-value can then be obtained on the basis of F-statistic distribution with respect to the degrees of freedom ($\nu_1 = n_1 - 1$ and $\nu_2 = n_2 - 1$); this will indicate whether H_0' should be rejected.

In summary, the unpaired t-test procedure is as follows:

(1) Calculate the F statistic using Equation 4.6 and determine whether H_0' should be rejected.
(2) If H_0' is rejected, use Equation 4.3 to calculate the t-statistic; otherwise, use Equation 4.4.
(3) Obtain the p-value on the basis of the t-statistic and the corresponding degrees of freedom to determine whether H_0 should be rejected.

4.4.3 *Variants of t-Test*

In addition to the classical t-statistics discussed above, several simplified forms are also available for the identification of differentially-expressed genes. For example, Golub et al. [114] have proposed a method called *neighborhood analysis*. To be specific, given the expression levels of gene g over all the experimental conditions, the following score is calculated:

$$P(g) = \frac{\mu_1(g) - \mu_2(g)}{\sigma_1(g) + \sigma_2(g)}, \tag{4.7}$$

where $\mu_1(g)$ and $\mu_2(g)$ are the means of the expression levels of gene g in classes 1 and 2, respectively, and $\sigma_1(g)$ and $\sigma_2(g)$ are the standard deviations of g in classes 1 and 2, respectively. Large values of $|P(g)|$ indicate a strong correlation between gene expression and class distinction, while a positive or negative $P(g)$ indicates that g is more highly expressed in class 1 or class 2, respectively. In [114], the 6,817 genes in the AML/ALL data set are sorted by their correlation value. The 25 most positively-correlated and the 25 most negatively-correlated genes are selected as the differentially-expressed genes.

In another example of a t-test variant, Pavlidis et al. [214] have adapted the *Fisher's discriminant criterion* (FDC) to define a score as:

$$F(g) = \frac{(\mu_1(g) - \mu_2(g))^2}{(\sigma_1^2(g) + \sigma_2^2(g))^2}. \tag{4.8}$$

Genes with higher score values are selected as differentially expressed genes. Equations 4.7 and 4.8 are similar to the t-statistic formula and considered to be variants of that method.

4.5 Non-Parametric Tests

The t-statistic and its variants start from the assumption that the data will follow a normal distribution. However, if this is not the case, the value of the these statistics may not represent the true degree of differential expression. As a result, using p-values obtained from the t-distribution as a test of gene expression may be meaningless in these instances. In fact, there are many sources of variability in a microarray experiment, and outliers are frequent. Thus, the distribution of intensities of many genes may not be normal in a real data set [70].

In this section, we will describe several *non-parametric methods* which do not place any assumptions on the observed data. These non-parametric methods do not rely on the estimation of parameters (such as the mean or the standard deviation) in describing the distribution of the variable of interest in the population.

4.5.1 *Classical Non-Parametric Statistics*

There are non-parametric equivalents of both the paired and unpaired t-tests. The *Wilcoxon sign-rank test* is the non-parametric equivalent of the paired t-test, while the the *Wilcoxon rank-sum test* (also called *Mann-Whitney test*) is the non-parametric equivalent of the unpaired t-tests [266]. As we have noted in the previous discussion, the unpaired t-test is actually a generalization of the paired case. Therefore, we will focus here on the Wilcoxon rank-sum test.

The Wilcoxon rank-sum test [305] organizes the observed data in value-ascending order. Each data item is assigned a rank corresponding to its place in the sorted list. These ranks, rather than the original observed values, are then used in the subsequent analysis. The major steps in applying the Wilcoxon rank-sum test are as follows:

(1) Merge all observations from the two classes and rank them in value-ascending order.
(2) Calculate the Wilcoxon statistics by adding all the ranks associated with the observations from the class with a smaller number of observations.
(3) Find the p-value associated with the Wilcoxon statistic from the Wilcoxon rank sum distribution table (see [141]).

The use of rank-based tests of this type is appropriate when the underlying distribution is far from normal. Moreover, the rank-sum test is much less sensitive to outliers and noise, typical characteristics of gene-expression data sets, than are parametric tests [70, 266]. An outlier will change the t-statistic value greatly but will have little impact on ranking. An example of the use of this type of test is found in [47], where Chambers et al. have applied the Wilcoxon sign-rank test to analyze the microarray data from a study of human cytomegalovirus infection.

Counterbalancing these benefits is the relative lack of sensitivity of the rank-sum tests in comparison with their parametric counterparts. Rank-sum p-values tend to be higher, increasing the difficulty of detecting real differences as statistically significant [10, 266]. If the sample sizes are large, the difference in sensitivity is minor. With the small sample sizes of typical microarray experiments, however, non-parametric tests have very little power to detect real differences in expression [10].

4.5.2   *Other Non-Parametric Statistics*

In addition to the classical rank-sum test, several other non-parametric statistics have also been proposed. For example, Ben-Dor et al. [21] use a *threshold number of misclassification* or TNoM score to select differentially-expressed genes. This method assumes that a differentially-expressed gene will exhibit significantly different values in the two classes and that the values can therefore be differentiated by a threshold number. Gene values which are more clearly separated by this threshold are more likely to arise from an up- or down-regulated gene. Given the expression values $\vec{g}$ of gene g over all the experimental conditions, the TNoM score is defined as follows:

$$TNoM(\vec{g}) = \min_{d,t} \sum_i 1\{l_i \neq \text{sign}(d \cdot (g_i - t))\}, \qquad (4.9)$$

where g_i is the expression level of g in the i-th experimental condition, l_i is the class label of the i-th condition. $d \in \{+1, -1\}$ is used to indicate the class label, and t is a threshold to separate the expression values of g. The term $\text{sign}(d \cdot (g_i - t))$ is called a "decision stump" which indicates the predicted class label based on g_i, d, and t. The rationale is that the sign of $d \cdot (g_i - t)$ is dependent on whether the expression level of gene g in condition i is greater than the threshold value t.

Equation 4.9 seeks the best decision stump for a given gene and then counts the classification errors this decision stump makes in differentiating known class labels. Fewer errors indicate that the threshold is more successful in differentiating the two classes, and, in turn, that it is more likely that the gene is up- or down-regulated. Like the classical non-parametric statistics, this method does not rely on any assumptions regarding the observed data.

4.5.3   *Bootstrap Analysis*

As discussed previously, a typical hypothesis test involves two major steps. First, an appropriate statistic is chosen and its value calculated from the observed data. Second, the p-value of the statistic is derived and a determination to reject or not to reject the null hypothesis is made on that basis. The first step of bootstrap analysis is similar to any of the classical parametric tests; for example, this analysis can begin with a calculation of the t-statistic. The major innovation is found in the second step. Rather than determining the p-value on the basis of the standard t-distribution (which is tabulated under the assumption of normal distribution), bootstrap analysis uses a resampling strategy to approximate the real distribution of the t-statistic.

In more detail, the bootstrap method constructs a large number of random data sets by resampling from the original data. That is, each data entry x_{ij} (the measurement of gene i under experimental condition j) is randomly assigned one of the measurements from the data set. The resulting data sets resemble the original data in their values. However, the correlation between genes and samples in the original data is completely disturbed through the randomization procedure.

The next step of the bootstrap method involves calculating the t-statistics for all the genes in each random data set and using the standard t-distribution to find the minimum p-value among the genes. The outcome of this process is an adjusted p-value for the original data set; this value is the proportion of resampled data sets in which the minimum p-value is less than or equal to the p-value of the real data set. In more formal terms, let p_i be the p-value of the i-th gene in the original data set, $p_j^{(b)}$ be the p-value of the j-th gene in the b-th random data set, and $\tilde{p}^{(b)} = min\{p_j^{(b)}\}$ be the minimum p-value in the b-th random data set. Then the adjusted p-value for the i-th gene in the real data set is

$$p_i' = \frac{\text{number of random data sets with } \tilde{p}^{(b)} \leq p_i}{\text{total number of random data sets}}.$$

Bootstrap analysis is based on the concept that, if the null hypothesis were true, then the real (observed) data set would exhibit characteristics similar to any of the randomized data sets. In other words, the value of the selected statistic T (and thus the p-value) calculated from the real data would appear as a typical value in the distribution of T (and p-value) from the randomized data sets. Conversely, if the value of T (and the p-value) from real data is "significantly abnormal," then we may be confident that the observed data are not formed by chance, and the null hypothesis should be rejected.

When compared to the other methods discussed in this chapter, bootstrap analysis has several significant advantages. As a non-parametric test, bootstrap analysis does not require that the data be normally distributed and is thus robust to noise and artifacts. Furthermore, bootstrap analysis is more sensitive and accurate than the classical non-parametric tests, as it is able to take into account the errors arising from "multiple testing" (discussed in detail in the next section). Finally, this method can be used with any statistical measure. That is, we can choose any statistic and evaluate its p-value using the resampling strategy. Therefore, bootstrap analysis is more appropriate for use with microarray data than either the t-test or classical non-parametric tests.

4.6 Multiple Testing

To select differentially-expressed genes, we usually apply the hypothesis test gene by gene. In practice, a microarray experiment typically involves thousands of genes. This means we have to repeatedly run the test for thousands of times. A problem with doing so many tests is that the number of false positives may be increased, a phenomenon called *multiple testing* in statistics.

Recall in a hypothesis test, a significance level α is usually specified before the test and the null hypothesis will be rejected if the p-value is smaller than α. Now consider we are facing thousands of genes. Even purely by chance, there could be some genes whose p-values appear to be smaller than α, but they are in fact *not* differentially expressed. In other words, we could make the Type I error and report false positives due to random effects. According to the definition of significance level (see Section 4.2.1), the probability of committing a Type I error is exactly α. Then the probability of not making a Type I error would be $1 - \alpha$.

Suppose we have N genes in the data set, the probability of make correct decisions for all genes is:

$$\text{Prob(globally correct)} = (1 - \alpha)^N, \tag{4.10}$$

and the probability of making at least one mistake is:

$$\text{Prob(wrong somewhere)} = 1 - (1 - \alpha)^N. \tag{4.11}$$

When α is small, the expected number of false positives is αN. For a very large N, the number of false positives may be large.

A possible approach to alleviating this problem is to control the *global* significance level, also known as the *family-wise error rate* (FWER). However, the FWER methods are often too conservative and result in too many false negatives. An alternative approach seeks to control the *false discovery rate* (FDR). The basic idea of this approach is to control the proportion of significant results that are in fact Type I errors. However, FDR assumes all the genes in the microarray experiment are independent, which is usually not true in reality. Instead, the *permutation-based correction* takes into consideration the possible correlations among genes and adjust the p-value based on the resampling theory. In this section, we will discuss these three approaches in detail. Moreover, we will present the *significance analysis of microarray data* (SAM) [289], which was specifically designed for microarray data. SAM integrates several features of FDR and permutation-based correction, and has become a very popular method for the identification of differentially expressed genes.

4.6.1   Family-Wise Error Rate

Before we describe the methods for controlling the error rate of multiple tests, let us first define several notations. Let N be the number of statistical tests performed, and $p_1, \ldots, p_N$ be the observed p-values for the individual tests. Suppose in R of the N tests, the individual null hypotheses are rejected, among which an unknown number of V decisions are actually false positives. We use α_s to denote the significance level specified for a single test; the null hypothesis will be rejected for the i-th test if $p_i < \alpha_s$. We use α_a to denote the probability of committing at least one false positive among all the hypotheses tested, i.e., Prob$(V > 0)$. Usually, the global-wise error rate α_a is also called the *family-wise error rate* (FWER) and the gene-wise error rate α_s is also called the *per-comparison error rate* (PCER) . As explained before, since the PCER tends to cause too many false positives when N increases, in general, most conventional multiplicity adjustments attempt to control the FWER.

4.6.1.1   *Šidák correction and Bonferroni correction*

The Šidák correction and Bonferroni correction consider the same question: to achieve a given global significance level α_a, what value of a gene-wise significance level α_s should be specified. According to our denotations, Formula 4.11 can be rewritten as

$$\alpha_a = 1 - (1 - \alpha_s)^N \tag{4.12}$$

or

$$\alpha_s = 1 - \sqrt[N]{1 - \alpha_a}. \tag{4.13}$$

This Formula is the *Šidák correction* for multiple comparisons [48]. It means that if we want to achieve a global significance level of α_a, we can set the significance level for each test as $1 - \sqrt[N]{1 - \alpha_a}$.

The *Bonferroni correction* [35, 36] approximates Formula 4.12 using the first two terms of the binomial expansion of $(1 - \alpha)^N$:

$$\alpha_a = 1 - (1 - \alpha_s)^N = 1 - (1 - N \cdot \alpha_s + \cdots) \approx N \cdot \alpha_s.$$

When α_s is small, $N \cdot \alpha_s$ approximates $1 - (1 - \alpha_a)^N$ well. In other words, when α_s is small, we need to set the significance level for each test as α_a/N (a simpler formula than that of Šidák correction) if we want to achieve a global significance level of α_a.

4.6.1.2 *Holm's step-wise correction*

While the Šidák correction and the Bonferroni correction are effective to avoid too many false positives, they may go to the other extremity: they may be too stringent to result in any positives at all. In our case, we may get none genes reported as up- or down-regulated. An alternative approach is Holm's sequential p-value adjustment [142], which takes the order of the observed p-values into account and adjusts more on smaller p-values than on larger p-values.

To be specific, Holm's step-wise correction procedure is as follows [142]:

(1) Choose the global significance level α_a.
(2) Order the genes according to their p-values in the ascending order.
(3) Compare the p-value (p_i) of the i-th gene in the ordered list with threshold $\tau_i = \frac{\alpha_a}{N-i+1}$.
(4) Report genes $1, \ldots, k$ as differentially expressed genes at the chosen α_a significance level, where $k = max_i\{p_i < \tau_i\}$ (the largest i for which $p_i < \frac{\alpha_a}{N-i+1}$).

4.6.2 *False Discovery Rate*

Although the Holm's step-wise correction is more relaxed than the Šidák correction and Bonferroni correction, it may still be too stringent and may result in too many false negatives. The reason is that all the methods in the previous subsection seek to control the probability of committing any single Type I error among all the N tests.

An alternative approach was proposed by Benjamini and Hochberg [23] to control the *false discovery rate* (FDR) instead of the FWER. To be specific, the FDR is defined as the expected proportion of false positives among the positive findings [10]:

$$FDR = E[\frac{V}{R}|R > 0] \cdot \text{Prob}[R > 0].$$

As a special case, the FDR is equal to the FWER if all the null hypotheses were true. However, this rarely happens in reality. In general, the more the number of hypotheses that are truly false, the smaller is the FDR. Therefore, control of FDR tends to be more relaxed than control of FWER at the same level of significance.

Similar to Holm's correction, the FDR correction is also a step-wise procedure as follows [23]:

(1) Choose the global significance level α_a.
(2) Order the genes according to their p-values in the ascending order.
(3) Compare the p-value (p_i) of the i-th gene in the ordered list with threshold $\tau_i = \frac{i}{N}\alpha_a$.
(4) Report those genes as differentially expressed genes if $p_i < \tau_i$.

Storey and Tibshirani [267] noted that an adjustment is only necessary when there are positive findings, i.e., there are cases when the null hypotheses are rejected. They proposed a modified version of the FDR, called the *positive discovery rate* (pFDR):

$$\text{pFDR} = E[\frac{V}{R}|R > 0].$$

Recall R and V are the number of positive findings and the number of false positives, respectively. Since the number of false positives is unknown, we have to estimate V in order to estimate the pFDR. This can be done through a permutation procedure, i.e., constructing B permutated data sets by changing the class labels of samples. Suppose an average number R^* of genes having the p-values smaller than threshold α_s over the B permutated data sets. We can assume there are no true positives in any permutated data set, and thus expect the number of false positives is R^*. Then a natural estimate of the pFDR is:

$$pFDR = \frac{R^*}{R}.$$

For more details of the pFDR, please refer to [267]. In Section 4.6.4, we will use SAM to illustrate the application of this method to estimate the false positive rate.

4.6.3   *Permutation Correction*

The FWER and FDR approaches control the global rate of false positives from different perspectives. However, neither of the approaches consider the possible correlation among data objects. For microarray data analysis, this problem is particularly important since genes are often highly correlated. For example, a group of genes may participate in the same pathway. In [304], Westfall and Young propose a step-down correction (the W-Y approach) that adjusts the p-value with the consideration of the possible correlation.

This correction procedure starts by permutating the class labels of samples. For each permutation, the p-values with respect to a chosen statistic, such as the t-statistic, of all genes are calculated and corrected using Holm's step-wise method discussed above. The permutation is repeated for a sufficiently large number of times. After the permutation procedure, the corrected p-value for a gene i will be the proportion of the times when the p-value for the real data set is greater than or equal to the p-value for a permutation:

$$p\text{-value for gene } i = \frac{\text{number of permutations for which } p_i^{(b)} \leq p_i}{\text{total number of permutations}},$$

where $p_i^{(b)}$ is the corrected p-value by Holm's step-wise method for permutation b, and p_i is the value of t-statistic for the real data set.

The main important advantage of the W-Y approach is that it fully takes into consideration all dependencies among genes. Disadvantages include the fact that it is an empirical process lacking the elegance of a more theoretical approach. Moreover, the label permutation process is extremely computationally intensive and thus prohibitively slow [83]. More details and an example of applying this method to microarray data can be found in [87]. In fact, this method is a refinement of the bootstrap analysis described before (see Section 4.5.3). The major difference between bootstrap and the W-Y method here is that W-Y method only permutates the *labels* while bootstrapping methods replace the values.

4.6.4 SAM: Significance Analysis of Microarrays

Tusher et al. [289] reviewed several approaches to adjusting p-values for multiple testing. They reported that the classical Bonferroni correction (see Section 4.6.1.1) was too stringent to find any differentially expressed genes. Although the permutation-based W-Y approach (see Section 4.6.3) defines "weak control" of the error rate and considers the correlation among genes, it is still too stringent for their data. The method of Benjamini and Hochberg [23] (see Section 4.6.2) relaxes the FWER and guarantees an upper bound for the FDR. However, this method assumes independent tests and the results tend to be too "granular" due to the limited number of permutations. In their experiment, they identified either zero or 300 significant genes depending on how the p-values was corrected.

To address the above challenges, Tusher et al. [289] proposed the Significance Analysis of Microarrays (SAM). Basically, SAM assigns a score to each gene according to its change in gene expression. Genes with scores greater than a threshold are considered as "potentially" significant. To control the false positives, SAM uses permutation of measurements to estimate the false discovery rate (pFDR) (see Section 4.6.2). The score threshold for genes is then adjusted iteratively according to the pFDR until a set of significant genes have been identified.

The SAM's score is very similar to the t-statistic:

$$d(i) = \frac{\overline{x}_1(i) - \overline{x}_2(i)}{s(i) + s_0} , \qquad (4.14)$$

where $\overline{x}_1(i)$ and $\overline{x}_2(i)$ are defined as the average levels of expression for gene i in classes 1 and 2, respectively. SAM assumes equal variance in groups 1 and 2 and applies the pooled variance to estimate the standard deviation $s(i)$:

$$s(i) = \sqrt{(\frac{\frac{1}{n_1} + \frac{1}{n_2}}{n_1 + n_2 - 2})(\sum_p (x_p(i) - \overline{x}_1(i))^2 \sum_q (x_q(i) - \overline{x}_2(i))^2)} ,$$

where $\sum_p$ and $\sum_q$ are the variances of the expression measurements in classes 1 and 2, respectively, and n_1 and n_2 are the numbers of measurements in classes 1 and 2, respectively.

Compared with the standard t-statistic, the SAM's score adds a "fudge" term, s_0, to the denominator. The rationale is that the variance $s(i)$ tends to be smaller at lower expression levels, which in turn, makes $d(i)$ dependent on the expression levels. However, to compare values of $d(i)$ across all genes, the distribution of $d(i)$ should be independent of the expression levels. This problem is especially troubling for microarray data since the number of samples in microarray experiments are typically limited and small standard errors can occur purely by chance.

To address this problem, SAM seeks to find a s_0 such that the dependence of $d(i)$ on $s(i)$ is as small as possible. In practice, this can be achieved by studying the relationship of $d(i)$ versus $s(i)$ using a sliding window across the data. An appropriate value of s_0 will be picked up such that the coefficient of variation of $d(i)$ is approximately constant as a function of $s(i)$. For details of implementation, please refer to the documentation accompanying the software package, SAM, at http://www-stat.stanford.edu/~tibs/SAM/.

Once the "fudge" term s_0 is estimated, the SAM's score $d(i)$ is applied for each gene. Now the problem is how to determine the threshold for significant genes and the expectation of false positives. SAM applies a permutation process to address this problem. This process proceeds by permuting the columns of the given data matrix, X, and assigning the first n_1 columns to class 1 and the remaining n_2 columns to class 2. A total number of B such permutations will be performed.

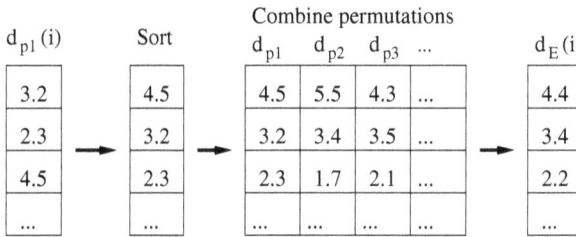

$d_{p1}(i)$	Sort	Combine permutations d_{p1} d_{p2} d_{p3} ...				$d_E(i)$
3.2	4.5	4.5	5.5	4.3	...	4.4
2.3	3.2	3.2	3.4	3.5	...	3.4
4.5	2.3	2.3	1.7	2.1	...	2.2
...	...	...	...	...	...	...

Fig. 4.5   An example of estimating the expected $d_E(i)$ from permutated data sets (Example is adapted from http://www.math.tau.ac.il/~nin/Courses/BioInfo04/SAM.ppt).

After the permutation process, SAM sorts the $d(i)$ values of the original data set in the descending order: $d(1) \geq d(2) \geq ... \geq d(N)$. Then for each permutated data set b, SAM also calculates and orders the scores: $d_b(1) \geq d_b(2) \geq ... \geq d_b(N)$. The expected order statistics $d_E(i)$ can be estimated by $d_E(i) = \sum_{b=1}^{B} d_b(i)/B$. For example, in Figure 4.5, the second column is the ordered scores of $d(i)$ for the original data set. The array shows the ordered scores of $d_b(i)$ for the permutated data sets, with each column corresponding to one permutation. The last column gives the expected score $d_E(i)$ by averaging the rows of the array. Intuitively, if a gene i has the SAM's score $d(i)$ substantially greater than its expected value $d_E(i)$, this gene is potentially significant. To identify these potential genes, $d(i)$ is usually plotted versus its expected values $d_E(i)$ (Figure 4.6). The solid line in the figure indicates the line for $d(i) = d_E(i)$, where the observed relative difference is identical to the expected relative difference. The dotted lines are drawn at a distance $\Delta = 1.2$ from the solid line.

Given a specific Δ, the procedure to declare significance is as follows: find the smallest $d(u)$ (the "upper cut point") such that $d(u) - d_E(u) \geq \Delta$,

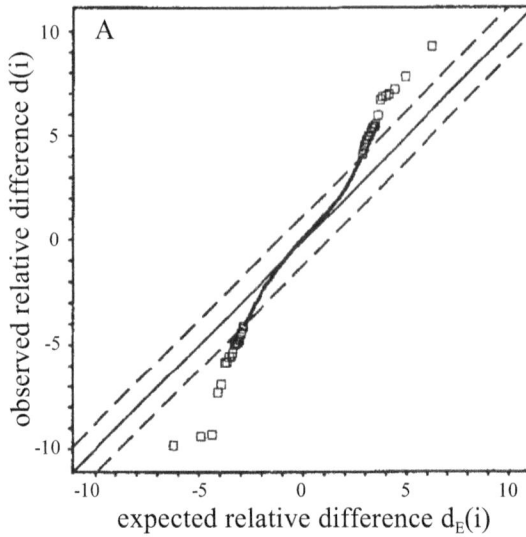

Fig. 4.6 The scatter plot of $d(i)$ vs. $d_E(i)$ to select potential significant genes (Figure is from [289]).

and report all the genes j such that $d(j) \geq d(u)$ as "significant positives;" similarly, find the largest $d(l)$ (called the "lower cut point") such that $d(l) - d_E(l) \leq -\Delta$, and report all the genes k such that $d(k) \leq d(l)$ as "significant negatives".

	d(i)
	8.6
t_1	4.6
	2.3
t_2	−0.6
	−1.2

$d_p(i)$			
4.5	5.5	4.3	7.9
3.2	4.4	3.5	1.6
2.3	2.7	2.1	0.1
1.1	0.4	1.0	−0.2
−0.8	−0.1	−1.0	−0.9

Fig. 4.7 An example of calculating the pFDR, which is adapted from http://www.math. tau.ac.il/~nin/Courses/BioInfo04/SAM.ppt).

The genes reported in the above procedure are considered as potentially significant genes. To estimate the error rate, SAM counts the number of genes which exceed the cut-offs d_u or d_l. Then the following equation is used to estimate the pFDR:

$$pFDR = \frac{\frac{1}{B}\sum_{b=1}^{B} C_b}{C},$$

where C_b is the number of potentially significant genes in the b-th permutated data set, while C is the number of potentially significant genes in the original data set. For example, in Figure 4.7, suppose the upper cut point t_1 is 4.6, and the lower cut point t_2 is -0.6. There are 2 entries in the permutation array above t_1 (5.5 and 7.9) and 3 entries below t_2 (-0.8, -1.0, and -0.9). Therefore, the average false positives in each data set is $(2 + 3)/4$. To estimate the pFDR, this number is further divided by 4, which is the number of potentially significant genes in the original data set. SAM usually tests a series of Δ and calculate the corresponding $pFDR$. Finally, the genes reported under the Δ with a satisfying $pFDR$ will be recognized as the differentially expressed genes. An overview of the SAM's method is summarized in Figure 4.8.

4.7 ANOVA: Analysis of Variance

In previous sections, we have described methods for analyzing differentially expressed genes in simple data sets with only two samples. In practice, microarrays are also being used to perform more complex experiments. For example, the Example Data Set D contains three groups of samples, and we are interested in identifying genes that were differentially expressed on one or more of the groups relative to the others. There are two possible ways the analysis can be performed [266]:

- A straight-forward method is to apply an unpaired t-test three times, to each pair of groups in turn; genes that are significant in one or more of the t-tests are then selected.
- An alternative method is to use a statistical test that compares all three groups simultaneously and reports a single p-value.

Stekel [266] noted that there are two problems with the first method. One is that by performing three tests, we increase the likelihood of false positives. This problem gets worse when the number of the groups increase; for example, with 10 groups, there would be 45 separate comparisons. The other problem is that each of these comparisons is not independent of the other, so it becomes very difficult to interpret the results.

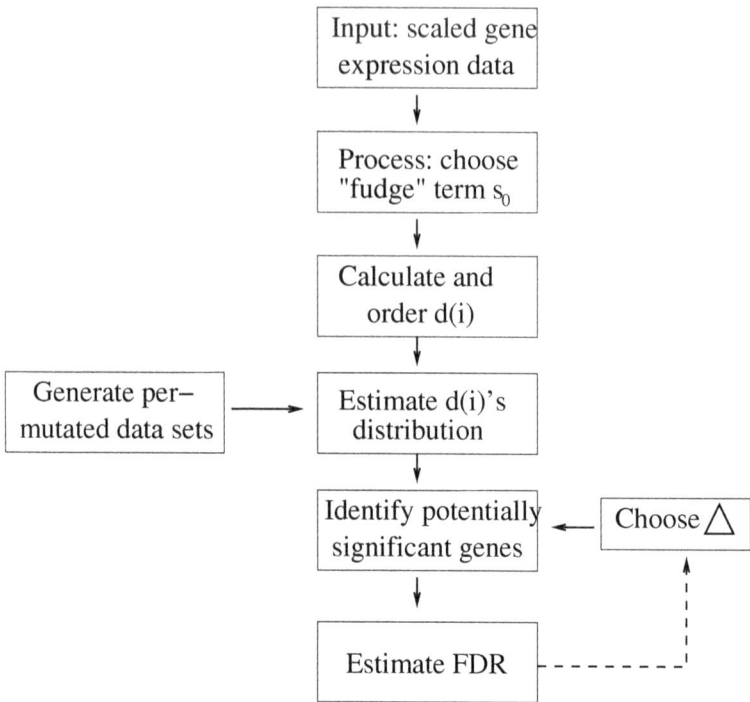

Fig. 4.8   Overview of the SAM's method, which is adapted from http://www.math. tau.ac.il/~nin/Courses/BioInfo04/SAM.ppt.

Due to the above problems, we usually adopt the second strategy for Example Data Set D. A typical approach taken by statisticians is the *Analysis of Variance* (ANOVA), which performs an analysis of the data with multiple groups, and returns a single p-value which suggests the level of significance whether one or more groups is different from others. If we assume the variance in gene expression comes from only one source, i.e., the different type of cancers the patients are suffering from, we actually perform the *one-way ANOVA*. Instead, if we consider the variance from multiple sources, e.g., the cancer type and the microarray experiment artifacts, we build more general ANOVA models which include multiple (correlated) factors and obtain one p-value for each of the factors separately. Such analysis is called the *two-way ANOVA* or *multifactor ANOVA*. In this section, we will present the applications of the one-way and two-way ANOVA to microarray data.

4.7.1 One-Way ANOVA

The one-way ANOVA regards the variance in a given data set comes from a single source. However, the variance can be divided into two parts. First, the measurements of each group vary around their mean, which forms the *within-group variance*. Second, the means of each group will vary around the overall mean of the data set, which forms the *inter-group variance*. The essential spirit of the one-way ANOVA is to study the relationship between the inter-group and the within-group variances.

Suppose the data set consists of k groups, and there are n_i measurements in each group i ($1 \leq i \leq k$). In the following, we will use x_{ij} to denote the j-th measurement in the i-th group, $\bar{x}_i$ to denote the mean of the measurements within the i-th group, and $\bar{\bar{x}}$ to denote the mean of all the measurements in the data set. The total variability of the data set is characterized by the *total sum of square*:

$$SS_{total} = \sum_{i=1}^{k} \sum_{j=1}^{n_i} (x_{ij} - \bar{\bar{x}})^2, \tag{4.15}$$

In fact, through simple mathematical transformations, Formula 4.15 can be rewritten as follows:

$$SS_{total} = \sum_{i=1}^{k} \sum_{j=1}^{n_i} (\bar{x}_i - \bar{\bar{x}})^2 + \sum_{i=1}^{k} \sum_{j=1}^{n_i} (x_{ij} - \bar{x}_i)^2. \tag{4.16}$$

The first term in above equation (called the *among group sum of squares*) characterizes the inter-group variance, while the second term (called the *error sum of squares* or *residual sum of squares*) describes the within-group variability. So Equation 4.16 can also be written as:

$$SS_{total} = SS_{group} + SS_{error} \tag{4.17}$$

This equation constitutes the basis of the ANOVA approach: the total sum of squares SS_{total} or the overall variability can be partitioned into the variability SS_{group} due to the difference between groups and the variability SS_{error} within groups.

Recall our problem is to determine whether a gene is differentially expressed on one or more of the groups relative to the others. Clearly, this problem can be formalized by the following hypotheses:

$H_0 : \mu_1 = \mu_2 = \cdots = \mu_k$
$H_a :$ There is at least one pair of means that are different from each other.

To compare the within-group variance with the inter-group variance, ANOVA uses the following F statistic:

$$F = \frac{MS_{group}}{MS_{error}} = \frac{SS_{group}/(k-1)}{SS_{error}/(N-k)}. \tag{4.18}$$

This statistic follows an F distribution with the degrees of freedom $\nu_1 = k - 1$ and $\nu_2 = N - k$ (N is the total number of measurements in the data set).

Dudoit et al. [86] performed a preliminary selection of significant genes based on the ratio of their between-group to within-group sums of squares, called the *signal-to-noise ratio*. For a gene j, this ratio is

$$BW(j) = \frac{\sum_i \sum_k I(y_i = k)(\overline{x}_{jk} - \overline{x}_j)^2}{\sum_i \sum_k I(y_i = k)(\overline{x}_{ji} - \overline{x}_{jk})^2},$$

where $\overline{x}_j$ and $\overline{x}_{jk}$ denote the average expression level of gene j across all tumor samples and across samples belonging to class k only, respectively. The user-specified number of k genes with the largest BW ratios are selected as significant genes.

4.7.2 Two-Way ANOVA

The idea behind two-way ANOVA is to build an explicit model about the multiple, possibly correlated sources of variance that affect the measurements, and then use the data to estimate the variance of each individual variable in the model. For instance, Kerr and Churchill [165, 166] proposed the following model to account for the multiple sources of variation in a microarray experiment:

$$log(y_{ijkg}) = \mu + A_i + D_j + V_k + G_g + (AG)_{ig} + (VG)_{kg} + \epsilon_{ijkg}$$

In this model, μ is the overall mean signal of the array, A_i is the effect of the ith array, D_j is the effect of the jth dye, V_k is the effect of the kth variety (such as the cancer type), G_g is the variation of the gth gene, $(AG)_{ig}$ is the effect of a particular spot on a given array, $(VG)_{kg}$ represents the interaction between the kth variety and the gth gene, and ϵ_{ijkg} represents the error term for array i, dye j, variety k and gene g. The error is assumed to be independent of y_{ijkg} and have a mean of zero. Finally, $log(y_{ijkg})$ is the measured log-ratio for gene g of variety k measured on array i using dye j.

To identify differentially regulated genes, the general Kerr-Churchill model can be modified as follows:

$$logR(g, s) = \mu + G(g) + \epsilon(g, s),$$

where $logR(g, s)$ is the measured log ratio for gene g and spot s, μ is the average log ratio over the whole array, $G(g)$ is a term for the differential regulation of gene g and $\epsilon(g, s)$ is a zero-mean noise term.

In this model, the mean log ratio μ can be estimated by

$$\hat{\mu} = \frac{1}{n \cdot m} \sum_{g,s} logR(g, s),$$

which is the average of the observed log ratios over all spots, where n is the number of genes and m is the number of replicates for each gene.

The effect of an individual gene g can be estimated by:

$$\hat{G}(g) = \frac{1}{m} \sum_{g} logR(g, s) - \hat{\mu}.$$

Clearly, the first term is the average log ratio over the replicated spots corresponding to gene g.

Using the estimates above, an estimate of the noise can be calculated as follows:

$$\hat{\epsilon}(g, s) = logR(g, s) - \hat{\mu} - \hat{G}(g).$$

Intuitively, the genes whose expression values are far deviated from the noise distribution are likely to be differentially expressed. A bootstrap analysis (see Section 4.5.3) can be performed to estimate the distribution of noise, and estimate the p-values of genes.

The advantage of two-way ANOVA is that it takes into consideration multiple sources of variance [82]. Thus, it is possible to distinguish interesting variations, such as gene regulation, from the experiment artifacts, such as differences caused by different dyes or arrays. Therefore, the two-way ANOVA is also popular for the normalization of microarray data. However, the application of the multifactor ANOVA requires very careful experimental design. In most cases, this requires repeating several microarrays with various mRNA samples, duplicating individual genes on multiple spots of a single chip, and swapping dyes if a multichannel technology is used, etc. In practice, due to the relatively expensive cost and intensive labor of microarray experiments, replicates of microarray are often very limited. Thus, the benefit of a full-scale ANOVA may only be received in the future when sufficient replicates are available.

4.8 Summary

In order to prepare the reader for the subsequent discussion of specific data-mining techniques, this chapter has offered an overview of the basic statistical approaches employed for the identification of differentially-expressed genes under various conditions. These approaches form the underpinning of many data-mining methods. The content of this chapter has largely been excerpted from a variety of publications and websites [10, 83, 266].

Chapter 5

Gene-Based Analysis

Contributor: Daxin Jiang

5.1 Introduction

The analysis of gene expression data sets is intended to identify co-expressed genes and coherent gene expression patterns. A group of *co-expressed genes* exhibits a common expression profile, while a *coherent gene expression pattern* (or, briefly, coherent pattern) characterizes the collective trend of the expression levels of a group of co-expressed genes. The coherent pattern can be viewed as a "template" to which the expression profiles of the corresponding co-expressed genes conform with only small divergences.

For example, Iyer's data set [147] records the expression profiles of 517 human genes with respect to a twelve-point time-series. In [147], Iyer et al. give a list of ten groups of co-expressed genes and the corresponding coherent gene expression patterns in the data set, which has been well accepted as the ground truth. Figure 5.1 presents two groups of co-expressed genes and their corresponding coherent patterns from the ground truth. The graphs in the upper portion of the figure show the expression profiles of the genes in these two groups. The profiles in each group appear to share a common trend, and this is shown in the lower row of the graph, which plot the point-wise median of the profiles. The error bars in the lower graphs indicate the standard deviations.

A variety of conventional and newly-developed clustering algorithms have been used to identify co-expressed genes and coherent patterns in microarray data. *Clustering* is the process of grouping data objects into a set of disjoint classes, called *clusters*. The objects within a class have high degree of similarity, while objects in separate classes are more dissimilar. In gene-based clustering, the genes are considered to be data objects, and experimental conditions (either samples or time points) are regarded as

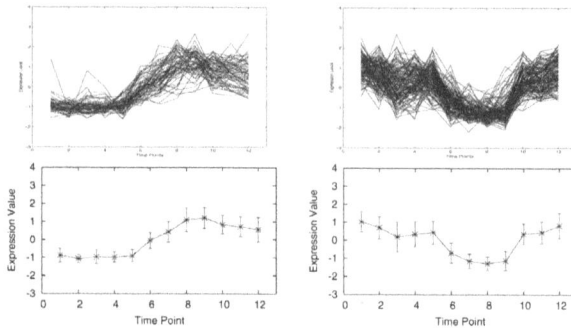

Fig. 5.1   Examples of co-expressed gene groups and corresponding coherent patterns.

attributes[1]. After the clustering process has been completed, each cluster can be regarded as a group of co-expressed genes, and the corresponding coherent pattern is simply the centroid of the cluster.

Previous studies have confirmed that clustering algorithms are useful in finding co-expressed gene groups and coherent patterns. The identified gene groups and patterns can further help to understand gene function, gene regulation and cellular processes. For example, a group of co-expressed genes often share similar cellular function. We may use the genes with known function to infer the function of other genes in the same group for which information has not been previously available [92, 277]. Furthermore, co-expressed genes in the same cluster are likely to be involved in the same cellular processes, and a strong correlation of expression patterns between those genes indicates co-regulation. Searching for common DNA sequences at the promoter regions of genes within the same cluster allows regulatory motifs specific to each gene cluster to be identified and *cis*-regulatory elements to be proposed [39, 277]. The inference of regulation through the clustering of gene expression data also gives rise to hypotheses regarding the mechanism of the transcriptional regulatory network [77].

In general, a clustering algorithm relies on some proximity measurement to evaluate the distance or similarity between a pair of data objects (i.e., genes) and seeks to optimize a specific object function. In this chapter, we will first introduce several proximity measures which have been widely used with microarray data. Three categories of clustering algorithms, *partition-based approaches*, *hierarchical approaches*, and *density-based approaches*, will be described in Sections 5.3, 5.4, and 5.5, respectively. In Section 5.6,

---

[1] In this chapter, we use the terms "objects" and "genes" exchangeably, and the terms "attributes", "features", and "experimental conditions" exchangeably.

we will present a novel approach which supports interactive user exploration of co-expressed genes and coherent patterns on the basis of their domain knowledge. Cluster validation techniques will be discussed in Section 5.7.

5.2 Proximity Measurement for Gene Expression Data

A *proximity measurement* measures the similarity (or distance) between two data objects. A data object O_i can be formalized as a numerical vector $\vec{O}_i = \{o_{ij}|1 \leq j \leq p\}$, where o_{ij} is the value of the jth feature for $\vec{O}_i$ and p is the number of features. The proximity between two objects O_i and O_j is measured by a *proximity function* of corresponding vectors $\vec{O}_i$ and $\vec{O}_j$.

5.2.1 *Euclidean Distance*

Euclidean distance is one of the most commonly-used methods to measure the distance between two data objects. The distance between objects O_i and O_j in p-dimensional space is defined as:

$$Euclidean(O_i, O_j) = \sqrt{\sum_{d=1}^{p} (o_{id} - o_{jd})^2}. \tag{5.1}$$

However, for gene expression data, the overall shapes of gene expression profiles are often of greater interest than the individual magnitudes of each feature. Euclidean distance does not score well for shift or scaled profiles. For example, in Figure 5.2(a), patterns A and B have exactly the same shape, but pattern B is shifted to a higher level. In Figure 5.2(b), patterns A and C demonstrate the same trend of rising and falling values, but pattern C is at twice the scale of pattern A. In both cases, it would be more appropriate to assign a distance score at or close to zero, considering the similarity of the overall shapes or trends between the expression profiles. In general, the distance function should be invariant to linear transformation, i.e., if $O_j = aO_i + b$ for $a, b \in R$, $distance(O_i, O_j)$ equals zero. However, Euclidean distance does not have such a property.

To solve this problem, a *standardization process* is usually performed before calculating the Euclidean distance. Each data object O_i is standardized to O_i' with $mean(O_i') = 0$ and $variance(O_i') = 1$ using $O_{ij}' = \frac{O_{ij} - \mu_i}{\sigma_i}$ where $1 \leq j \leq p$, $\mu_i = (\sum_{d=1}^{p} o_{id})/p$ and $\sigma_i = \sqrt{\frac{1}{p}\sum_{d=1}^{p}(o_{id} - \mu_i)^2}$. After the standardization process, the distance between two data objects O_i and O_j

is defined as

$$distance(O_i, O_j) = Euclidean(O'_i, O'_j). \qquad (5.2)$$

We can easily prove the above distance is invariant to linear transformation of patterns. Figure 5.2(c) shows that the standardized profiles of A, B and C completely overlap with each other; thus, the distance between any pair of these profiles is zero.

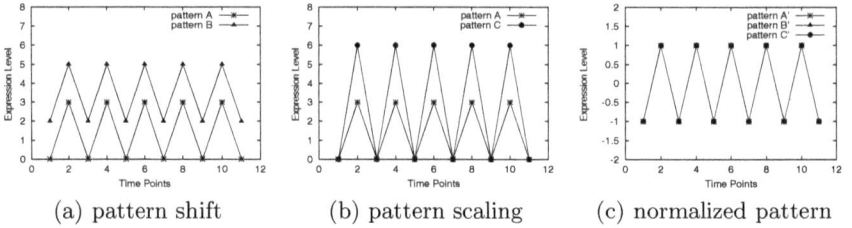

(a) pattern shift   (b) pattern scaling   (c) normalized pattern

Fig. 5.2   Examples of shifting and scaled profiles and the effect of the standardization process.

5.2.2   Correlation Coefficient

5.2.2.1   Pearson's correlation coefficient

In contrast to Euclidean distance, which measures the distance (dissimilarity) between two patterns, *Pearson's correlation coefficient* measures the extent to which two patterns are similar with each other. Given two data objects O_i and O_j, Pearson's correlation coefficient is defined as

$$Pearson(O_i, O_j) = \frac{\sum_{d=1}^{p} (o_{id} - \mu_i)(o_{jd} - \mu_j)}{\sqrt{\sum_{d=1}^{p} (o_{id} - \mu_i)^2}\sqrt{\sum_{d=1}^{p} (o_{jd} - \mu_j)^2}}, \qquad (5.3)$$

where μ_i and μ_j are the means for $\vec{O}_i$ and $\vec{O}_j$, respectively. The value of Pearson's correlation coefficient ranges between -1 and 1 with a higher value indicating stronger similarity. From a statistical view, each data object can be regarded as a random variable with p observations. Pearson's correlation coefficient measures the similarity between two profiles by calculating the linear relationship of the distributions of the two corresponding random variables. The definition indicates that Pearson's correlation coefficient is invariant to linear transformation. We can, additionally, prove that $Pearson(O_i, O_j) = Pearson(O'_i, O'_j)$, where O'_i and

O'_j are the standardized objects of O_i and O_j, respectively. We can further prove that $Euclidean(O'_i, O'_j) = \sqrt{2p}(\sqrt{1 - Pearson(O'_i, O'_j)})$. This equation discloses the consistency between Pearson's correlation coefficient and Euclidean distance after data standardization; if a pair of data objects O_i, O_j has a higher correlation than pair O_l, O_k ($Pearson(O'_i, O'_j) > Pearson(O'_l, O'_k)$), then pair O_i, O_j has a smaller distance than pair O_l, O_k ($Euclidean(O'_i, O'_j) < Euclidean(O'_l, O'_k)$), and vice versa. Thus, we can expect the effectiveness of a clustering algorithm to be equivalent, regardless of whether Euclidean distance or Pearson's correlation coefficient is chosen as the proximity measure.

Pearson's correlation coefficient is widely used and has proved effective as similarity measure for gene expression data. However, empirical study has shown that Pearson's correlation coefficient is not robust to outliers [133] and may generate *false positives*, i.e., assigning a high similarity score to a pair of dissimilar patterns. If two gene profiles have a common high peak or valley at a single experimental condition, the correlation will be dominated by this condition, although the profiles at remaining conditions may be not similar at all. In Figure 5.3(a), profiles A and B have a high correlation coefficient when all experimental conditions are considered. However, removing the single outlier from the feature set (see Figure 5.3(b)) will result in a negative correlation. In practice, such outlier may occur because of experimental errors.

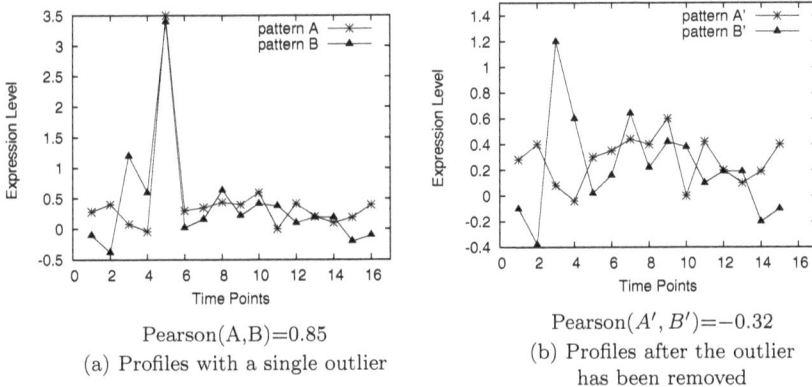

Pearson(A,B)=0.85
(a) Profiles with a single outlier

Pearson(A', B')=−0.32
(b) Profiles after the outlier has been removed

Fig. 5.3 An example of profiles with a single outlier.

5.2.2.2 *Jackknife correlation*

The issues identified above are addressed by the *jackknife correla-tion* [91, 133] which is defined as

$$Jackknife(O_i, O_j) = min\{\rho_{ij}^{(1)}, \ldots, \rho_{ij}^{(l)}, \ldots, \rho_{ij}^{(p)}\},$$

where $\rho_{ij}^{(l)}$ is the Pearson's correlation coefficient of data objects O_i and O_j with the lth attribute deleted. Use of the jackknife correlation avoids the "dominance effect" of single outliers. More general versions of the jackknife correlation that are robust to more than one outlier can similarly be derived. However, the generalized jackknife correlation, which would involve the enumeration of different combinations of features to be deleted, is computationally costly and is rarely used.

5.2.2.3 *Spearman's rank-order correlation*

Another issue with Pearson's correlation coefficient is that it assumes an approximately Gaussian distribution of the points and may not be robust for non-Gaussian distributions [77]. For example, in Figure 5.4, profiles A and B show a highly concordant trend in expression level across the experi-mental conditions. However, the Pearson's correlation coefficient of profiles A and B is not very high. An alternative measure, *Spearman's rank-order correlation coefficient* was suggested by D'Haeseleer [77]. The rank correla-tion is derived by replacing the numerical expression level o_{id} with its rank r_{id} among all time points. For example, $r_{id} = 3$ if o_{id} is the third high-est value among o_{ik}, where $1 \leq k \leq p$. Spearman's correlation coefficient does not require the assumption of Gaussian distribution and is also more robust against outliers than Pearson's correlation coefficient. However, as a consequence of ranking, a significant amount of information present in the data is lost. Therefore, on average, Spearman's rank-order correlation coefficient may not perform as well as Pearson's correlation coefficient.

5.2.3 **Kullback-Leibler Divergence**

The *Kullback-Leibler divergence* (abbreviated K-L divergence and also called the *relative entropy*) is an information-theoretic approach to mea-suring the distance between two gene expression profiles. In general, the relative entropy between two probability mass functions $u(w)$ and $v(w)$ over

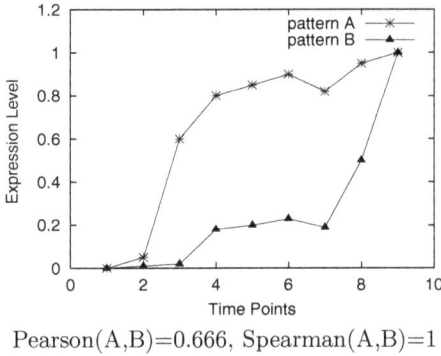

Pearson(A,B)=0.666, Spearman(A,B)=1

Fig. 5.4   An example of two profiles with high rank correlation but low linear correlation.

the random variable W is defined as [64]:

$$KL(u\|v) = \sum_{w \in W} u(w)log\frac{u(w)}{v(w)} \ .$$

Given a random variable W with a true distribution u, the K-L divergence $KL(u\|v)$ measures the inefficiency of assuming the distribution of W as v.

As with Pearson's correlation coefficient, the K-L divergence also regards each gene expression profile O_i as a random variable with p observations. To apply the K-L divergence, a profile O_i is converted to its probability mass function by calculating the fractional contribution of the expression level at each experimental condition to the sum of expression levels at all conditions; i.e., $u_i(w) = \frac{O_{iw}}{\sum_{d=1}^{p} O_{id}}$. The K-L divergence always takes non-negative values, and is zero if and only if $u = v$.

Kasturi et al. [160] noted that the K-L divergence has several important and useful properties. These are: i) convergence in the K-L sense implies convergence in the L_1 norm sense but no proof is known for the reverse; ii) the χ^2 statistic is twice the first term in the Taylor expansion of the K-L divergence; and iii) $KL(u\|v)$ is convex in the pair (u, v). They used the K-L divergence in conjunction with an unsupervised self-organizing map algorithm (see Section 5.3.2). The clustering results from two gene expression data sets were found to be superior to those obtained with the hierarchical clustering algorithm using the Pearson's correlation coefficient [160].

5.3   Partition-Based Approaches

Partition-based algorithms divide a data set into several mutually-exclusive subsets based on certain clustering assumptions (e.g., there are k clusters in the data set) and optimization criteria (e.g., minimize the sum of distances between objects and their cluster centroids). We can further divide the partition-based methods into four sub-categories: *the K-means algorithm and its variations* [133, 188, 227, 258, 277], *the Self-Organizing Map (SOM) and its extensions* [132, 171, 271, 282], *graph-theoretical algorithms* [22, 124, 251, 313], and *model-based algorithms* [102, 111, 194, 320].

5.3.1   *K-means and its Variations*

The K-means algorithm [188] is a typical partition-based clustering method. Given a pre-specified number K, the algorithm partitions the data set into K disjoint subsets which optimize the following objective function:

$$E = \sum_{i=1}^{K} \sum_{O \in C_i} |O - \mu_i|^2, \qquad (5.4)$$

where O is a data object in cluster C_i and μ_i is the centroid (mean of objects) of C_i. Thus, the objective function E tries to minimize the sum of the squared distances of objects from their cluster centers.

Algorithm 5.1   : K-means

Initialization:
    Randomly or manually select K data objects as initial cluster centroids;
Iteration:
    Step 1: **for** each data object O_i assign O_i to the cluster with the nearest centroid
    Step 2: **for** each cluster C_j recalculate the centroid μ_j as the mean of all data objects $O \in C_j$
    Repeat steps 1 and 2 until no more changes occur, or the amount of change falls below a pre-defined threshold.

The K-means algorithm (see Algorithm 5.1) is simple and fast. The time complexity of K-means is $O(l * k * n)$, where l is the number of iterations, k is the number of clusters, and n is the total number of objects. In [277],

Tavazoie et al. applied the K-means algorithm on the gene expression data collected by Cho et al. [54]. They found that each cluster contains a significant portion of genes with similar functions. Furthermore, by searching the upstream DNA sequences of genes within the same cluster, they extracted eighteen motifs which are promising candidates for novel *cis*-regulatory elements. However, the K-means algorithm also has several drawbacks as a gene-based clustering algorithm. First, the number of gene clusters in a gene expression data set is usually unknown in advance. To detect the optimal number of clusters, users usually run the algorithms repeatedly with different values of k and compare the clustering results. For a large gene expression data set which contains thousands of genes, this extensive parameter fine-tuning process may not be practical. Second, gene expression data typically contain significant noise. The K-means algorithm forces each gene into a cluster, which may cause the algorithm to be sensitive to noise [252, 258].

Recently, several new algorithms [133, 227, 258] have been proposed to overcome the drawbacks of the K-means algorithm. We call them *variations of the K-means algorithm*, since, in essence, they also are intended to minimize the overall divergence of objects from their cluster centers (see Formula 5.4). One common feature of these algorithms is that they use some thresholds to control the coherence of clusters. For example, Ralf-Herwig et al. [227] introduced two parameters γ and ρ, where γ is the maximal similarity between two separate cluster centroids, and ρ corresponds to the minimal similarity between a data point and its cluster centroid. In [133], Heyer et al. constrained the clusters to have a diameter no larger than a threshold. Motivated by this approach, Smet et al. [258] proposed a more efficient algorithm in which data object x is assigned to cluster c if the assignment has a higher probability than a threshold. Compared with cluster models in [133] and [227], the model in [258] is more adaptive to various data structures, since the clusters are not rigidly bounded by the radius or diameter threshold.

The clustering process of the above approaches extracts from the data set all clusters with qualified coherence. This feature obviates the need for users to input the number of clusters. In addition, with the coherence control, outliers may only end up with trivial clusters, i.e., clusters with very few members. Thus, they will not compromise the significant groups of co-expressed genes.

The K-means algorithm and its variations require the initial stipulation of some global parameters such as the number of clusters or a coherence threshold. The clustering process is like a "black box;" users input the data set and the parameter values, and the clusters are generated. There is no intensive interaction between the user and the mining procedures. While this simplifies processing, they are not sensitive to the local structures of the data set and provide no opportunity to exploit user domain knowledge of the data set.

5.3.2   *SOM and its Extensions*

The Self-Organizing Map (SOM) was developed by Kohonen [171] on the basis of a single layered neural network. The data objects are presented as input one at a time. The output neurons are organized with a simple neighborhood structure such as a two-dimensional $p * q$ grid (see [271] for a visual picture of this structure). Each output neuron of the neural network is associated with a p-dimensional reference vector, where p is the dimensionality of the input data objects. In the learning process, each input data point is "mapped" to the output neuron with the "closest" reference vector (the winning node, see Algorithm 5.2). Reference vectors (output nodes), which are in some neighborhood of the winning node, are updated by moving them toward the input pattern. For SOM learning each data object acts as a training sample which directs the movement of the reference vectors towards the denser areas of the input vector space. As a result, reference vectors are trained to fit the distributions of the input data set. When the training is complete, clusters are identified by mapping all data points to the output neurons.

Tamayo et al. [271] applied the SOM algorithm in a study of hematopoietic differentiation. The expression patterns of $1,036$ human genes were mapped to a $6 * 4$ SOM. After the clustering process, the genes were organized into biologically-relevant clusters that suggested novel hypotheses about hematopoietic differentiation. For example, in the reported result in [271], Cluster 15 captured 154 genes involved in the "differentiation therapy" which is part of the standard treatment for patients with acute promyelocytic leukemia. Among the 154 genes, some showed unexpected regulation patterns. This provides interesting insights into the poorly-understood mechanism of differentiation.

One of the remarkable features of SOM is that it allows users to impose a partial structure on the clusters, and arranges similar patterns as neighbors in the output neuron map. This feature facilitates easy visualization and interpretation of the clusters and thus partly supports the explorative

---

Algorithm 5.2   : Self-Organizing Map

Initialization: Randomly initialize reference vectors m_i for all output neurons

Iteration:

Repeat the following steps until termination condition is met. Termination conditions: the maximum number of iterations has been reached or the amount of changes in reference vectors m_i is less than some predefined threshold.

Step 1: Randomly select a data object x and find the "winning" neuron c with the closest reference vector m_c defined by

$$\forall i, \|x - m_c\| \leq \|x - m_i\|.$$

Step 2: Update the reference vector m_i using

$$m_i = m_i + \mu \cdot h_{c,i} \cdot (x - m_i),$$

where μ is the learning rate and $h_{c,i}$ is the "neighborhood function" which determines both a neighborhood of winning node c and the factors of changes for reference vectors in that neighborhood. $h_{c,i}$ is based on the pre-defined topological relationships between output neurons. Both μ and $h_{c,i}$ are gradually reduced in each iteration.

---

analysis of gene expression patterns.

However, as with the K-means algorithm, SOM requires a user to pre-specify the number of clusters, something which is typically unknown in the case of gene expression data. Moreover, as pointed out by Herrero et al. [132], if the data set is abundant with irrelevant data points, such as genes with invariant patterns, SOM will produce an output in which these data points will populate the vast majority of clusters. The most interesting patterns will be missed, since they are collapsed into only a few clusters.

Recently, several new algorithms [132, 192, 282] have been proposed based on the SOM algorithm. These algorithms can automatically determine the number of clusters and dynamically adapt the map structure to the data distribution. For example, Herrero et al. [132] extended the SOM by a binary tree structure. At first, the tree only contains a root node connecting two neurons. After a training process similar to that of the SOM algorithm, the data set is segregated into two subsets. The neuron with less coherence is then split into two new neurons. This process is repeated level by level until all the neurons in the tree satisfy some coherence threshold. Other examples of SOM extensions are Fuzzy Adaptive Resonance Theory (Fuzzy ART) [282] and supervised Network Self-Organized Map (sNet SOM) [192]. In general, they provide some approaches to measuring the coherence of a neuron (e.g, *vigilance criterion* in [282] and *grow parameter* in [192]). The output map is adjusted by splitting existing neurons or adding new neurons to the map until the coherence of each neuron in the map satisfies a user-specified threshold.

SOM is an efficient and robust clustering technique. A hierarchical structure can also be built on the basis of SOM; one example is SOTA (Self Organizing Tree Algorithm) [80]. Moreover, by systematically controlling neuron splitting, SOM can easily adapt to the local structures of the data set. However, the current versions of SOM require that the splitting process be controlled by a user-specified coherence threshold, something which is difficult to identify with gene expression data.

5.3.3   *Graph-Theoretical Approaches*

Given a data set X, we can construct a *proximity matrix* P, where $P[i,j] = proximity(O_i, O_j)$, and a weighted graph $\mathcal{G}(V,E)$, called a *proximity graph*, where each data point corresponds to a vertex. For some clustering methods, each pair of objects is connected by an edge, with weight assigned according to the proximity value between the objects [251, 308]. For other methods, proximity is mapped only to either 0 or 1 on the basis of some threshold, and edges only exist between objects i and j, where $P[i,j]$ equals 1 [22, 124]. Graph-theoretical clustering techniques are explicitly presented in terms of a graph, thus converting the problem of clustering a data set into such graph theoretical problems as finding the minimum cut or maximal cliques in the proximity graph $\mathcal{G}$.

5.3.3.1   *HCS and CLICK*

Hartuv et al. [124] proposed an algorithm *HCS* (Highly Connected Subgraph) which recursively splits a graph G into a set of "highly connected" components along the minimum cut. Each highly-connected component is considered to be a cluster. In graph theory, a *cut* S of a graph G consists of a set of edges whose removal results in a disconnected graph. The *edge-connectivity* $k(G)$ of G is defined as the minimum number of edges whose removal disconnects G. A cut S of G is called a *minimum cut* if and only if $|S| = k(G)$. In [124], a graph (or subgraph) G with $n > 1$ vertices is called "highly connected" if $k(G) > \frac{n}{2}$. At each recursion of HCS, the operation MINCUT(G) returns a minimum cut C which separates the graph G into two subgraphs H and $\overline{H}$. The recursion continues until both H and $\overline{H}$ are highly connected (see Figure 5.5).

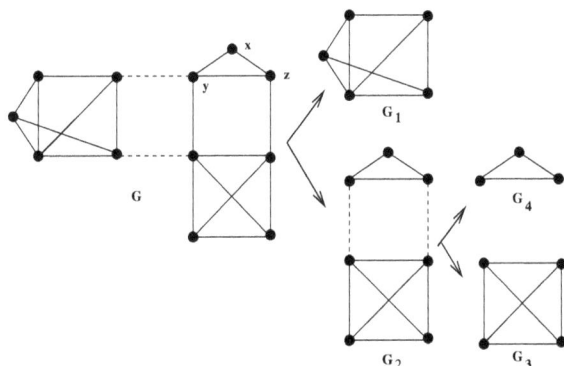

Fig. 5.5 An example of applying the HCS algorithm to a graph. Minimum cut edges are denoted by broken lines. Reprinted from [124], Copyright 1999, with permission from Elsevier.

Motivated by HCS, Shamir et al. presented the algorithm *CLICK* (CLuster Identification via Connectivity Kernels) in [251]. CLICK makes the probabilistic assumption that, after standardization, pair-wise similarity values between elements (in the same or different clusters) will be normally distributed. Under this assumption, the weight ω_{ij} of an edge (i, j) is defined as the probability that vertices i and j are in the same cluster. The clustering process of CLICK iteratively finds the minimum cut in the proximity graph and splits the data set recursively into a set of connected components from the minimum cut. CLICK also takes two post-pruning steps to refine the clustering results. The *adoption step* handles the remaining singletons and updates the current clusters, while the *merging step* iteratively merges two clusters with similarity exceeding a predefined threshold.

In [251], the authors compared the clustering results of CLICK as applied to two public gene expression data sets with those of GENECLUSTER [271] (a SOM approach) and Eisen's hierarchical approach [92] (will be discussed in Section 5.4.1). In both cases, the clusters obtained by CLICK demonstrated better quality in terms of cluster homogeneity and separation. However, CLICK has the potential of going astray and generating highly unbalanced partitions which separate a handful of outliers from the remaining data objects. Furthermore, in gene expression data, two clusters of co-expressed genes may significantly intersect. In such situations, CLICK is unlikely to properly split the two clusters, which are likely to be reported as one highly-connected component.

5.3.3.2   *CAST: Cluster affinity search technique*

In [22], Ben-Dor et al. introduced the concept of a *corrupted clique graph* data model. The input data set is assumed to come from the underlying cluster structure by "contamination" by random errors caused by the complex process of microarray experimentation. Specifically, it is assumed that the true clusters of the data points can be represented by a *clique graph H*, a disjoint union of complete sub-graphs in which each clique corresponds to a cluster. The similarity graph G is derived from H by flipping each edge/non-edge with probability α. Therefore, clustering a data set is equivalent to identifying the original clique graph H from the corrupted version G with as few flips (errors) as possible.

Ben-Dor et al. presented in [22] both a theoretical algorithm and a practical heuristic called *CAST* (Cluster Affinity Search Technique, see Algorithm 5.3). CAST takes as input a real, symmetric, n-by-n similarity matrix S $(S(i,j) \in [0,1])$ and an *affinity threshold t*. The algorithm searches the clusters individually, with the cluster currently being searched denoted by C_{open}. Each element x has an *affinity value* $a(x)$ with respect to C_{open}, as $a(x) = \sum_{y \in C_{open}} S(x,y)$. An element x has a high affinity value if it satisfies $a(x) \geq t|C_{open}|$; otherwise, x has a low affinity value. CAST alternates between adding high-affinity elements to the current cluster and removing low-affinity elements from it. When the process stabilizes, C_{open} is considered a complete cluster, and this process continues with each new cluster until all elements have been assigned to a cluster.

Algorithm 5.3 : Clustering Affinity Search Technique (CAST)

Initializations:

$C \leftarrow \emptyset$ //The collection of closed clusters

$C_{open} \leftarrow \emptyset$ //The constructed cluster

$U \leftarrow \{1, \ldots, n\}$ //Elements not yet assigned to any cluster

$a(\cdot) \leftarrow 0$ //Reset the affinity

Iterations:

while $(U \cup C_{open} \neq \emptyset)$ do

  Select the element u with maximal affinity in U.

  if $(a(u) \geq t|C_{open}|)$ //u is of high affinity

    $C_{open} \leftarrow C_{open} \cup \{u\}$ //Insert u into C_{open}

    $U \leftarrow U - \{u\}$ //Remove u from U

    for all x in $U \cup C_{open}$ do

      $a(x) = a(x) + S(x,u)$ //Update the affinity

    end for

  else //No high affinity elements outside C_{open}

    Select the vertex v with the minimal affinity in C_{open}.

    if $(a(v) < t|C_{open}|)$ //v is of low affinity

      $C_{open} \leftarrow C_{open} - \{v\}$ //Remove v from C_{open}

      $U \leftarrow U \cup \{v\}$ //Insert v into U

      for all x in $U \cup C_{open}$ do

        $a(x) = a(x)\text{-}S(x,v)$ //Update the affinity

      end for

    else //C_{open} is clean

      $C \leftarrow C \cup C_{open}$ //Close the cluster

      $C_{open} \leftarrow \emptyset$ //Start a new cluster

      $a(\cdot) \leftarrow 0$ //Reset affinity

    end if

  end if

end while

Return the collection of clusters C.

The affinity threshold t of the CAST algorithm is the average similarity between the objects within a cluster. CAST specifies the desired cluster quality through t and applies a heuristic searching process to identify qualified clusters one at a time. Therefore, CAST does not depend on a user-defined number of clusters and deals effectively with outliers. Nevertheless, CAST has the usual difficulty of determining a "good" value for the global parameter t.

5.3.4   *Model-Based Clustering*

Model-based clustering approaches [102, 111, 194, 320] provide a statistical framework to model the cluster structure of gene expression data. The data set is assumed to come from a finite mixture of underlying probability distributions, with each component corresponding to a different cluster. The goal is to estimate the parameters $\Theta = \{\theta_i \mid 1 \leq i \leq k\}$ and $\Gamma = \{\gamma_r^i \mid 1 \leq i \leq k, 1 \leq r \leq n\}$ that maximize the mixture likelihood $L_{mix}(\Theta, \Gamma) = \prod_{r=1}^{n} \sum_{i=1}^{k} \gamma_r^i f_i(x_r|\theta_i)$, where n is the number of data objects, k is the number of components, x_r is a data object (i.e., a gene expression pattern), $f_i(x_r|\theta_i)$ is the density function of x_r of component C_i with some unknown set of parameters θ_i (*model parameters*), and γ_r^i (*hidden parameters*) represents the probability that x_r belongs to C_i. Usually, the parameters Θ and Γ are estimated by the EM algorithm [69]. The EM algorithm iterates between Expectation (E) steps and Maximization (M) steps. In the E step, hidden parameters Γ are conditionally estimated from the data with the current estimated Θ. In the M step, model parameters Θ are estimated so as to maximize the likelihood of complete data, given the estimated hidden parameters. When the EM algorithm converges, each data object is assigned to the component (cluster) with the maximum conditional probability.

Several early studies, including [102, 111, 320], impose a model of multivariate Gaussian distributions on gene expression data. Although the Gaussian model works well for gene-sample data (where the expression levels of genes are measured under a collection of samples), it may not be effective for time-series data (where the expression levels of genes are monitored over a continuous series of time points). This occurs because the Gaussian model treats the time points as unordered, static attributes and ignores the inherent dependency of the gene expression levels over time.

To better describe the gene expression dynamics in time-series data, several new models have been introduced, including [16, 186, 230, 244]. In [16], each gene expression profile was modeled as a *cubic spline* and each time point influences the overall smooth expression curve. Luan et al. [186] independently developed a similar model, called *B-splines*. In [230], Ramoni et al. assumed that the time-series follow an *autoregressive model*, where the value of the series at time t is a linear function of the values at several previous time points. Schliep et al. [244] proposed a restricted *hidden Markov model* to account for the dependencies in time-series data.

An important advantage of model-based approaches is that they provide an estimated probability γ_r^i that data object x_r will belong to cluster C_i. However, gene expression data are typically "highly-connected" [153]; there may be instances in which a single gene has a high correlation with two or more different clusters. Thus, the probabilistic feature of model-based clustering is particularly suitable for gene expression data. However, model-based clustering relies on the assumption that the data set fits a specific distribution, which may often not be the case. The modeling of gene expression data sets, in particular, is an ongoing effort by many researchers, and there is currently no well-established model to represent gene expression data. Yeung et al. [320] studied several kinds of commonly-used data transformations and assessed the degree to which three gene expression data sets fit the multi-variant Gaussian model assumption. The raw values from all three data sets fit the Gaussian model poorly, and there is no uniform rule to indicate which transformation would best improve this fit.

5.4 Hierarchical Approaches

In contrast to partition-based clustering, which attempts to directly decompose the data set into a set of disjoint clusters, hierarchical clustering generates a hierarchical series of nested clusters which can be graphically represented by a tree, called a *dendrogram*. The branches of a dendrogram not only record the formation of the clusters but also indicate the similarity between the clusters. By cutting the dendrogram at some level, we can obtain a specified number of clusters. By reordering the objects such that the branches of the corresponding dendrogram do not cross, the data set can be arranged with similar objects placed together. Hierarchical clustering algorithms can be further divided into either *agglomerative* approaches or *divisive* approaches, based on the formation of the hierarchical dendrogram.

5.4.1 *Agglomerative Algorithms*

Agglomerative algorithms (which take a bottom-up approach) initially regard each data object as an individual cluster, and, at each step, merge the closest pair of clusters until all the groups are merged into one cluster. Different measures of *cluster proximity*, such as single link, complete link and minimum-variance [84, 161], are used to derive various merge strategies.

Eisen et al. [92] applied an agglomerative algorithm called *UPGMA* (for Unweighed Pair Group Method with Arithmetic Mean) and adopted a method to graphically represent the clustered data set. In this method, each cell of the gene expression matrix is colored according to the measured fluorescence ratio, and the rows of the matrix are re-ordered based on the hierarchical dendrogram structure and a consistent node-ordering rule. After clustering, the original gene expression matrix is represented by a colored table (see Figure 5.6), where large contiguous patches of color represent groups of genes that share similar expression patterns over multiple conditions.

Hierarchical clustering not only groups together genes with similar expression pattern but also provides a natural way to graphically represent the data set. The graphic representation gives users a thorough inspection of the whole data set so that the users can obtain an initial impression of the distribution of data. Eisen's method is much favored by many biologists and has become one of the most widely-used tools in gene expression data analysis [6, 7, 92, 147, 219].

However, as pointed out in previous studies [15, 271], traditional agglomerative clustering algorithms may not be robust to noise. They often base merging decisions on local information and never trace back to ensure that poor decisions made in the initial steps are corrected later. In addition, hierarchical clustering results in a dendrogram, with no guidance on cutting the dendrogram to derive clusters. Given a typical gene expression data set with thousands of genes, it is unrealistic to expect users to manually inspect the entire tree.

To render the traditional agglomerative method more robust to the noise, Šášik et al. [238] proposed a novel approach called *percolation clustering*. In essence, percolation clustering adopts a statistical bootstrap method to merge two data objects (or two subsets of data objects) when they are significantly coherent with each other. In [15], Bar-Joseph et al. replaced the traditional binary hierarchical tree with a k-ary tree, where each non-leaf node was allowed to have at most k children. A heuristic algorithm was also presented to construct the k-ary tree, which reduced susceptibility to noise and generated an optimal order for the leaf nodes. These two approaches increase the robustness of the derived hierarchical tree. However, neither of them indicates how to cut the dendrogram to

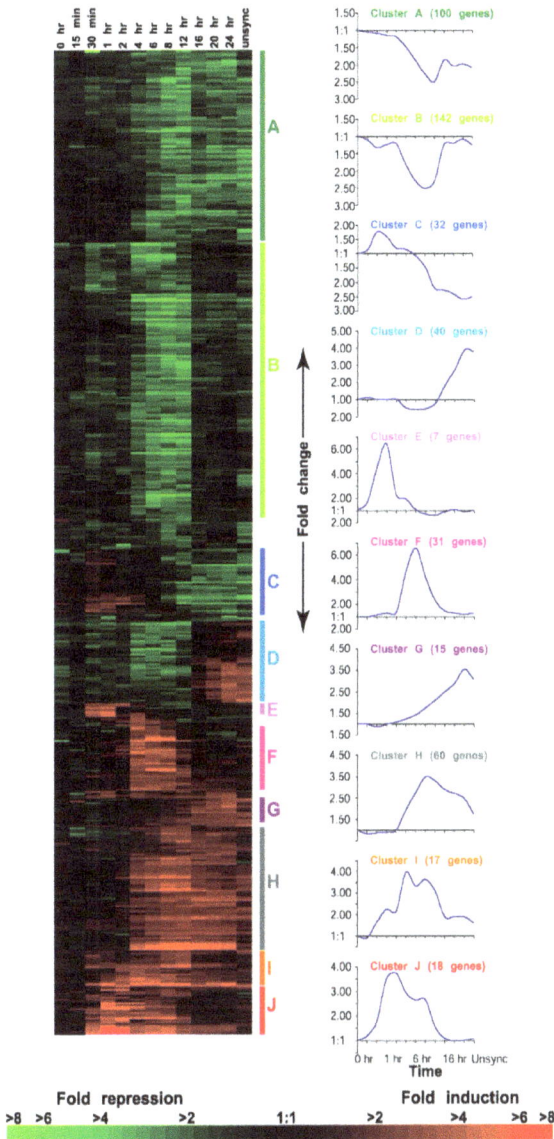

Fig. 5.6   An image showing the different classes of gene expression profiles. Five hundred and seventeen genes whose mRNA levels changed in response to serum stimulation were selected. This subset of genes was clustered hierarchically into groups on the basis of the similarity of their expression profiles, using the procedure of Eisen et al. [92]. The expression pattern of each gene in this set is displayed here as a horizontal strip. For each gene, the ratio of mRNA levels is represented by a color, according to the color scale at the bottom. The graphs show the average expression profiles for the genes in the corresponding "cluster" (indicated by the letters A to J and color coding). Reprinted with permission from [147]. Copyright 1999 AAAS.

obtain meaningful clusters.

Seo et al. [248] developed an interactive tool, called the *Hierarchical Clustering Explorer (HCE)*, to help users derive clusters from the dendrogram. HCE visualizes the dendrogram by setting the distance from the root to an internal node N according to the coherence between the two children N_1 and N_2 of N. That is, the more coherent are N_1 and N_2, the more distant is N from the root. A user can select how to cut the dendrogram horizontally by dragging the *"minimum similarity bar."* However, the HCE system is only a visualization tool that facilitates inspection of the dendrogram and, like the methods discussed above, does not provide guidance on cutting the dendrogram.

5.4.2   *Divisive Algorithms*

Divisive algorithms (i.e., top-down approaches) start with a single cluster which contains all the data objects. The algorithm splits clusters iteratively until each cluster contains only one data object or a certain stop criterion is met. The various divisive approaches are primarily distinguished by the manner in which clusters are split at each step.

5.4.2.1   *DAA: Deterministic annealing algorithm*

Alon et al. [7] used a divisive approach, called the *deterministic-annealing algorithm (DAA)* [233, 234] to split genes. First, two initial cluster centroids C_j, $j = 1, 2$, were randomly defined. The expression pattern of gene k was represented by a vector $\vec{g}_k$, and the probability of gene k belonging to cluster j was determined according to a two-component Gaussian model:

$$P_j(\vec{g}_k) = exp(-\beta|\vec{g}_k - C_j|^2)/\sum_j exp(-\beta|\vec{g}_k - C_j|^2).$$

The cluster centroids were recalculated by $C_j = \sum_k \vec{g}_k P_j(\vec{g}_k)/\sum_k P_j(\vec{g}_k)$, and the *EM algorithm* [69] (discussed in Section 5.3.4) was then applied to solve P_j and C_j. When $\beta = 0$, there was only one cluster, $C_1 = C_2$. Increasing β in small increments to a threshold resulted in two distinct, converged centroids. The entire data set was recursively split until each cluster contained only one gene.

5.4.2.2 *SPC: Super-paramagnetic clustering*

Super-Paramagnetic Clustering (SPC), proposed by Blatt et al. in [31], is based on the physical properties of an inhomogeneous ferromagnetic model. *SPC* first transforms the data set into a *distance graph* where each vertex corresponds to a data object. Two vertices V_i and V_j in the graph are connected by an edge if and only if their corresponding objects O_i and O_j satisfy the *K-mutual-neighbor criterion*; i.e., O_j is one of the K nearest objects to O_i, and vice versa. Moreover, an edge in the distance graph is associated with a weight $J_{ij} > 0$, with a smaller Euclidean distance (see Equation 5.1) between O_i and O_j associated with a greater weight.

The clustering process of *SPC* is equivalent to the partitioning of a weighted graph. Given a specific temperature T, an *correlation function* is estimated between each pair of neighboring vertices, based on the edge weight J_{ij} and the temperature T. Two vertices with high correlation will be assigned to the same cluster. At very low temperatures, all objects belong to a single cluster. When T is increased, clusters subdivide until, at very high temperatures, each object forms a single cluster. The clusters found at all temperatures form a dendrogram.

Getz et al. [110] applied *SPC* to a yeast cell-cycle expression data set [263]. Figure 5.7 illustrates the dendrogram generated by *SPC*. Three out of the eleven identified clusters corresponded to known phases of the cell cycle, while other clusters revealed features that had not been previously identified and may serve as the basis of future experimental investigations. In [109], the authors compared the performance of *SPC* with that of an agglomerative hierarchical algorithm (average linkage) [123] and *SOM* [171]. Some of the advantages of *SPC* are its robustness against noise and initialization, a clear signature of cluster formation and splitting, and an unsupervised self-organized determination of the number of clusters at each resolution.

In summary, hierarchical clustering not only groups together genes with similar expression pattern but also provides a natural means to graphically represent the data set. The graphic representation allows users to make a thorough inspection of the entire data set and thus obtain an initial impression of the distribution of data. Eisen's method, discussed above, is favored by many biologists and has become the most widely-used tool in gene expression data analysis [6, 7, 92, 147, 219]. As noted earlier, though, the conventional agglomerative approach suffers from a lack of robustness [271], and a small perturbation of the data set may greatly change the structure

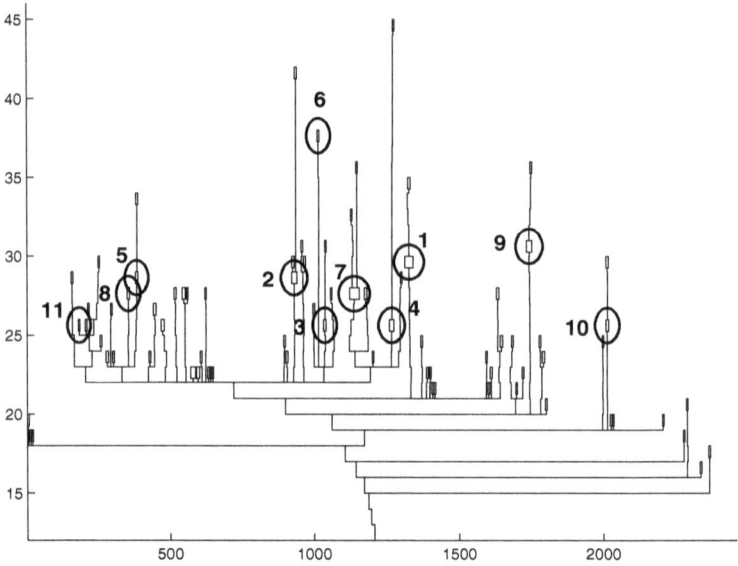

Fig. 5.7   Dendrogram of genes generated by SPC for yeast cell-cycle expression data (Figure from [110]).

of the hierarchical dendrogram. Another drawback of the hierarchical approach is its high computational complexity. To construct a "complete" dendrogram (each leaf node corresponds to one data object and the root node corresponds to the entire data set), the clustering process should engage in $\frac{n^2-n}{2}$ merging (or splitting) steps. The time complexity for a typical agglomerative hierarchical algorithm is $O(n^2 logn)$ [149]. Furthermore, for both agglomerative and divisive approaches, the "greedy" nature of hierarchical clustering prevents the refinement of the previous clustering steps. If a "bad" decision is made in the initial steps, it can never be corrected in the following steps.

5.5   Density-Based Approaches

Density-based approaches describe the distribution of a given data set by the "density" of data objects. The clustering process involves a search of the "dense areas" in the object space [95].

5.5.1 *DBSCAN*

The DBSCAN algorithm introduced by Ester et al. [95] is grounded on a density-based notion of clusters. To measure the "density" of data objects, DBSCAN first defines the *Eps-neighborhood* of an object p as a set of objects $N_{eps}(p)$ such that the distance between p and each object q in $N_{eps}(p)$ is smaller than a user-specified threshold *Eps*. Intuitively, an object p has a "high" density (and thus is called a *core object*) if $N_{eps}(p) \geq MinPts$, where $MinPts$ is a user-specified threshold.

An object p is called *density-connected* to an object q with respect to *Eps* and *MinPts* if there is an object o such that both p and q are *density-reachable* from o (see Figure 5.8(a)). The *density-reachability* from q to p with respect to *Eps* and *MinPts* is described by two situations. First, p is called *directly density-reachable* from q if $p \in N_{eps}(q)$ and q is a *core object*. A more general situation is given as follows: p is *density-reachable* from q if there is a chain of objects $p_1, \ldots, p_n$, $p_1 = q$, $p_n = p$ such that p_{i+1} is *directly density-reachable* from p_i ($1 \leq i \leq n-1$) (see Figure 5.8(b)).

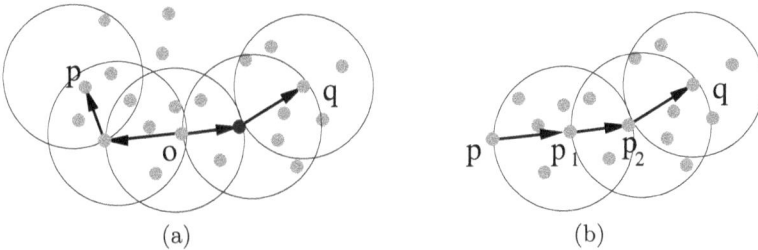

Fig. 5.8   (a) Density connectivity and (b) density reachability in DBSCAN. Figures are adapted from [95] with permission from ACM.

Given the two thresholds *Eps* and *MinPts*, DBSCAN defines a *cluster* C as a non-empty set of data objects satisfying the following two conditions: (1) $\forall p, q$: if $p \in C$ and q is *density-reachable* from p with respect to *Eps* and *MinPts*, then $q \in C$ (maximality); and (2)$\forall p, q \in C$, p is *density-connected* to q with respect to *Eps* and *MinPts* (connectivity). *Noise* is those objects which fall outside all clusters.

The clustering process of *DBSCAN* scans the data set only once and reports all the clusters and noise. For each data object o, DBSCAN checks the Eps-Neighborhood $N_{eps}(o)$ of o. If $N_{eps}(o)$ contains more than $MinPts$ data objects (i.e., o is a core object), DBSCAN creates a new cluster and then iteratively retrieves all data objects which are density-reachable from

o with respect to Eps and $MinPts$. If o is a border object, no points are density-reachable from o, and DBSCAN visits the next point in the data set.

5.5.2 *OPTICS*

While DBSCAN is able to discover clusters with arbitrary shape and is quite efficient for large data sets, the algorithm is very sensitive to input parameters. It may generate very different clustering results from slightly differences in parameter settings [11]. Building on DBSCAN, Ankerst et al. [11] introduced the algorithm *OPTICS*. OPTICS does not generate clusters explicitly but instead creates an ordering of the data objects and illustrates the cluster structure of the data set. In essence, this ordering contains information that is equivalent to the clustering of DBSCAN, with a wide range of parameter settings.

Given the parameters Eps and $MinPts$, DBSCAN categorized the objects in a data set into *core objects*, *border objects*, and *noise* on the basis of their density. OPTICS introduces two additional concepts to provide a more precise description of object density. First, the *core-distance* of an object o is the smallest Eps' value that determines o to be a core object; that is, core-distance$(o)=\min\{Eps'| \ |N_{Eps'}(o)| \geq MinPts\}$. If o is not a core object, the core-distance of o is undefined. Second, the *reachability-distance* of an object p with respect to another object o is the greater of the core-distance of o or the distance between o and p. In other words, reachability-distance$(p,o)=\max($core-distance$(o),$distance$(o,p))$. If o is not a core object, the reachability-distance of p with respect to o is undefined. For example, in Figure 5.9, assume that the parameters $Eps = 4mm$ and $MinPts = 5$, and the core-distance Eps' of object o is 2mm. Let $d(o, p_1)$ and $d(o, p_2)$ denote the distance between objects o and p_1 and the distance between o and p_2, respectively. According to the definition, the reachability-distance of p_1 with respect to o is Eps' since $d(o, p_1) < Eps'$, while the reachability-distance of p_2 with respect to o is $d(o, p_2)$ since $Eps' < d(o, p_2)$.

The OPTICS algorithm generates an ordering of objects by scanning the data set once. The algorithm maintains a queue in which data objects are sorted in ascending order according to their reachability-distances. The ordering of objects is simply the sequence in which the data objects are extracted from the queue. Once the ordering of the objects and their reachability-distances are obtained, clusters can be extracted using this information. Figure 5.10 shows a reachability plot for a simple two-

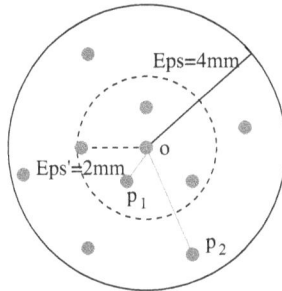

Core-distance(o)=2mm, Reachability-distance(p_1,o)= Core-distance(o)=2mm, and Reachability-distance(p_2,o)=d(o,p_2).

Fig. 5.9 Core-distance and reachability-distance in OPTICS (Figure is adapted from [11] with permission from ACM).

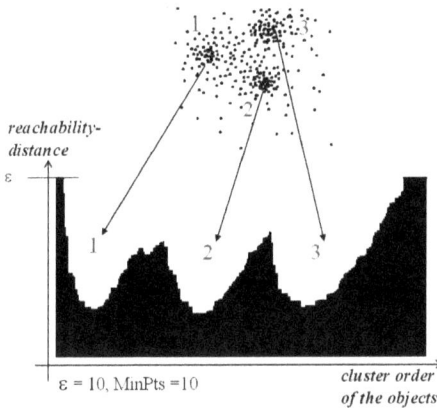

Fig. 5.10 Cluster ordering in OPTICS (Figure is from [11] with permission from ACM).

dimensional data set, where each valley in the plot corresponds to one cluster.

5.5.3 DENCLUE

Unlike the local density measures employed by DBSCAN and OPTICS, DENCLUE [134] measures object density from a global perspective. Data objects are assumed to "influence" each other, and the density of a data object is the sum of influence functions from all data objects in the data set. The incorporation of a variety of *influence functions* allows DENCLUE to be a generalization of many partition-based, hierarchical, and density-based clustering methods.

Given x and y as data objects in a p-dimensional space F^p, the *influence function* of y on x is a function $f_B^y : F^p \to R_0^+$, which is defined in terms of a basic influence function f_B:

$$f_B^y(x) = f_B(x, y). \tag{5.5}$$

The influence function can be arbitrarily set. In practice, it can be defined on the basis of a distance measure, such as the Euclidean distance function (see Section 5.2.1). For example, the basic influence function can be a *square wave influence function*,

$$f_{Sqaure} = \begin{cases} 0 \text{ if } d(x, y) > \sigma \\ 1 \text{ otherwise} \end{cases}, \tag{5.6}$$

or a *Gaussian influence function*,

$$f_{Guass} = e^{-\frac{d(x,y)^2}{2\sigma^2}}, \tag{5.7}$$

where $d(x, y)$ is the Euclidean distance between x and y, and σ is a parameter. Figure 5.11 illustrates the effect of different influence functions.

The *density function* at an object $x \in F^p$ is defined as the sum of influence functions of all data objects. Given n data objects, $D = \{x_1, \ldots, x_n\} \subset F^p$, the density function at x is defined as

$$f_B^D = \sum_{i=1}^{n} f_B^{x_i}(x). \tag{5.8}$$

For example, the density function based on the Gaussian influence function (Formula 5.7) is

$$f_{Gauss}^D(x) = \sum_{i=1}^{n} e^{-\frac{d(x,x_i)^2}{2\sigma^2}}. \tag{5.9}$$

Based on the definition of density function $f_B^D(x)$, the *gradient* of $f_B^D(x)$ is defined as

$$\nabla f_B^D(x) = \sum_{i=1}^{n} (x_i - x) \cdot f_B^{x_i}(x). \tag{5.10}$$

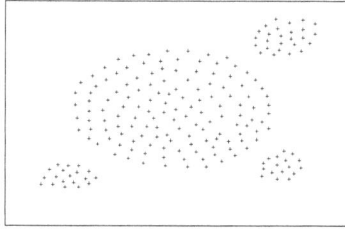

(a) An example data Set

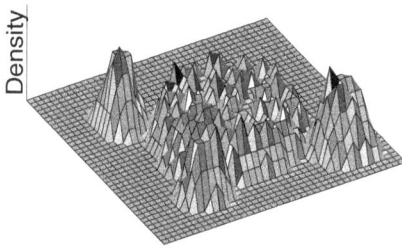

(b) Square Wave function (c) Gaussian function

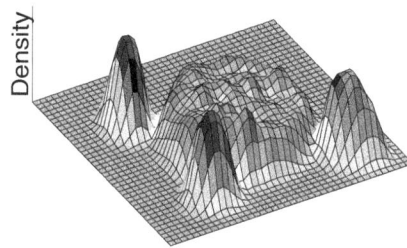

Fig. 5.11 Examples illustrating various influence functions of DENCLUE (Figures are from [134] with permission from ACM).

Where the Gaussian influence function is used, the gradient is defined as

$$\nabla f_{Gauss}^{D}(x) = \sum_{i=1}^{n}(x_i - x) \cdot e^{-\frac{d(x,x_i)^2}{2\sigma^2}}. \tag{5.11}$$

The gradient of the density function at an object x indicates the direction along which a denser area can be reached. In other words, if we start from x and "climb" along the direction of the gradient of its density function, we will finally reach the local maxima of the density function, or the *density attractor*. Formally stated, an object x is *density-attracted* to a density attractor x^*, if and only if there exists a series of points in the p-dimensional space such that $x^0 = x$ and

$$x^i = x^{i-1} + \delta \cdot \frac{\nabla f_B^{D}(x^{i-1})}{\|\nabla f_B^{D}(x^{i-1})\|}. \tag{5.12}$$

Based on the above notations, DENCLUE defines both *center-defined clusters* and *arbitrary-shaped clusters*. A center-defined cluster with respect to parameter ξ for a density attractor x^* is a subset $C \subseteq D$, with $x \in C$ being density-attracted by x^* and $f_B^D(x^*) \geq \xi$. Objects $x \in D$ are called *outliers* if they are density-attracted by a local maximum x_o^* with $f_B^D(x_o^*) < \xi$. An arbitrary-shape cluster with respect to ξ for the set of density-attractors X is a subset $C \subseteq D$, where 1) $\forall x \in C$, $\exists x^* \in X$: $f_B^D(x^*) \geq \xi$, x is density-attracted to x^*; and 2) $\forall x_1^*, x_2^* \in X$: $\exists$ a path $Q \subset F^p$ from x_1^* to x_2^* with $\forall q \in Q$: $f_B^D(q) \geq \xi$.

DENCLUE rests on a solid mathematical foundation and allows a compact mathematical description of arbitrarily-shaped clusters. Moreover, by assigning different influence functions, DENCLUE is a generalization of many partition-based, hierarchical, and density-based clustering methods. Furthermore, DENCLUE has good clustering properties for data sets with high dimensionality and large amounts of noise. The computational efficiency of DENCLUE is significantly higher than some influential algorithms, such as DBSCAN (by a factor up to 45) [134]. However, the method requires careful selection of the density parameter σ and the noise threshold ξ, as the setting of such parameters may significantly influence the quality of the clustering results [120]. Moreover, DENCLUE outputs all clusters at the same level. Therefore, it cannot support an exploration of hierarchical cluster structures which exploits users' domain knowledge.

To the best of our knowledge, the density-based approaches described above have not been directly applied to gene expression data for cluster analysis. But the basic idea of density-based clustering has been incorporated into the approach discussed below in Section 5.6.

5.6   GPX: Gene Pattern eXplorer

The various existing approaches discussed in this chapter have proved to be useful in identifying co-expressed gene groups and coherent patterns. However, the specific characteristics of gene expression data and special requirements arising from the domain of biology still pose challenges to the effective clustering of gene expression data [152, 153, 154].

A microarray data set typically contains multiple groups of coexpressed genes and their corresponding coherent patterns. As a general observation, *there is usually a hierarchy of coexpressed genes and coherent patterns in a typical gene expression data set*. For example, as shown in Figure 5.12, a group of co-expressed genes S taken from Iyer's data set [147] can be split into two subgroups S_1 and S_2, and S_2 can be further split into two sub-

subgroups S_{21} and S_{22}. The expression profiles of genes within each smaller subgroup become increasingly more uniform and the patterns more coherent when compared with the higher-level groups. Therefore, these groups of coexpressed genes form a hierarchy. At the upper levels of the hierarchy, large groups of genes generally conform to "rough" coherent expression patterns. At lower hierarchical levels, these larger groups are broken into smaller subgroups. Those smaller groups of coexpressed genes conform to "fine" coherent patterns; these patterns inherit some features from the "rough" patterns and add some distinct characteristics.

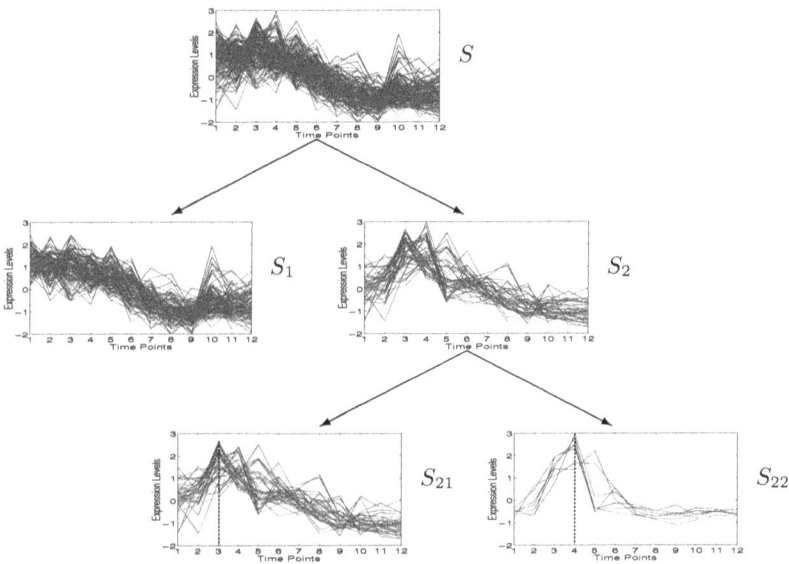

Fig. 5.12   The hierarchy of a coexpressed gene group.

The apparent simplicity of this organization is complicated by *the lack of a rigorous definition or objective standard to unambiguously identify co-expressed gene groups*. The interpretation of "co-expression" often depends on knowledge from domain experts. Typically, three situations may happen in the analysis of gene expression data:

- Biologists can often bring some prior knowledge to the analysis of a microarray data set. For example, some genes are known to be closely related in function, while some genes are known not to stay in the same cluster. If such prior knowledge is integrated into the clustering process, the mining results may be substantially improved.

- A microarray experiment often involves thousands of genes. However, only a small subset (perhaps several hundreds) of those genes may play important roles in the underlying biological processes. In an initial examination, biologists may browse through the "rough" patterns in the data set. They may then choose several patterns of particular interest and decompose them into "finer" patterns in further analysis. In other words, biologists may have different requirements for the coherence of different parts of the data set.
- The domain knowledge of biologists is typically incomplete. That is, the functions of many genes in a data set are still unclear, and there could be various hypotheses regarding the functions of those genes. For example, in Figure 5.12, the subset of genes S_2 can be split into two sub-subsets S_{21} and S_{22}. The genes in S_{21} and S_{22} exhibit similar expression profiles. The critical difference is that the genes in S_{21} are up-regulated at the third time point, while the expression levels of genes in S_{22} peak at the fourth time point. Two hypotheses could explain this phenomenon. It is possible that the genes in S_{22} are up-regulated by the genes in S_{21}. If this is the case, it is meaningful to split S_2 into S_{21} and S_{22}. Alternatively, the genes in S_{21} and S_{22} may both be up-regulated by some common factors, and the genes in S_{21} have responded more quickly than the genes in S_{22}. In this case, the genes in S_{21} and S_{22} may have similar functions, and it would be appropriate not to split S_2. Given such uncertainties, biologists would prefer an exploratory tool which can illustrate the possible options for partitioning the data set and assist in evaluating the range of hypotheses based on the underlying data structure.

The discussion of this issue leads us to pose the following question: *how can we allow biologists to interactively unfold the hierarchy of groups of co-expressed genes and derive the corresponding coherent patterns?* Flexibility is key, as various users may wish to explore the structure of a data set using criteria appropriate to their research goals and background knowledge.

Another challenge of analyzing microarray data is the high connectivity of gene expression data, which complicates the determination of appropriate cluster boundaries. Gene expression data typically include many genes which fall between the groups of co-expressed genes. These "intermediate" genes build "bridges" between co-expressed gene groups. An example taken from yeast expression data ($CDC28$ [263]) is shown in Figure 5.13. The two genes in the first row have very different expression profiles and thus cannot belong to the same co-expressed gene group. However, in the same data set, we can find a series of genes in which each gene is quite similar to its predecessor; such a series is illustrated in the lower rows of Figure 5.13.

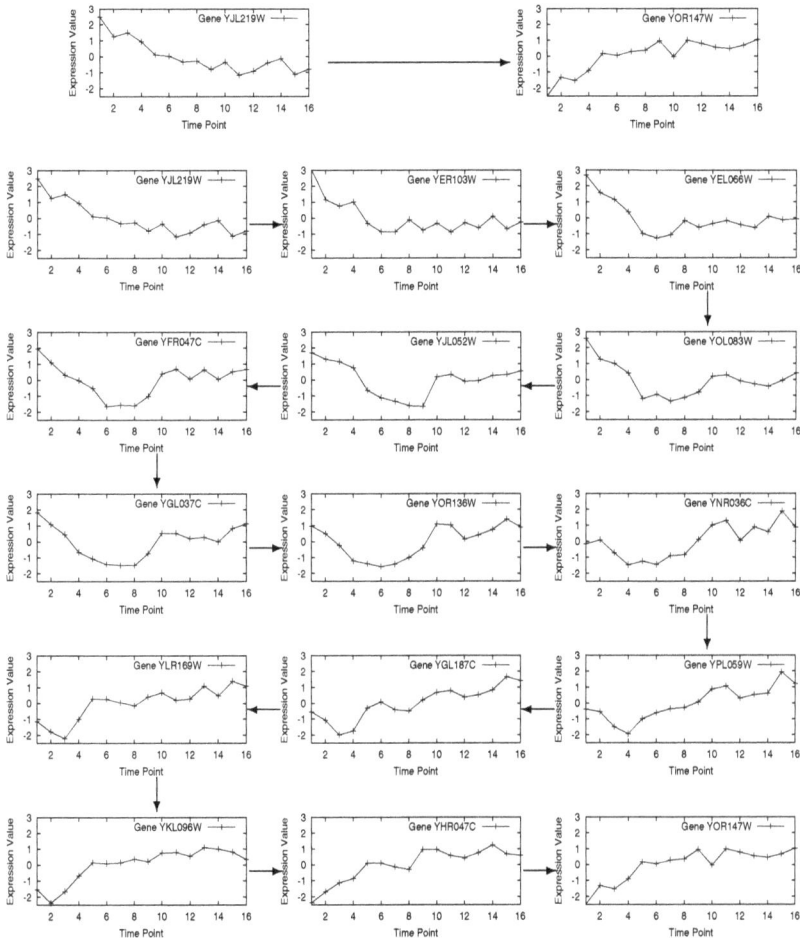

Fig. 5.13   The gradual change from one expression profile to a complete different profile.

The biological role of "intermediate" genes can be different. On the one hand, some may participate in several cellular processes and, thus, should be classified into multiple clusters. On the other hand, the majority of intermediate genes may not be involved in any biological processes of interest and, thus, do not belong to any clusters. In other words, these intermediate genes are simply noise. For example, in [54], only 416 out of 6220 monitored transcripts were recognized as five cell-cycle regulated clusters, while the remaining 5, 804 located around the clusters were considered to be noise. Among the 416 cell-cycle regulated genes, 22 belong to multiple cell cycle phases. The large amount of intermediate genes poses a big challenge: *gene expression data are often highly connected, and it is difficult to determine the borders between clusters.* Most methods make the decision by force and may fall in one of the following two situations:

- The data set is decomposed into numerous small clusters. Some clusters will consist of groups of co-expressed genes, while many clusters will be made up of intermediate genes. Since there is no absolute standard, such as size or compactness, with which to rank the resulting clusters, it may require significant user effort to distinguish meaningful from those trivial clusters. This situation is illustrated in Figure 5.14(a).
- The data set is decomposed into several large clusters, each of which contains both co-expressed genes and many intermediate genes. However, the heavy representation of intermediate genes may lead to the skewing of cluster centroids. These "warped" centroids do not accurately represent the coherent patterns in the groups of co-expressed genes. This situation is exemplified in Figure 5.14(b).

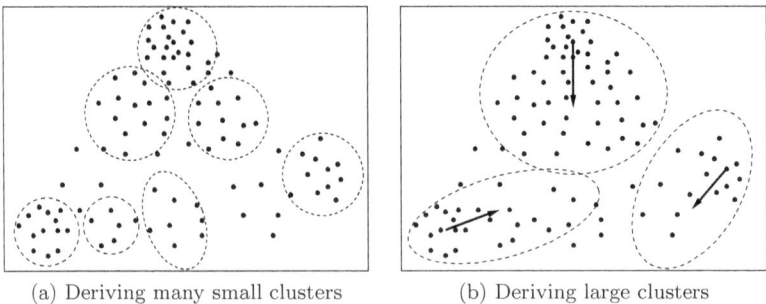

(a) Deriving many small clusters   (b) Deriving large clusters

Fig. 5.14   Handling intermediate genes. (a) Deriving many small clusters and (b) deriving large clusters.

In the remainder of Section 5.6, we will explore new ways to address the challenges associated with mining co-expressed genes and coherent patterns from two-dimensional gene expression data. A framework of *interactive exploration* has the potential to overcome many of the drawbacks of conventional methods for the analysis of microarray data. This Gene Pattern eXplorer (GPX) approach[2] supports exploration by users as guided by their domain knowledge and accommodates disparate user requirements for varying degrees of coherence in different parts of the data set. The interactive exploration framework consists of three major computational components: a *density-based model* to describe the coherence relationship between genes, an *attraction tree structure* which summarizes these coherence relationships in a manner which facilitates interactive exploration, and a *coherent pattern index graph* to give users ranked indications of the existence of coherent patterns. This approach also adopts a novel strategy for handling intermediate genes. Users first identify coherent patterns and then determine the borders of co-expressed gene groups on the basis of the distance between a gene and the coherent patterns. An intermediate gene is thus allowed to participate in more than one cluster. The results of an extensive performance study conducted with both synthetic and real-world gene expression data sets have indicated that this approach is effective and scalable in mining gene expression data.

5.6.1 The Attraction Tree

To enable the interactive exploration of coherent gene expression patterns, information regarding the coherence among genes must be extracted and organized. GPX employs a novel density-based method for constructing an attraction tree structure. Once the attraction tree is built, the original data set will no longer need to be referenced. We will first describe the measure of the distance between two genes and the definition of the density of genes. The structure of the attraction tree will then be explicated.

5.6.1.1 The distance measure

For gene expression data, people are usually interested in the overall shapes of expression profiles instead of the absolute magnitudes. The commonly used Euclidean distance does not work well for scaling and shifting profiles [296]. Hence, the *Pearson's correlation coefficient* is used to measure the similarity between two expression patterns (see Section 5.2 for details of Euclidean distance and Pearson's correlation coefficient).

---

[2]With permission from IEEE.

Given an object O, we *standardize* the object to O' such that each O' has a mean of 0 and a variance of 1 over all the attributes. Then, the similarity and distance between the data objects are defined respectively as

$$similarity(O_i, O_j) = d_P(O'_i, O'_j) \qquad (5.13)$$

and

$$distance(O_i, O_j) = d_E(O'_i, O'_j), \qquad (5.14)$$

where $d_P(O'_i, O'_j)$ and $d_E(O'_i, O'_j)$ are the Pearson's correlation coefficient and the Euclidean distance between O'_i and O'_j, respectively. After the normalization, the similarity definition and the distance definition are consistent, i.e., given objects O_1, O_2, O_3 and O_4, $similarity(O_1, O_2) > similarity(O_3, O_4)$ implies $distance(O_1, O_2) < distance(O_3, O_4)$.

5.6.1.2 *The density definition*

The density of a data object O reflects the distribution of the other objects in O's neighborhood. The radius-based measure defines the density of O as the number of data objects within O's ε-neighborhood, the density of O, where ε is a radius parameter specified by the user. In such a definition, every neighbor of O is treated equally, no matter what the real distance between O and the neighbor is. One drawback of such a measure is that it is insensitive to the internal structure within a radius ε. For example, in Figure 5.15(a), intuitively, object O_2 should have a higher density than object O_1. However, under the definition of the radius-based density, their density measures are the same.

As an alternative measure, the *k-nearest neighbor density* (KNN) uses

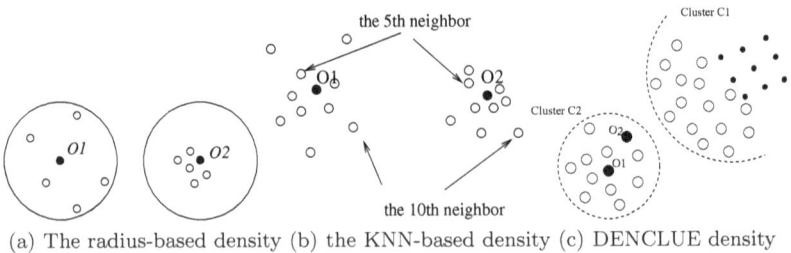

(a) The radius-based density (b) the KNN-based density (c) DENCLUE density

Fig. 5.15   The drawbacks of some density measures.

the distance between an object and its kth nearest neighbor; in this definition, a smaller distance, indicates a higher density. That solves the problem
in Figure 5.15(a). However, a global value k may not fit all the objects. For
example, in Figure 5.15(b), object O_2 should have a higher density than
object O_1 does. However, when $k = 5$ and $k = 10$, both objects have an
exactly same density value.

DENCLUE [134] (see Section 5.5.3) defines an *influence function* to
describe the influence between two objects. For example, the *Gaussian
influence function* can be defined as

$$f(O_i, O_j) = e^{-\frac{d(O_i,O_j)^2}{2\sigma^2}},$$

where $d(O_1, O_2)$ is the distance between objects O_1 and O_2 and σ is a
parameter. Given a data set $\mathcal{D}$, the density of an object O is the sum
of influences from all the objects in the data. That is, $density(O) = \sum_{O_j \in \mathcal{D}} f(O, O_j)$.

DENCLUE takes a "global" view of the neighborhood of a data object
O, i.e., all the data objects in the data set is considered to contribute to
the density of O. This avoids the problem of picking the parameter values
such as ε and k. However, counting the influence from related far away
data objects may corrupt the "real" cluster structure. For example, in Figure 5.15(c), object O_1 is the medoid of cluster C_2 and should have a higher
local density than O_2. However, according to the definition by DENCLUE,
O_2 has a higher density than O_1 since it is close to a neighboring large
cluster C_1.

Basically, we want to measure the density of an object O as the sum
of the influences from objects within its own cluster, while ignoring the
contribution of objects in other clusters. In [251], Sharan et al. use an *EM
algorithm* [69] (discussed in Section 5.3.4) to estimate the average pairwise
similarity $\bar{S}$ between data objects within the same cluster. They assume
that both the pairwise similarity S_1 between data objects within the same
cluster and the pairwise similarity S_0 between data objects in different
clusters follow normal distribution, if the Pearson's correlation coefficient
is used as the similarity measure and the number of data objects in the
data set is large enough. In this section, we will use this method, and,
consequently, modify the density definition formulated by DENCLUE as

$$f(O_i, O_j) = e^{-\frac{distance(O_i,O_j)^2}{2\sigma^2}} \tag{5.15}$$

$$density(O) = \sum_{O_j \in \mathcal{D}, similarity(O,O_j) \geq \bar{S}} f(O, O_j) \qquad (5.16)$$

where $similarity(O, O_j)$ and $distance(O_i, O_j)$ are defined by Equations 5.13 and 5.14, and $\bar{S}$ is the estimated average pairwise similarity between data objects within the same cluster. Objects O_i and O_j are called *neighbors* if $similarity(O_i, O_j) \geq \bar{S}$. The *EM* algorithm [69] is used to estimate the value of $\bar{S}$. We will address the determination of an appropriate value for parameter σ in Section 5.6.2.3.

5.6.1.3 *The attraction tree*

Based on the density definition, a gene O_i is "influenced" by its neighbors. The direction of the *united influence* from all neighbors of O_i is an $m-$dimensional vector determined by

$$I(O_i)^{(d)} = \sum_{O_j \in \mathcal{D}, similarity(O_i,O_j) \geq \bar{S}} \frac{1}{f(O, O_j)} O_j^{(d)} \quad (1 \leq d \leq m), \quad (5.17)$$

where m is the number of attributes of object O_j, and $O_j^{(d)}$ is the d-th attribute of O_j.

Intuitively, the direction of the united influence on object O_i indicates the dense region in O_i's neighborhood. If O_i moves toward the direction of $I(O_i)$, O_i is likely to reach an area with higher density. In particular, if an object O_i has a higher density than all of its neighbors, O_i is a *local maximum*. There are two special cases of a local maximum. In the first case, if an object O_i has no neighbors at all, O_i is a *noise object*. In the second case, if an object O_i has a higher density than any other object in the data set, O_i is the *global maximum*, denoted by O_{max}. A data object O_i is "attracted" by its *attractor* O_j (denoted by $O_i \rightarrow O_j$) according to the following definition:

$$A(O_i) = \begin{cases} O_i & \text{if } O_i = O_{max} \\ \arg \max_{O_1 \in A_1} similarity(O_1, O_i) & \text{if } O_i \text{ is a noise object} \\ \arg \max_{O_2 \in A_2} similarity(O_2, I(O_i)) & \text{if } O_i \text{ is not a local} \\ & \quad \text{maximum} \\ \arg \max_{O_1 \in A_1} similarity(O_1, I(O_i)) & \text{otherwise.} \end{cases}$$

$$(5.18)$$

Algorithm 5.4   : Construct Attraction Tree

Input: A set $\mathcal{D}$ of n objects and the number of nearest neighbors K
Output: The attractor of each object $O_i \in \mathcal{D}$
Proc ConstructTree()
  (1) Use EM algorithm to estimate the average pairwise similarity
     within a cluster $\bar{S}$;
  (2) for each object O_i
    (2.1) determine O_i's neighbors
       $N(O_i) = \{O_j | similarity(O_i, O_j) \geq \bar{S}\}$;
    (2.2) order $N(O_i)$ according to the similarity with O_i
       in descending order;
    (2.3) calculate the density of O_i and the direction of the united
       influence $I(O_i)$;
  end for
  (3) let $local\_maxima\_set = \emptyset$;   //record local maxima's
  (4) for each object O_i // find attractor
    (4.1) let $attractor(O_i) = O_i$, $A(O_i) = \emptyset$, $k = 1$ and $j = 1$;
    (4.2) repeat
        let O_j be the jth element in $N(O_i)$;
        if $(density(O_j) > density(O_i))$
          or $((density(O_j) == density(O_i))$ and $(j < i))$
        then $A(O_i)\cup = \{O_j\}$, $k++$
        end if
        let j++;
        until $(k == K)$ or $(j == ||N(O_i)||)$
    (4.3) if $A(O_i) \neq \emptyset$
       then select $attractor(O_i)$ according to Equation 5.18
       else $local\_maxima\_set\cup = \{O_i\}$
       end if
  end for
  (5) for each object $O_i \in local\_maxima\_set$ //handle local maxima's
    find $A(O_i)$ and $attractor(O_i)$ as described in Section 5.6.1.3
  end for
end Proc

In the above definition, A_1 is the set of local maximums O_1 such that O_1 has a higher density than O_i, while A_2 is the set of O_i's neighbors O_2 such that O_2 has a higher density than O_i. The attraction from an object to another (i.e., $O_i \to O_j$) forms a partial order. Given any data object $O_i \neq O_{max}$, we can recursively trace the attractor of O_i until we reach O_{max}. Therefore, we can derive an *attraction tree* T where each node corresponds to an object O_i such that

$$Parent(O_i) = \begin{cases} nil & \text{if } O_i = O_{max} \\ A(O_i) & \text{otherwise.} \end{cases}$$

The weight of each edge $e(O_i, O_j)$ is defined as the similarity between O_i and O_j.

To locate the attractor of object O, all neighbors of O must be searched. However, for a large and highly-connected data set, this operation can be expensive. As an approximation, we only keep the K nearest objects $O_j \in A(O)$ with respect to O. There are two extreme settings for K. When $K = 1$, the attractor of O becomes the nearest higher-density neighbor of O. At the other extreme, when $K = \infty$, the approximation is ignored, and all neighbors of O are exhaustively searched to find the attractor of O (see Algorithm 5.4).

Intuitively, if K is set to a small number, we need only search a relatively small neighborhood for the attractor. As a result, an object O_i may first be attracted to a "lower-level" local maximum O_{i1}, which covers a relatively small neighborhood A_1 (Figure 5.16); while O_{i1} will then be attracted to a "upper-level" local maximum O_{i2}, which covers a larger neighborhood A_2. However, when we assign a large number of K, we will need to search a larger neighborhood for the attractor. In this situation, the object O_i is more likely to be directly attracted to the "upper-level" local maxima O_{i2}. That is, a smaller K favors a more detailed local structure, and, on average, a data object is attracted to the root of the attraction tree through more intermediate steps. In contrast, a large K "shortcuts" the attractor paths between a data object and the root and tends to delineate rough structures.

5.6.1.4 *An example of attraction tree*

To illustrate the concept of the attraction tree, let us consider a synthetic data set $\mathcal{D}$ as represented in parallel coordinates in Figure 5.17. This data set contains three coherent patterns, P_1, P_2, and P_3. We denote the groups of objects which conform to coherent pattern P_i as G_i and represent the

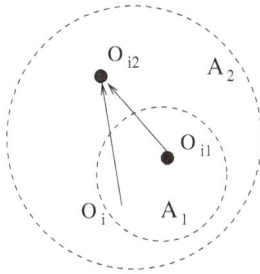

Fig. 5.16 The effect of the number of attractors K.

objects conforming to P_1, P_2, and P_3 by the solid blue, green, and red lines, respectively. There is also some noise in data set $\mathcal{D}$, and this is represented by the dotted cyan lines. Suppose O_1, O_2, and O_3 are the objects with locally maximal density in G_1, G_2, and G_3, respectively. In the resulting attraction tree, other objects in G_1, G_2, and G_3 will be attracted (directly or indirectly) to O_1, O_2, and O_3. Thus, O_1, O_2, and O_3 become the roots of attraction subtrees T_1, T_2, and T_3, respectively, where T_1, T_2, and T_3 contain all the objects of G_1, G_2, and G_3.

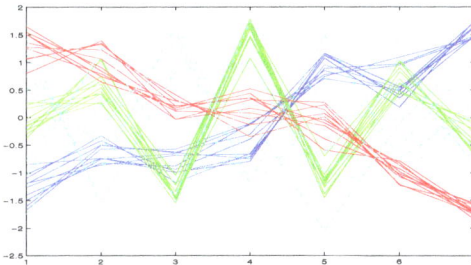

Fig. 5.17 A simplified synthetic gene expression data set.

Figure 5.18(a) is the attraction tree ($K = 1$) for $\mathcal{D}$. In this example, the density of O_2 is greater than that of both O_1 and O_3. Thus, O_2 has the globally maximal density and becomes the root of the tree. O_1 and O_3 are attracted to O_2 and become the roots of subtrees. Figure 5.18(b) is the attraction tree for $\mathcal{D}$ with $K = \infty$, and it reveals a structure similar to that in 5.18(a). However, objects in 5.18(b) tend to be attracted directly to upper-level attractors, giving that tree a flatter structure.

This example demonstrates two characteristics of the attraction tree structure. First, the attraction tree is *self-closed*. A group of objects con-

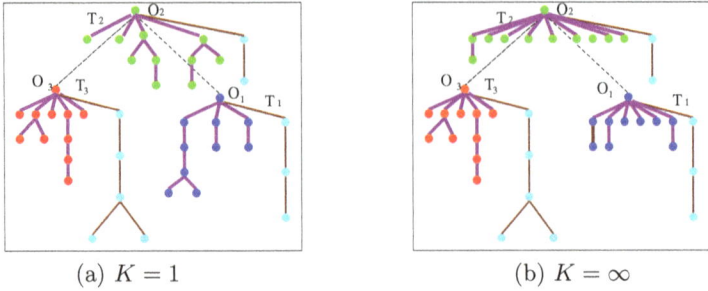

(a) $K = 1$ (b) $K = \infty$

Fig. 5.18   The attraction tree for the data set in Figure 5.17.

forming to the same coherent pattern forms an attraction subtree. Objects
conforming to different coherent patterns are not mixed in the same attraction subtree. Second, the attraction tree is *robust to noise*. The root of each
attraction subtree has the locally maximal density and represents the coherent pattern for that attraction subtree. Objects matching the coherent
pattern stay connected with each other, while noise objects are connected
either to the roots of the subtrees or to each other. A child O_j of a subtree root O_i must conform to the same coherent pattern as O_i if the edge
(O_i, O_j) has a high weight. Otherwise, O_j must be a noise object. Even in
a noisy environment, the density of noise will still be relatively lower than
that of the co-expressed objects. Therefore, the attraction tree structure
will remain stable, and the representative of coherent patterns will not be
perturbed by the presence of noise.

5.6.2   *Interactive Exploration of Coherent Patterns*

We now describe the application of the concepts defined in Section 5.6.1
to the interactive exploration of coherent patterns. The goal is to plot a
coherent pattern index graph, where the genes are ordered into an *index list*
such that the genes sharing a coherent pattern stay close to each other in
the list. Each gene is assigned a coherent pattern index value such that, if
there is a consecutive sublist of genes sharing a coherent pattern, the first
gene in the sublist has a significantly high index value and the following
genes has a low index value. For example, the coherent pattern index graph
for the synthetic data set in Figure 5.17 is shown in Figure 5.19. In the
coherent index graph, a sharp pulse strongly indicates the existence of a
coherent pattern. Such pluses can guide users in deriving coherent patterns

and their corresponding co-expressed genes. Users can recursively examine the selected subsets of co-expressed genes as well as their sub-patterns in depth.

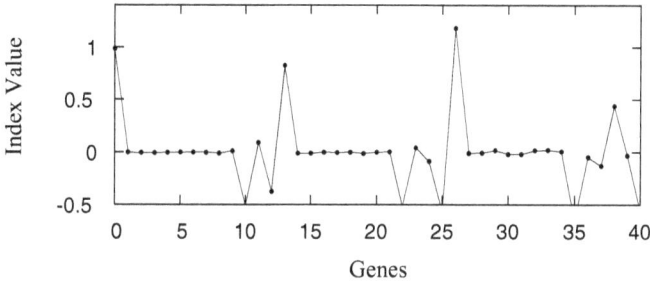

Fig. 5.19   The coherent pattern index graph for the data set in Figure 5.17.

5.6.2.1   *Generating the index list*

Ordering genes into a list allows us to plot the genes and examine the probability of each to be a "leader" in a group of co-expressed genes in a two-dimensional space. An ordered *index list* can be generated based on the following observations:

(1) In the attraction tree, the edges connecting a pair of objects O_1 and O_2 conforming to the same coherent pattern P have heavy weights (represented by thick purple lines in Figure 5.18). Genes connected by those edges should remain in close proximity in the list.

(2) The edges connecting a pair of intermediate (noise) objects O_1 and O_2 or connecting a pattern-correlated object and an intermediate (noise) object have moderate weights (represented by thin brown lines in Figure 5.18). Genes connected by those edges should also remain in proximity in the list but should be more separated than those addressed in case 1.

(3) The edges connecting a pair of objects O_1 and O_2 conforming to different coherent patterns P_1 and P_2 have light weights (represented by dashed yellow lines in Figure 5.18). Genes connected by those edges should be widely separated in the list.

Algorithm 5.5   : The Gene-Ordering Algorithm

```
Proc ordering(AttractionTree root)
    processedVertices.add(root)
    for each child ch of root do edgeHeap.insert(edge(root,ch))
    while ( !edgeHeap.isEmpty() ) do
        currentEdge = edgeHeap.extract();
        currentVertex = currentEdge.endVertex
        processedVertices.add(currentVertex)
        for each child ch of currentVertex do
            edgeheap.insert(edge(currentVertex,ch))
        end for
    end while
end Proc
```

On the basis of these observations, the gene-ordering algorithm (see Algorithm 5.5) is developed. This algorithm maintains a list, called *processedVertices*, to record the visiting order of the nodes in the attraction tree T. Starting from the root of T, all the edges connecting the root with its children are put into a heap, where the edges are sorted in descending order of weight. We then iteratively extract the edge with the highest weight from the heap. At this point, the start vertex of the edge must have been processed since, otherwise, the edge could not have been put into the heap. We put the end vertex of the edge *currentVertex* into the list *processedVertices* and put all the edges connecting *currentVertex* and its children into the *edgeHeap*. The loop continues until all of the edges in the tree have been visited. The resulting *processedVertices* is the *index list* of the genes.

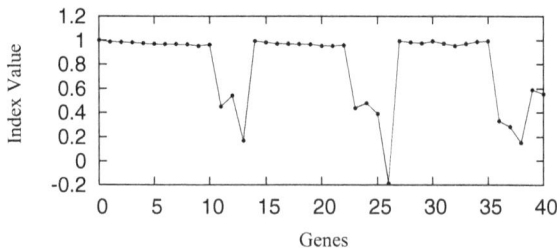

Fig. 5.20   The similarity curve for Figure 5.17.

Figure 5.20 shows the objects in the index list derived from the attraction tree in Figure 5.18. For each object, the similarity value plotted in the figure is the similarity between the object and its parent in the attraction tree. The similarity curve appears to be divided into three level terraces separated by two valleys. Each valley corresponds to the edge connecting different attraction subtrees. Since such edges have significantly lower weights than other edges, our search strategy does not allow the nodes in subtree T_2 to be visited before all the nodes in subtree T_1 have been visited. Similarly, the visit to subtree T_3 cannot start until the visit to subtree T_2 is finished.

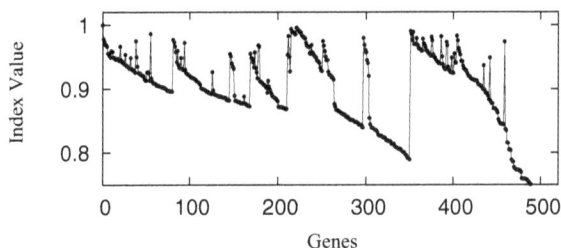

Fig. 5.21   The similarity curve for Iyer's data set.

While the similarity curve is informative, it is not always effective, especially with large data sets. For example, the similarity curve shown in Figure 5.21 is messy. This is because the similarity curve cannot distinguish co-expressed genes from a pair of similar intermediate genes formed by chance. To solve this problem, we design the coherent pattern index graph, where the beginning of a potential coherent pattern will be indicated clearly.

5.6.2.2 *The coherent pattern index and its graph*

As previously noted, in the construction of an attraction tree and index list, co-expressed genes are located in subtrees and thus are arranged as neighbors in the index list. This structure becomes the genesis of the coherent pattern index. In an index list, we may observe a subsequence S of consecutive genes which are more coherent to their parents in the attraction tree than are the genes preceding subsequence S. This configuration strongly suggests that S is the starting segment of a group of co-expressed genes.

The above observation leads us to focus on the recognition of *probes*, short subsequences of genes which appear at the beginning of a group of co-expressed genes. In a similarity curve, the similarity between a gene and its parent is plotted. For a gene g_i in an index list $g_1 \cdots g_n$, let $Sim(g_i)$ be g_i's similarity value in the similarity curve. $Sim(g_i) = 0$ if $(i < 1)$ or $(i > n)$. Let p be the minimum size of probe. For each gene g_i in the index list $g_1 \cdots g_n$, we define the coherent pattern index $CPI(g_i)$ as follows:

$$CPI(g_i) = \sum_{j=1}^{p} Sim(g_{i+j}) - \sum_{j=0}^{p-1} Sim(g_{i-j}). \qquad (5.19)$$

Intuitively, a high coherent pattern index value indicates a strong potential that a given gene is the start of a group of co-expressed genes. The graph plotting the coherent pattern index values with respect to the index list is called the *coherent pattern index graph*. The valleys in the similarity curve correspond to the sharp pulses in the coherent pattern index graph. In particular, from the above definition, the first $(p - 1)$ genes in the index list always generate the first sharp pulse. Figure 5.22 is the coherent pattern index graph derived from Figure 5.21 with $p = 5$. The coherent pattern index graph clearly indicates the existence of coherent patterns.

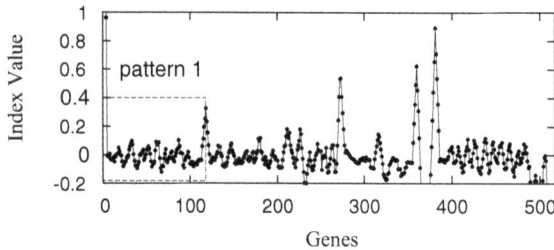

Fig. 5.22   The coherent pattern index graph for Iyer's data set.

5.6.2.3 *Drilling down to subgroups*

Figure 5.22 clearly indicates that there are five major coherent patterns in the data set. However, *can we further investigate the groups of co-expressed genes conforming to the coherent patterns and identify subgroups of co-expressed genes that conform to any sub-patterns?*

Suppose a user accepts the five major coherent patterns reported by the system and clicks on the corresponding peaks in the coherent pattern index graph. The system will split the attraction tree T for the entire data set into five exclusive attraction subtrees. Each subtree corresponds to one coherent pattern, and the genes conforming to that coherent pattern are gathered in that subtree. The original data set is thus partitioned into five subsets.

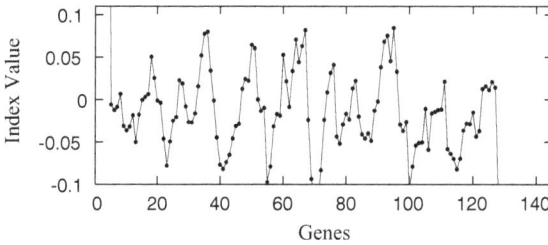

Fig. 5.23   The coherent pattern index graph for a subset of genes in Iyer's data set.

The user may now select the first subset of genes $\mathcal{D}_1$ (indicated by the dashed box in Figure 5.22) in order to move to a finer level of analysis. Figure 5.23 shows the local coherent pattern index graph for the selected subset of genes. It should be noted that Figure 5.23 is not simply a higher-resolution extracted from Figure 5.22. Rather, $\mathcal{D}_1$ is assembled from the attraction tree such that genes conforming to the coherent pattern are selected. The attraction tree, index list and coherent pattern index graph are then generated, with only the genes in the selected subset considered. The user can specify local parameters (e.g., σ) for computing the influence and density in the subset of genes.

According to the influence function (Equation 5.15), a smaller σ will boost the relative influence of a gene on its neighborhood. A detailed discussion of the effect of σ on the influence calculation can be found in [134]. We use the standard deviation of the pairwise distance between genes to determine the value of σ. When the data set is split into smaller subsets, the standard deviation will decrease.

With the help of index graphs, users can recursively explore the coherent patterns until satisfying results are achieved. The coherent patterns and the corresponding co-expressed gene groups form a hierarchical tree T, where each node on T at level k contains a subset of genes G_i^k conforming to the same coherent pattern P_i^k. We select the expression profile of the root object of the attraction subtree with respect to G_i^k as the representative of the coherent pattern P_i^k.

As the final step of this approach, the following method is adopted to report groups of co-expressed genes from the identified coherent patterns in T. First, for each coherent expression pattern P_j, all the genes in the data set are ordered according to their similarity to P_j. Those genes with a similarity value greater than $\bar{S}$, the estimated average similarity between genes within the same cluster, are assigned to cluster G_j. Second, P_j is adjusted to the centroid of the genes in G_j. The above two steps repeat until the assignments of genes do not change. Finally, P_j and G_j are returned as a coherent expression pattern and the corresponding group of co-expressed genes. In particular, genes similar to more than one coherent pattern will participate in multiple clusters, while genes not similar to any coherent pattern are considered as noise and filtered out.

5.6.3   *Experimental Results*

GPX was tested on both real-world and synthetic data sets and the results were compared with several other approaches [152].

Iyer's Data

Iyer et al. [147] monitored the expression levels of $8,600$ distinct human genes during a twelve-point time-series of serum stimulation. Those genes whose expression levels significantly changed during the time-series were selected for cluster analysis. Only 517 genes survived this significance test; other genes were filtered out. In other words, Iyer's data set contains 517 data objects with twelve attributes. In [147], the authors gave a list of ten co-expressed gene groups and the corresponding coherent patterns. We adopted this as the ground truth for the following experiments.

Spellman's Data

Spellman et al. [263] reported the genome-wide $6,220$ mRNA transcript levels during the cell cycle of the budding yeast *S. cerevisiae* synchronized by three independent methods. From these data sets, we have selected the *cdc*15 time-series since it contains the largest number of cell cycles and the most coherent expression patterns. From the $6,200$ genes monitored, 800 were found to be cell-cycle-dependent. The expression levels of those 800 peak at one of the following five phases: the early M/G_1 phase, the G_1 phase, the S phase, the G_2 phase, or the M phase. All of the cell-cycle correlated genes, together with their peaking phases, are listed at http://genome-www.stanford.edu/cellcycle/. From this set, we filtered out those genes which lack more than one-third of the measured expression values. The remaining 747 genes naturally form five co-expressed gene groups and conform to five coherent patterns. We used this data set to test the ability of the GPX approach and other algorithms to detect these five cell-cycle correlated patterns.

5.6.3.1 *Interactive exploration of Iyer's data and Spellman's data*

Figure 5.24 illustrates the exploration process. The pulses selected to split the (sub) data sets are marked in the corresponding coherent pattern index graphs. Initially, the coherent pattern index graph for the entire data set indicates five "major" coherent patterns. A user may accept this indication and ask the system to split the data set accordingly. The user may consider some "major patterns" to be already biologically meaningful, and require no further division into "finer patterns" (for example, the second and the fourth patterns in the second row of the figure). Other major patterns may still need further investigation.

The system generates the coherent pattern index graphs for the remaining subsets (i.e., the first, third, and fifth subsets in the second row of the figure). Each of the coherent pattern index graphs for these subsets contains multiple significant pulses. The highest of these peaks is generally a good guide to further splitting the data set and exploring the finer patterns. The user can ask the system to split the subset according to the highest peak in the graph. If the result is unacceptable, the user can easily "roll back" to the previous level and split the data set on the basis of the second highest peak. In our experiment, we assumed that the user chooses the highest peak in each subset and splits the data set accordingly. The hierarchy extends to the third level. Such an interactive exploration can be conducted recursively, until the user is satisfied with the resulting coherent patterns and co-expressed genes.

The exploration process of Sepllman's data is illustrated in Figure 5.25. Again, users can explore the data structure level by level with the guidance of the coherent pattern index graph until satisfying results have been achieved.

5.6.3.2 *Comparison with other algorithms*

We compared the coherent patterns identified by the GPX approach with the ground truth and with the results produced by six other methods: two classical partition-based approaches, *K-means* and *SOM*; two graph-

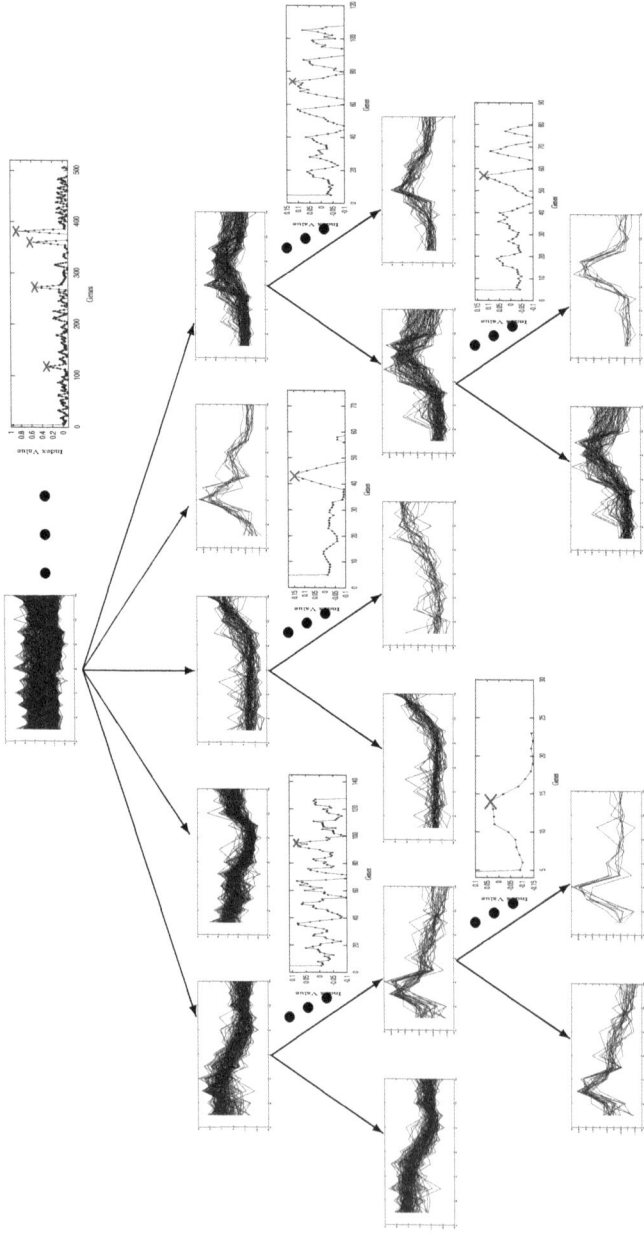

Fig. 5.24   The hierarchy of co-expressed gene groups in Iyer's data

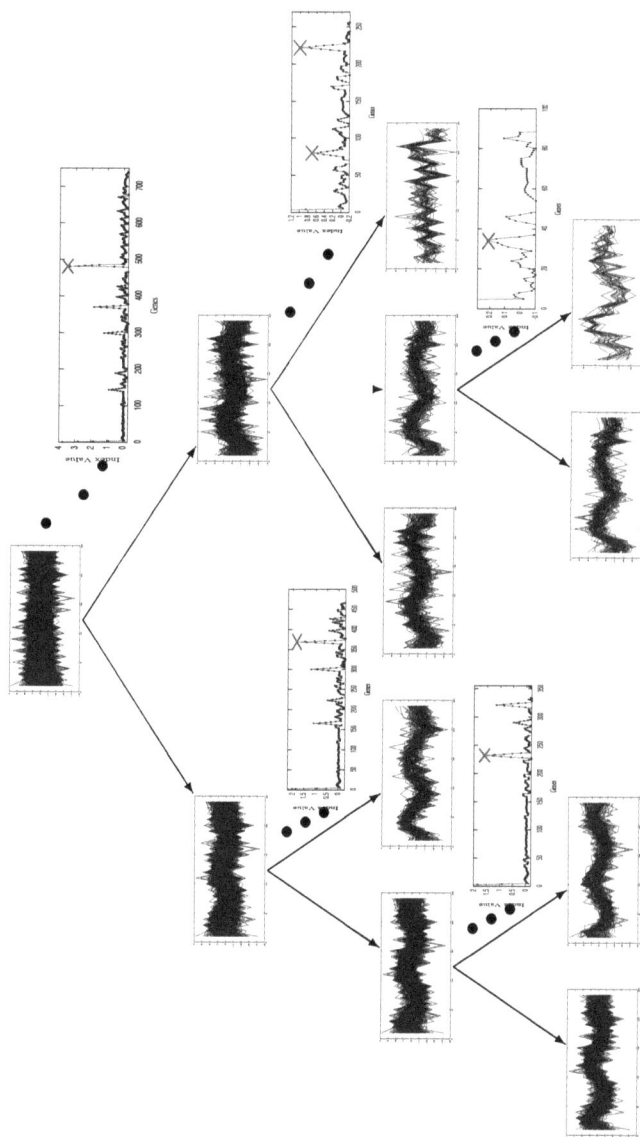

Fig. 5.25   The hierarchy of co-expressed gene groups in the Spellman's data set.

based approaches, *CAST* (Cluster Affinity Search Technique) [22] and *CLICK*(Cluster Identification via Connectivity Kernels) [251]; a clustering algorithm newly developed for gene expression data, *ADAPT* (Adaptive quality-based clustering) [258] and a hierarchical approach *SOTA* (Self Organizing Tree Algorithm) [132].

We implemented the *K-means* and *SOM* algorithms and set the number of clusters as equal to the number of coherent patterns in the ground truth. *CAST* was implemented according to the algorithm described in [22]. The program was run with a wide range of settings for parameter t (the *affinity threshold*) and the result best matches the ground truth was selected. *CLICK* was downloaded from http://www.cs.tau.ac.il/~rshamir/expander/expander.html. We accepted the default parameter setting in the software. *ADAPT* has a web interface at http://www.esat.kuleuven.ac.be/~thijs/Work/Clustering.html. We set the minimum number of genes in a cluster as five and accepted the default value of 0.95 as the minimum probability of genes belonging to a given cluster. *SOTA* has a web interface at http://gepas.bioinfo.cnio.es/cgi-bin/sotarray. We accepted the default parameter settings suggested by the web site.

We used the ground-truth patterns provided in [147, 263] as the domain knowledge for GPX to guide the growth of the hierarchical tree of co-expressed genes and coherent patterns. That is, we split the nodes on the tree on the basis of the corresponding pattern index graphs until the tree fully captures the domain knowledge. The coherent patterns were then returned as described in Section 5.6.2.3. To make a fair comparison, for those algorithms with default parameter values, such as CLICK, ADAPT, and SOTA, we simply adopted the default values; for those algorithms without default parameter values, such as K-means, SOM and CAST, we tuned the parameters as follows: for K-means and SOM, we set the parameter, i.e., the number of clusters K, as the number of coherent patterns in the ground truth; for CAST, we increased the parameter t with a small step (0.01) within its full range, i.e., from 0 to 1, and ran the algorithm repeatedly. The result which best "matches" the ground truth was picked.

Suppose $\{P_1, \ldots, P_n\}$ is the set of coherent patterns in the ground truth and $\{\tilde{P}_1, \ldots, \tilde{P}_m\}$ is the set of coherent patterns identified by a particular mining method. For each pattern P_i in the ground truth, we identified the pattern $\tilde{P}_j$ in the mining results which most closely resembles P_i, and called $\tilde{P}_j$ the "match" for P_i. Table 5.1 lists the similarity between the ground-truth patterns for Iyer's data set and the corresponding "matches"

Table 5.1   Coherent patterns discovered in Iyer's data set by different approaches.

Pattern	GPX (10)	Kmeans (10)	SOM (10)	ADAPT (11)	CLICK (7)	CAST (9)	SOTA (50)
1	**0.998**	**0.973**	**0.983**	**0.956**	0.884	**0.955**	**0.962**
2	**0.996**	**0.950**	**0.992**	**0.911**	**0.991**	0.887	**0.936**
3	**0.993**	**0.910**	0.872	**0.993**	**0.994**	**0.997**	**0.947**
4	**0.995**	**0.996**	**0.989**	**0.984**	0.883	**0.968**	**0.955**
5	**0.964**	0.882	0.716	0.868	0.886	0.855	**0.962**
6	**0.940**	**0.965**	0.764	**0.989**	**0.970**	**0.984**	**0.972**
7	**0.972**	0.880	0.892	**0.976**	**0.990**	0.719	**0.988**
8	**0.995**	**0.963**	0.917	**0.997**	**0.914**	**0.999**	**0.958**
9	**0.907**	**0.910**	0.848	0.824	0.844	0.800	**0.940**
10	**0.987**	**0.930**	**0.983**	**0.981**	**0.976**	**0.996**	**0.960**

identified by the various tested methods. We call a ground-truth pattern P_i is "accurately identified" if there exists a matching pattern $\tilde{P}_j$ such that the similarity between P_i and $\tilde{P}_j$ is greater than 0.9. The patterns accurately identified by each algorithm are indicated in a bold font. The numbers in parenthesis in the second row indicate the number of coherent patterns *returned* by each method. Not every pattern returned by one of these algorithms necessarily matches a ground-truth pattern.

The comparison of the patterns reported by each approach with the ground truth indicates that:

- GPX and *SOTA* were the only two approaches that accurately identified all ten ground-truth patterns. However, *SOTA* over-estimated the number of patterns; it reported 50 coherent patterns with 80% false positives, while GPX system reported ten patterns with zero false positives.

- Many of the methods failed to identify ground-truth patterns 5 and 9. These ground-truth patterns are shared by a small number of co-expressed genes and are similar to patterns 3 and 6, respectively. Therefore, most clustering algorithms merged them into patterns 3 and 6, respectively.

- All methods other than the GPX approach and *CLICK* split the genes sharing ground-truth coherent pattern 2 into several smaller subsets. This is because the coherence of the genes in this group is much weaker than that of other clusters.

The full range of algorithms was also tested on Spellman's data set. Again, the GPX approach was the only one to accurately identify all the

Table 5.2   Coherent patterns discovered in Spellman's data set by different approaches.

Pattern	GPX (7)	Kmeans (5)	SOM (5)	ADAPT (21)	CLICK (9)	CAST (19)	SOTA (99)
1	**0.901**	**0.928**	0.194	0.884	0.855	**0.900**	**0.938**
2	**0.970**	**0.976**	**0.972**	**0.972**	**0.978**	**0.970**	**0.968**
3	**0.980**	**0.950**	0.552	**0.953**	**0.970**	0.888	**0.940**
4	**0.901**	0.773	0.437	0.796	**0.984**	0.888	**0.961**
5	**0.945**	**0.965**	**0.964**	**0.962**	**0.978**	**0.956**	**0.956**

ground-truth patterns with a low false positive rate (see Figure 5.25, and Table 5.2). Ground-truth patterns 1 and 4 in this data set are difficult to identify. Pattern 1 corresponds to the genes peaking at the early $M/G1$ phase, which is an intermediate phase between the M (pattern 5) and $G1$ (pattern 2) phases. Some tested approaches assigned the genes following pattern 1 to either pattern 2 or pattern 5. Pattern 4 is a "weak" pattern which is conformed to by a small number of genes. Some approaches cannot effectively adapt to different cluster granularities, and thus fail to identify pattern 4.

As mentioned before, the interpretation of coherent patterns and co-expressed genes depends on the domain knowledge. Users may have different requirements for cluster granularity in different part of the data set. The GPX approach addresses this challenge by adopting an interactive approach and supporting flexible exploration incorporating the domain knowledge of users.

5.6.4   *Efficiency and Scalability*

We tested the efficiency and scalability of the GPX method using synthetic data sets of various sizes. In fact, GPX proceeds in two steps. First, in the *pre-processing step*, we normalize the data objects and calculate the pairwise distance between data objects. In the *exploration step*, we construct the attraction tree structure and generate the pattern index graph to support interactive exploration. Figure 5.26(a) and (b) illustrate the computation time for the pre-processing and exploration steps. As indicated there, GPX is scalable with respect to the number of genes, and the computation time is dominated by the pre-processing step.

(a) Pre-processing step

(b) Exploration step

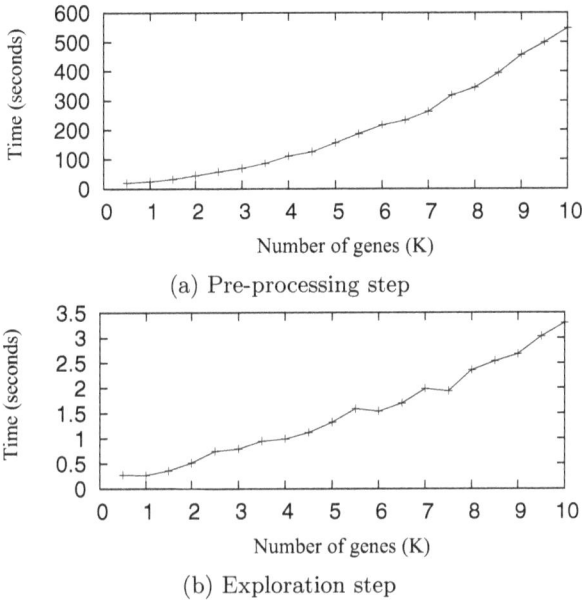

Fig. 5.26   The scalability of the two steps in our algorithm.

5.7   Cluster Validation

The previous sections have reviewed a number of clustering algorithms which partition a data set on the basis of particular clustering criteria. However, different clustering algorithms, or even the use of a single clustering algorithm with different parameters, generally result in different sets of clusters. Therefore, it is important to compare various clustering results and select the one that best fits the "true" data distribution. Cluster validation is the process of assessing the quality and reliability of the cluster sets derived from various clustering processes.

Cluster validity is assessed on three general bases. First, the quality of clusters can be measured in terms of the degree of their *homogeneity* and *separation*. By definition, objects within one cluster are assumed to be similar to each other, while objects in different clusters are dissimilar. The second aspect relies on a given "ground truth" of the clusters. The "ground truth" may come from domain knowledge, such as known function families of genes. Cluster validation is based on the extent of agreement between the clustering results and this "ground truth." The third aspect of cluster validity focuses on the reliability of the clusters, or the likelihood that the cluster structure has not been formed by chance. In this section, we will discuss these three aspects of cluster validation.

5.7.1 Homogeneity and Separation

The homogeneity of a cluster is defined by some measure which quantifies the similarity of data objects in the cluster cluster C. For example,

$$H_1(C) = \frac{\sum_{O_i, O_j \in C, O_i \neq O_j} Similarity(O_i, O_j)}{||C|| \cdot (||C|| - 1)}.$$

This definition represents the homogeneity of cluster C by the average pairwise object similarity within C. An alternate definition evaluates the homogeneity with respect to the "centroid" of the cluster C, i.e.,

$$H_2(C) = \frac{1}{||C||} \sum_{O_i \in C} Similarity(O_i, \bar{O}),$$

where $\bar{O}$ is the "centroid" of C. Other definitions, such as the representation of cluster homogeneity via maximum or minimum pairwise or centroid-based similarity within C can also be useful and perform well under certain conditions. Cluster separation is analogously defined from various perspectives to measure the dissimilarity between two clusters C_1, C_2. For example,

$$S_1(C_1, C_2) = \frac{\sum_{O_i \in C_1, O_j \in C_2} Similarity(O_i, O_j)}{||C_1|| \cdot ||C_2||}$$

and

$$S_2(C_1, C_2) = Similarity(\bar{O}_1, \bar{O}_2),$$

where $\bar{O}_1$ and $\bar{O}_2$ are the centroids of C_1 and C_2, respectively.

Since these definitions of homogeneity and separation are based on the similarity between objects, the quality of C increases with higher homogeneity values within C and lower separation values between C and other clusters. Once we have defined the homogeneity of a cluster and the separation between a pair of clusters, for a given clustering result $\mathcal{C} = \{C_1, C_2, \dots, C_K\}$, we can define the homogeneity and the separation of $\mathcal{C}$. For example, Sharan et al. [251] used definitions of

$$H_{ave} = \frac{1}{N} \sum_{C_i \in \mathcal{C}} ||C_i|| \cdot H_2(C_i)$$

and

$$S_{ave} = \frac{1}{\sum_{C_i \neq C_j} ||C_i|| \cdot ||C_j||} \sum_{C_i \neq C_j} (||C_i|| \cdot ||C_j||) S_2(C_i, C_j)$$

to measure the average homogeneity and separation for the set of clustering results $\mathcal{C}$.

5.7.2   Agreement with Reference Partition

If the "ground truth" of the cluster structure of the data set is available, we can test the performance of a clustering process by comparing the clustering results with the "ground truth." Given the clustering results $\mathcal{C} = \{C_1, ..., C_p\}$, we can construct a $n * n$ binary matrix C, where n is the number of data objects, $C_{ij} = 1$ if O_i and O_j belong to the same cluster, and $C_{ij} = 0$ otherwise. Similarly, we can build the binary matrix P for the "ground truth" $\mathcal{P} = \{P_1, ..., P_s\}$. The agreement between $\mathcal{C}$ and $\mathcal{P}$ can be disclosed via the following values:

- n_{11} is the number of object pairs (O_i, O_j), where $C_{ij} = 1$ and $P_{ij} = 1$;
- n_{10} is the number of object pairs (O_i, O_j), where $C_{ij} = 1$ and $P_{ij} = 0$;
- n_{01} is the number of object pairs (O_i, O_j), where $C_{ij} = 0$ and $P_{ij} = 1$;
- n_{00} is the number of object pairs (O_i, O_j), where $C_{ij} = 0$ and $P_{ij} = 0$.

Some commonly used validation indices [119, 259] have been defined to measure the degree of similarity between $\mathcal{C}$ and $\mathcal{P}$:

$$\text{Rand index: } Rand = \frac{n_{11} + n_{00}}{n_{11} + n_{10} + n_{01} + n_{00}},$$

$$\text{Jaccard coefficient: } JC = \frac{n_{11}}{n_{11} + n_{10} + n_{01}},$$

$$\text{Minkowski measure: } Minkowski = \sqrt{\frac{n_{10} + n_{01}}{n_{11} + n_{01}}}.$$

The *Rand index* and the *Jaccard coefficient* measure the extent of agreement between $\mathcal{C}$ and $\mathcal{P}$, while *Minkowski measure* illustrates the proportion of disagreements to the total number of object pairs (O_i, O_j), where O_i, O_j belong to the same set in $\mathcal{P}$. It should be noted that the *Jaccard coefficient* and the *Minkowski measure* do not (directly) involve the term n_{00}. These two indices may be more effective in gene-based clustering because a majority of pairs of objects tend to be in separate clusters, and the term

n_{00} would dominate the other three terms in both accurate and inaccurate solutions. Other methods are also available to measure the correlation between the clustering results and the "ground truth" [119]. Again, the optimal index selection is application-dependent.

5.7.3　*Reliability of Clusters*

While a validation index can be used to compare different clustering results, this comparison will not reveal the reliability of the resulting clusters; that is, the probability that the clusters are not formed by chance. In the following subsection, we will review some representative approaches to measuring the significance of the derived clusters.

5.7.3.1　*P-value of a cluster*

In [277], Tavazoie et al. mapped the genes in each resulting cluster to the 199 functional categories in the Martinsried Institute of Protein Sciences function classification scheme (MIPS) database. For each cluster, P-values were calculated to measure the statistical significance for functional category enrichment. The authors used the hyper-geometric distribution to calculate the probability of observing at least k genes from a functional category within a cluster of size n:

$$P = 1 - \sum_{i=0}^{k-1} \frac{\binom{f}{i} \binom{g-f}{n-i}}{\binom{g}{n}},$$

where f is the total number of genes within a functional category and g is the total number of genes within the genome. Since the expectation of P within the cluster would be higher than 0.05%, the authors regarded clusters with P-values smaller than $3 * 10^{-4}$ as significant. Jakt et al. [150] integrated the assessment of the potential functional significance of both gene clusters and the corresponding postulated regulatory motifs (common DNA sequence patterns), and developed a method to estimate the probability (P-value) of finding a certain number of matches to a motif in all of the gene clusters. A smaller probability indicates a higher significance of the clustering results.

5.7.3.2 *Prediction strength*

In [321], Yeung et al. proposed an approach to the validation of gene clusters based on the idea of "prediction strength." Intuitively, if a cluster of genes formed with respect to a set of samples (attributes) has possible biological significance, then the expression levels of the genes within that cluster should also be similar to each other in "test" samples that were not used to form the cluster. Yeung et al. proposed a specific *figure of merit (FOM)* to estimate the predictive power of a clustering algorithm. Suppose $C_1, \ldots, C_k$ are the resulting clusters based on samples $1, \ldots, (e - 1), (e + 1), \ldots, m$, and sample e is left out to test the prediction strength. Let $R(g, e)$ be the expression level of gene g under sample e in the raw data matrix. Let $\mu_{C_i}(e)$ be the average expression level in sample e of the genes in cluster C_i. The *figure of merit* with respect to e and the number of clusters k is defined as

$$FOM(e, k) = \sqrt{\frac{1}{n} * \sum_{i=1}^{k} \sum_{x \in C_i} (R(x, e) - \mu_{C_i}(e))^2}.$$

Each of the m samples can be left out in turn, and the *aggregate figure of merit* is defined as $FOM(k) = \sum_{e=1}^{m} FOM(e, k)$. The FOM measures the mean deviation of the expression levels of genes in e relative to their corresponding cluster means. Thus, a small value of FOM indicates a strong prediction strength, and therefore a high level reliability of the resulting clusters. Levine et al. [179] proposed another figure of merit $\mathcal{M}$ based on a resampling scheme. This scheme assumes that the cluster structure derived from the entire data set should be able to "predict" the cluster structure of subsets of the full data. $\mathcal{M}$ measures the extent to which the clustering assignments obtained from the resamples (subsets of the full data) agree with those from the full data. A high value of $\mathcal{M}$ against a wide range of resampling indicates a reliable clustering result.

5.8 Summary

In this chapter, we have reviewed a series of approaches to gene-based analysis, focusing in particular on gene-based clustering techniques. The intent of the clustering process is the identification of groups of highly co-expressed genes from the mass of noisy gene expression data. Such clusters provide a useful basis for further investigation of gene function and gene regulation.

Early efforts to apply conventional clustering algorithms, such as the K-means [277], SOM [271], and hierarchical approaches [92], have progressed to the development of algorithms specifically designed to address the particular challenges of gene-based clustering. Experimental studies of such clustering algorithms as CLICK [251] and CAST [22] developed specifically for gene-expression data have noted improved performance. However, each of these algorithms is based on some particular clustering criteria and/or assumptions regarding data distribution, so the performance of each may vary greatly with different data sets. For example, the general-purpose K-means or SOM algorithms may outperform the more tailored approaches if the target data set contains few outliers and the number of clusters in the data set is known. In contrast, CAST or CLICK may be a more appropriate choice for use with a very noisy gene expression data set in which the number of clusters is unknown.

There are a number of characteristics of gene expression data which are not addressed by even these more targeted algorithms. In particular, gene expression data is characterized by high connectivity through a large number of "intermediate" genes, hindering the identification of clear boundaries between clusters. In this chapter, we have described an advanced approach, called GPX [152, 153], which provides an interactive framework for exploring coherent patterns in gene expression data. Unlike many other clustering approaches, GPX does not start by partitioning the data set into clusters of co-expressed genes. Rather, it prompts users with a menu of potential coherent patterns. The arbitrary setting of cluster borders which weakens other methods is thus avoided, and enhanced performance is obtained even with data sets containing many "intermediate" genes. Experimental studies have indicated that GPX can identify most of the coherent patterns in a data set with higher accuracy than other comparable clustering methods.

Chapter 6

Sample-Based Analysis

Contributors: Daxin Jiang, Chun Tang, and Xian Xu

6.1 Introduction

One of the most exciting areas to which microarray technology has been applied is the challenge of deciphering complex diseases such as cancer. Since most tumors exhibit unique expression patterns, gene expression data are often referred to as "signatures" or "portraits" [57]. The simultaneous monitoring of large-scale gene expression levels enables the identification of cell types which share these common expression patterns. As summarized by Chung et al. [57], previous studies have shown that DNA microarrays can help investigators to develop expression-based classifications for many types of cancer, including breast [117, 129, 220, 260, 291, 303], brain [187, 223], ovary [151, 297, 302], lung [19, 27, 105], colon [7, 183, 329], kidney [270], prostate [78, 175, 256, 302], gastric [136], leukemia [98, 139, 318], and lymphoma [6, 235, 254].

In these studies, samples are taken from two or more groups of individuals with heterogeneous phenotypes, pathologies, or clinical outcomes. These samples are hybridized to microarrays in an effort to find a small number of genes which are strongly correlated with the groups of individuals. These genes are often called *informative genes* [114], since they may help biomedical researchers to understand disease mechanisms. They can also be used to resolve levels of heterogeneity among cells that are not apparent by eye and to provide a more accurate prognosis and prediction of response to therapy [57]. Figure 6.1 illustrates the mechanism of using gene expression profiles to distinguish individuals with different phenotypes.

This approach is very promising, since tumors of different types (such as malignant and benign tumors) can be very difficult to distinguish by conventional morphological, histological, clinical, or pathological means. An examination of their gene expression patterns, or signatures, offers much

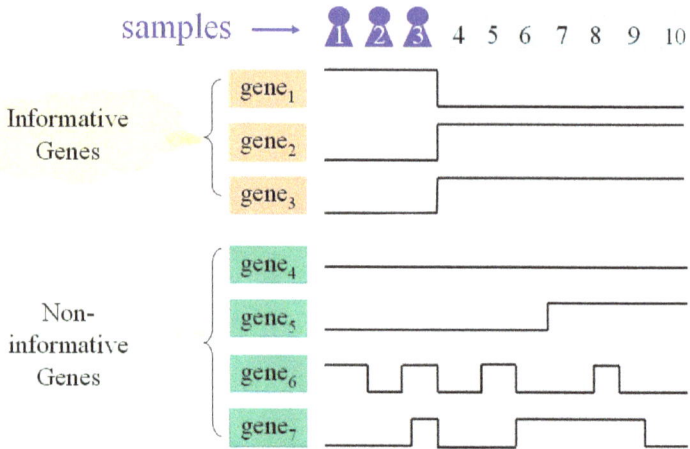

Fig. 6.1   A simplified illustration of the mechanism of "expression signature." The polylines show the expressions profiles in an example gene expression data set. The first three samples belong to one phenotype and the rest of the samples to the second phenotype. The first three genes are "informative genes" which exhibit expression profiles strongly correlated to the phenotype structure. For example, the first gene exhibits a "high" expression level in all individuals with phenotype 1 and a "low" expression level in all individuals with phenotype 2. The expression levels of the full set of informative genes form the "expression signature." For example, the "signature" of phenotype 1 is "($gene_1$ =high, $gene_2$ =low, $gene_3$ =low)."

better potential for accurate discrimination. For example, Golub et al. [114] demonstrated the feasibility of differentiating two types of leukemia, acute lymphocytic leukemia (ALL) and acute myelocytic leukemia (AML), on the basis of gene expression information. Distinguishing ALL from AML is critical for successful treatment, but no single test is currently sufficient to establish the diagnosis. In [114], Golub et al. chose 50 informative genes and built a model which generated predictions for 29 of the 34 testing samples with 100% accuracy.

In another example, Staudt and colleagues [6, 235] conducted a systematic characterization of gene expression in diffuse large B-cell lymphoma (DLBCL) malignancies using DNA microarrays. DLBCL is clinically heterogeneous; 40% of patients respond well to current therapy and have prolonged survival, whereas the remainder succumb to the disease. Although a combination of clinical parameters is currently used to assess a patient's risk profile, these prognostic variables are considered to be proxies for the underlying cellular and molecular variation within DLBCL [6]. From the gene expression patterns of hundreds of DLBCL samples, Staudt and colleagues identified three subtypes: a germinal center B cell-like (GCBL) subtype, an activated B cell-like (ABL) subtype, and a third subtype (T3) lacking high expression of either the GCBL- or the ABL-defining genes. They found that the GCBL subtype was associated with relatively good clinical outcomes, while the ABL subtype displayed an inferior clinical outcome.

The two cases summarized above exemplify two of the fundamental problems of sample-based analysis, termed *class discovery* and *class prediction* by Golub et al. [114]. Class prediction refers to the assignment of tumors to known classes, such as the classification of a new leukemia case as ALL or AML. Class discovery refers to the identification of new cancer classes or subtypes of tumors, such as the automatic differentiation of the GCBL, ABL, and T3 varieties without prior knowledge of these subtypes.

The process of class prediction is often recognized as a typical "supervised learning" problem in the field of pattern recognition. Numerous methods for classifying multidimensional data objects into *known* classes have been presented in the literature of machine learning. In general, these methods proceed in two phases. First, in a *training phase*, the algorithm searches the hypothesis space and finds the target function that is (approximately) most consistent with the given training samples. In a *testing phase*, the algorithm uses the learned target function to predict the class label of new observations. When these methods are applied to classifying samples, the training data consist of a set of samples with known class labels (such as cancerous/non-cancerous) which may have been obtained via morphological or clinical methods. The goal is to predict the class label for a given new sample based on its expression profile.

In contrast to class prediction, class discovery is a classical "unsupervised learning" problem, in the sense that no pre-defined class labels and training samples are provided. In other words, the tumor samples are automatically partitioned into several groups exhibiting greater within-group than between-group similarities. The clustering techniques discussed in Chapter 5 are a clear fit for this task. That discussion pertained to gene-based analysis, in which genes are considered as data objects and samples (or time points) serve as attributes. The situation is reversed in sample-based analysis, where samples are treated as data objects, while the genes play the role of attributes.

A specific characteristic of microarray data is that *the number of samples in a microarray experiment is typically by far smaller than that of the genes*; this is known as the *"large p, small n"* problem in statistics.

Consequently, both conventional "supervised learning" or "unsupervised learning" methods may not be effective for sample analysis. In this chapter, we will first explore in Section 6.2 why this "large p, small n" problem affects the mining results and how the dimensionality of the data may be reduced. In Sections 6.3 and 6.4, we will describe several representative approaches to class prediction and class discovery. Methods for validating the mining results will be offered in Section 6.5.

6.2   Selection of Informative Genes

Microarray data are often extremely asymmetric in dimensionality. At one extreme, a microarray data set usually contains thousands or even tens of thousands of genes. At the other extreme, the number of samples is usually no more than a few hundreds. Such extreme asymmetry between the dimensionality of genes and samples presents several challenges to conventional clustering and classification methods, which were generally designed to process a large number of data objects with relatively few attributes. Some of the challenges these methods encounter in processing asymmetric microarray data sets are listed below:

- *Proximity measure.* Most methods for sample-based analysis rely on some proximity function (see Section 5.2) to measure the distance or similarity between a pair of samples. However, when the number of attributes is very high, some proximity measures, such as the Euclidean distance, may become meaningless. That is, the distance of an object to its nearest neighbor approaches the distance to the farthest neighbor [25].
- *Overfitting.* The problem of "overfitting" occurs when an algorithm adapts to the training samples too exactly, losing sufficient ability to generalize in the prediction of new samples. In consequence, while the classification of the training examples may be perfect, the accuracy of predication with test samples drops dramatically. The large number of features characteristic of microarray data sets may render the predictive algorithm prone to overfitting to the limited number of training samples [257].
- *Multiplicity.* A parallel examination of a large volume of genes may incorrectly identify some as being differentially expressed between different samples when, in fact, these differences may be due to random vari-

ation [10, 266]. Multiple testing was discussed in detail in Section 4.6.

- *Curse of dimensionality.* The term "curse of dimensionality" was coined by R. Bellman [20]. In the context of sample-based analysis, it refers to the exponential growth of the hypothesis space with respect to the number of features. In general, a clustering or classification method needs to search the feature space to find a solution. Given the large number of genes and the exponential growth of the search space, the efficiency of learning will drop dramatically [266].

These difficulties suggest the appropriateness of reducing the data dimensionality to improve the effectiveness and efficiency of clustering and classification methods. In fact, many genes in a microarray experiment are irrelevant to the problem under study and need not be considered in the clustering or classification process. For example, a study of a typical biological process, such as a determination of the differences between tumor samples, seldom involves more than a few dozen genes [308].

Determining which genes to be used in the clustering or classification procedure is essential for the success of sample-based analysis. Gene selection is necessary not only to reduce data dimensionality but also to identify those genes that are closely related to the cell types. In general, the approaches to gene selection can be categorized as supervised and unsupervised, depending on whether the class labels of the samples are given a priori. In the following subsections, we will discuss these methods in more detail.

6.2.1 Supervised Approaches

6.2.1.1 Differentially expressed genes

Given a set of samples with known class labels, the simplest strategy to reduce the number of genes is to select those that are differentially expressed across the different classes. In Chapter 4, we have discussed a series of statistical approaches used to test whether an individual gene is differentially expressed across two or more groups. Based on these tests, those genes that allow rejection of the null hypothesis with a certain significance level will be used as features in the sample-based analysis.

As genes expressed without significant variation between classes are likely to be irrelevant, this approach is conceptually sound. However, Amaratunga and Cabrera [10] have noted that it still faces the following problems in practice:

- *Granularity of filtering.* In any specific biological system, the number of genes needed to fully characterize a macroscopic phenotype is usually unclear. As a result, excessive filtering of genes may lead to loss of most of the information necessary for accurate classification, while insufficient filtering may retain significant noise.
- *Correlation among genes.* The data set may include a set of genes that together acts as a classifier, while each individual gene in the set does not. Unfortunately, differentially-expressed genes are usually individually identified without considering the inherent correlation among genes.
- *False positives.* Hypothesis testing may result in false positives, where the identification of a differentially-expressed gene in the training samples cannot be reproduced in the testing data. The large number of genes in a microarray experiment predisposes this data to the problem of false positives. Multiple testing and false positives have been discussed in detail in Section 4.6.

To exploit the inherent correlation among genes, an alternative strategy for gene selection is to consider multiple genes simultaneously instead of testing them individually. Several methods address the problem of false positives by incorporating gene selection into the classification procedure itself. These methods build classifiers on the basis of the groups of genes under consideration, and the group of genes which achieves the highest prediction accuracy as measured by cross-validation (see Section 6.5) will be considered to be the informative genes.

The following subsections present three novel approaches to gene selection. The first, *gene pairs* [32], evaluates how well a pair of genes in combination distinguishes two experimental classes. The second approach, *virtual genes* [311, 312], employs the inherent correlation among n ($n \geq 2$) genes to predict the class labels. The third approach [181, 182] considers the combined discriminative power of a subset of genes by integrating a genetic algorithm (the gene selection process) with a KNN algorithm (the classification process, see Section 6.3.2.1) to achieve a better set of informative genes.

6.2.1.2   Gene pairs

Bo and Jonassen [32] showed that a pair of genes in combination may separate two classes well. To evaluate the discriminative power of gene pairs, Hotelling's t-statistic (the multivariate form of the t-test) is used to score the genes. In [32], the authors presented two methods to search for highly discriminative gene pairs; these are a naïve "all pairs" method and the heuristic "greedy pairs" method.

The all-pairs method individually examines the Hotelling's t-statistic for each pair of genes and selects those with the highest scores. The selected gene pairs must be exclusive in order to preclude an inaccurate biasing effect by one exceptionally high-scoring gene on several low-scoring companion genes. This is accomplished by selecting the gene pair with the highest t-statistic score and then removing from consideration all other pairs containing either of those genes.

Clearly, this method is computationally very expensive when applied to a microarray data set with thousands of genes. An alternative "greedy pairs" method ranks all genes on the basis of individual t-statistic measurements rather than considering combinations of genes. The gene g_i with the highest individual t-statistic is identified, and the gene g_j that together with g_i maximizes the paired t-score is then selected. These two genes are then removed from the gene set, and the procedure is repeated with the remainder of the set until a desired number of gene pairs has been obtained.

In [32], these methods for selecting gene pairs were tested on leukemia [114] and colon cancer [7] data sets and were then compared with other two methods, *individual ranking* and *forward selection* (see [32]). The effectiveness of gene selection was measured indirectly by the prediction accuracy of three standard classifiers: Fisher's linear discriminant (LDA), diagonal linear discrimination (DLDA), and k-nearest neighbor (KNN). LDA and DLDA will be discussed in Section 6.3.1 and KNN algorithms in Section 6.3.2.1. Bo and Jonassen [32] compared the gene sets selected by the all-pairs and greedy-pairs methods with those generated by several standard methods which select genes individually. These latter methods include the t-statistic [310] (see Section 4.4), a variant of the t-statistic introduced by Golub et al. [114] (see Section 4.4.3), an adaption by Pavlidis et al. of Fisher's discriminant criterion [214] (see Section 4.4.3), the TNoM score introduced by Ben-Dor et al [21] (see Section 4.5.2), and the signal-to-noise ratio developed by Dudoit et al. [86] (see Section 4.4.3). In some cases, the all-pairs and greedy-pairs methods outperformed the standard methods by a large margin.

6.2.1.3   *Virtual genes*

In [311, 312], Xu and Zhang investigated the correlation among genes as a potential predictor of the class labeling of samples. Genes are well-known to interact with each other through gene regulatory networks, and thus it is intuitively reasonable that the inherent correlation among genes may disclose the complex mechanism that distinguishes the samples. For example, Figure 6.2 shows the expression profiles of genes H09719 and L07648 across the samples in Alon's data set [7]. Samples in the left region of the vertical bar (shown in the dark gray color) in Figure 6.2 are cancerous, while samples in the right region are normal. It is evident that both genes express randomly across the two classes of samples, so neither gene can serve as a definitive discriminator between the two sample classes. However, a closer observation of the figure reveals that the expression levels of gene H09719 are consistently higher than those of gene L07648 in cancerous samples, while, for all but one of the normal samples, the expression levels of L07648 are higher than those of H09719. This suggests that, in some cases, the correlation among gene expression values may be more effective than their absolute magnitudes in distinguishing samples.

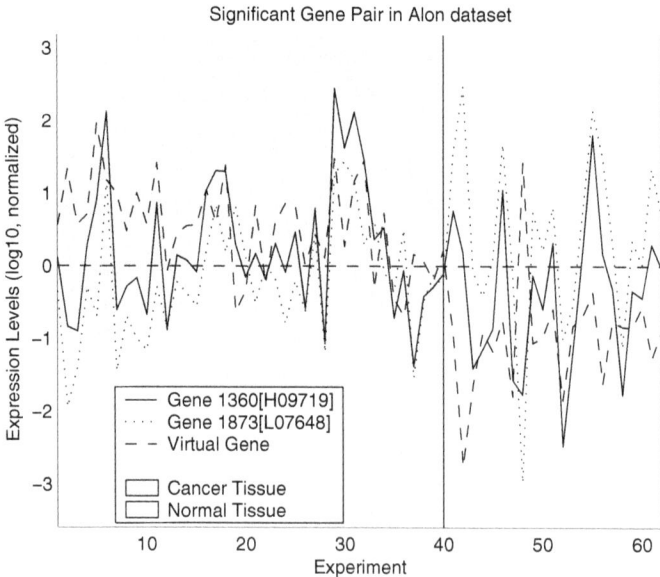

Fig. 6.2   A pair of genes from Alon's data set [7] serves as a better predictor of class labels than does a single gene.

Based on the above observation, Xu and Zhang [311, 312] proposed the concept of *virtual genes* as features to classify samples. In essence, virtual genes are linear combinations of actual genes taken from the given microarray data set. However, unlike typical feature-extraction approaches such as PCA (see Section 6.2.2.1), the virtual-gene approach is supervised, and the linear combination carries clear biological meaning.

Using the virtual-gene approach, a given gene expression data set can be represented by $\mathcal{E} = (G, S, L, E)$, where G is the set of genes, S is the set of samples, L is the set of class labels of the samples in S, and E is the an $n \times m$ real-valued data matrix where each cell $e_{ij} \in E$ is the expression level of $g_i \in G$ under $s_j \in S$. A *virtual gene* is defined as a triplet $VG = (G_v, W, b)$ where $G_v \subseteq G$, W is a matrix of size $|G_v| \times 1$, and b is a parameter. The *virtual gene expression* VE of a virtual gene VG is a linear combination of the expression profiles of G_v, i.e., $VE(VG) = W' \times E_v + b$, where W' is the transpose of W and E_v is the projected submatrix of E on the subset of genes G_v.

In the definition of virtual genes, the parameters W and b can be determined by applying FLD (Fisher's linear discriminant) to maximize linear separability between sample classes. The discriminative power of a virtual gene expression with respect to the sample classes can be measured using a single-gene-based score such as the t-statistic (see Section 4.4). In particular, the *pairwise virtual gene* is a special case of virtual genes in which the number of genes involved is limited to two. In this case, only the correlations between a pair of genes are considered. Through this constraint, the computation can be carried out efficiently. The empirical studies reported in [311, 312] indicate this pairwise virtual-gene method performed well when applied to three publicly-available data sets.

Exhaustive examination of all pairwise virtual genes requires $O(n^2)$ computation, where n is the number of genes. For a large number of genes, exhaustive search of all pairs of genes becomes inefficient. Such exhaustive search may also incorporate unwanted noise, since not all pairs of genes bear biological meaning. For example, the relative expression levels of genes that are expressed in different locations in a cell, in different biological processes, or without biological interactions may not be biologically significant at all. Ideally, only the pairs of genes with some biological interaction shall be examined. In [311, 312], the authors adopted a clustering strategy to find the biologically related genes. Genes are first clustered into groups on the basis of their expression profiles, and only those pairs of genes which occur within the same clusters are examined. This strategy assumes that gene clusters may roughly correspond to some biological pathway and that genes within a given cluster are more likely to be biologically related. This windowing process also substantially improves the efficiency of the gene selection algorithm.

As already noted, application of the pairwise virtual gene selection algorithm begins by clustering genes on the basis of their expression levels. The virtual gene expression value for each pair of genes in a cluster is then calculated and a single-gene-based discriminative score (the t-statistic) is derived. All the scores are stored in the *pair_score* matrix. In the next step of the algorithm, the virtual gene with the highest score is selected, and the scores for other virtual genes are modified according to two parameters. First, the scores of other virtual genes that share genes with the selected gene are degraded by a constant α ranging between 0 and 1. This dampens the effect of a single dominant salient gene. In the extreme case, when α is set to 0, all other virtual genes sharing constituent genes with a selected virtual gene will be ignored. The second parameter affecting selection of virtual genes is $\beta \in [0,1]$, which controls the likelihood that virtual genes in the same gene cluster will be selected. Different gene clusters usually correspond to different regulative processes in a cell. Therefore, choosing genes from different clusters broadens the spectrum of the selected gene set. In the extreme case, when $\beta = 0$, only one virtual gene will be selected from each gene cluster. After updating the pairwise scores, the algorithm loops back to find the virtual gene with the next-highest score. This process repeats until k virtual genes have been selected.

In [311, 312], the effectiveness of pairwise virtual gene selection was compared with several other gene selection methods, including the single-gene-based t-statistics [32], the S2N score [7], and the Hotelling's t-statistic [32]. Testing involved application of these methods to three public gene expression data sets, a colon cancer data set [7], a leukemia data set [114], and a multi-class cancer data set [229]. The experimental results showed that the classifier using pairwise virtual genes had a higher prediction accuracy than those built on other methods. Additionally, the virtual-gene approach was robust to initial clustering results and various types of classifiers.

6.2.1.4   Genetic algorithms

In [181, 182], Li et al. proposed an approach which combines a *genetic algorithm* and a k-nearest neighbor (KNN) method to identify genes that can jointly discriminate between different classes of samples. The genetic algorithm is essentially a heuristic method to search complex high-dimensional space [140], while the KNN method is a classical classification method. In this section, we will describe the genetic algorithm. For details of the KNN method, please refer to Section 6.3.2.1.

Genetic algorithms were inspired by several interesting observations in evolutionary biology. As we described in Section 2.6, organisms with different genotypes may have different phenotypes, which more or less fit into their environments. Analogously, in genetic algorithms, the particular problem to be solved can be viewed as the "environment," while the possible solutions to the problem form the population of "individuals." An individual is characterized by a "genotype" which corresponds to the parameters of the solution. The fitness of the individual is determined by its ability to solve the problem.

Algorithm 6.1   : Genetic Algorithm

---

(1) Initialize p number of individuals $\mathcal{P}$ with random parameters.

(2) For each $P_i \in \mathcal{P}$, compute $Fitness(P_i)$.

(3) While $max\{Fitness(P_i)\} < \tau$ do
   Create a new generation $\mathcal{P}'$

   (a) *Select*: Probabilistically select $(1-r)p$ members of $\mathcal{P}$ to $\mathcal{P}'$. The probability $Pr(P_i)$ is related to its fitness score:

   $$Pr(P_i) = \frac{Fitness(P_i)}{\sum_{j=1}^{p} Fitness(P_j)},$$

   (b) *Crossover*: Probabilistically select $\frac{r \cdot p}{2}$ pairs of individuals from $\mathcal{P}$, according to $Pr(P_i)$ above. For each pair of individuals, produce two offspring and add to $\mathcal{P}'$,
   (c) *Mutate*: Choose m percent of the members of $\mathcal{P}'$ with uniform probability. For each individual, randomly change one parameter,
   (d) *Update*: $\mathcal{P} \leftarrow \mathcal{P}'$, and
   (e) *Evaluate*: for each $P_i \in \mathcal{P}$, compute $Fitness(P_i)$.

(4) Return the individual from $\mathcal{P}$ that has the highest fitness.

As first expounded in Darwin's *Theory of Natural Selection*, all organisms reproduce themselves, and those which are well-adapted to their environment will survive and reproduce better than others. The metaphor carries through to genetic algorithms, where the most fit individuals (solutions) are selected to produce offspring both asexually and sexually. With asexual reproduction, the offspring inherit the "genomes" (parameters) of the parent, with possible changes arising from random mutation. With sexual reproduction, the "genomes" of two individuals are cross-recombined to produce a new individual. In other words, the parameters of the two solutions are reshuffled to derive a new solution.

A general genetic algorithm is described in [197]. Within a population $\mathcal{P}$ of p individuals, $Fitness(\cdot)$ is a function that assigns a fitness score to an individual, τ is a threshold specifying the termination criterion, and r and m are the rates for cross-over and mutation, respectively. The major steps of a genetic algorithm are outlined in Algorithm 6.1.

Li et al. [181, 182] implemented a genetic algorithm using the prediction accuracy of the KNN method as the fitness function. This approach was tested on several microarray data sets, including lymphoma [6], colon [7], and leukemia [114] data. They found that the majority of samples were correctly classified using the genes reported by the genetic algorithm.

6.2.2   Unsupervised Approaches

Unsupervised approaches reduce the dimensionality of microarray data without training samples. *Principal component analysis* (PCA) [157] is a classical method which projects the original data along a few directions in an attempt to capture the major variations in the data. However, the results obtained through PCA are often difficult to interpret. "Gene shaving" [126] is a PCA-based approach developed specifically for microarray data. After "shaving" the gene dimension for several iterations, the algorithm reports a set of informative gene groups. These approaches will be discussed in more detail in the following subsections.

6.2.2.1   PCA: Principal component analysis

Principal component analysis (PCA) [157] is a classical technique designed to reduce the dimensionality of a given data set by projecting it to a new set of coordinates (the principal components, or PCs) that capture the variance of the data set as much as possible. The PCs are uncorrelated and orthogonal to each other. In particular, they are ordered so that the kth PC has the kth largest variance among all PCs. For example, as shown in Figure 6.3, the first principal component (PC1) is obtained by drawing a solid line that best describes the direction of variation. Any variation that is not captured by that first PC can be captured by the second (orthogonal)

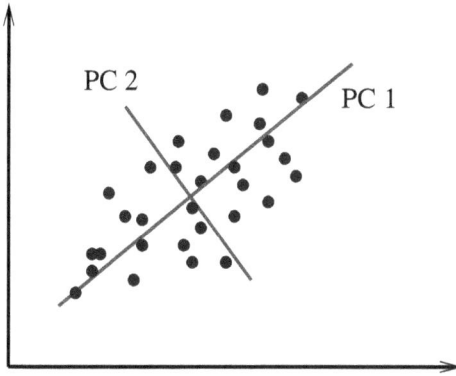

Fig. 6.3   PCA is applied to an example data set. The first principal component (PC1) captures the major variance in the data set, while the second principal component (PC2) represents the "residual" variance. Note PC1 is orthogonal to PC2.

principal component (PC2).

The PCs are the *eigenvectors* with the corresponding eigenvalues (ranked from largest to smallest) of the *covariance matrix* of the original data. An *eigenvector* of a matrix A is defined as a vector z such that:

$$Az = \lambda z,$$

where λ is a real value called *eigenvalue*. For instance, the matrix:

$$A = \begin{bmatrix} 2 & 1 \\ 3 & 4 \end{bmatrix}$$

has the eigenvalues $\lambda_1 = 5$ and $\lambda_2 = 1$ and the eigenvectors $z_1 = \begin{bmatrix} 1 \\ 3 \end{bmatrix}$ and $z_2 = \begin{bmatrix} -1 \\ 1 \end{bmatrix}$.

It can be verified that:

$$Az_1 = \begin{bmatrix} 2 & 1 \\ 3 & 4 \end{bmatrix} \begin{bmatrix} 1 \\ 3 \end{bmatrix} = 5 \cdot \begin{bmatrix} 1 \\ 3 \end{bmatrix} = \lambda_1 z_1,$$

and

$$Az_2 = \begin{bmatrix} 2 & 1 \\ 3 & 4 \end{bmatrix} \begin{bmatrix} -1 \\ 1 \end{bmatrix} = 1 \cdot \begin{bmatrix} -1 \\ 1 \end{bmatrix} = \lambda_2 z_2.$$

The first several PCs identified through PCA typically capture most of the variation in the original data and are therefore the most useful in reducing the dimensionality of the data. In contrast, the last few PCs are often assumed to contain only the residual "noise" in the data and are thus discarded. For instance, the data points in Figure 6.3 lie mainly along a single direction, although each particular data point has two coordinates. In this example, the first principal component is sufficient to capture most of the variance present in the data, and the second principal component may be discarded. To transform the data points from the original high-dimensional space into the new coordinate system defined by PCs, we need only multiply each original data point by the matrix of the eigenvectors.

Although PCA is effective in reducing the dimensionality of data while still capture the salient variations, it is not ideal for sample-based analysis. Each principal component is a linear combination of all the original features, which makes it difficult to identify those features which are most important [9, 126]. As previously noted, finding informative genes is as important as sample classification. Furthermore, the direction of the PCs is determined solely based on the variance of the data, with no consideration of the classes of the data points, so they may not necessarily be the most "informative" directions [83].

6.2.2.2   *Gene shaving*

Gene shaving [126] seeks to identify groups of genes which serve as the informative genes to classify samples. A group of genes must satisfy the following criteria: 1) the genes in each cluster behave in a similar manner, which suggests similar or related function among genes; 2) the cluster centroid shows high variance across the samples, which indicates the potential of this cluster to distinguish sample classes; and 3) the groups are as much uncorrelated between each other (which encourages seeking groups of different specification) as possible.

The gene shaving algorithm performs an iterative principal component analysis. The major steps of the algorithm are outlined in Algorithm 6.2.

Algorithm 6.2 : Gene-Shaving Algorithm

---

(1) Computes the first principal component of the given data set X.

(2) Shaves off a proportion α (typically 10%) of the rows having the smallest inner product with the first principal component.

(3) Repeats steps 1 and 2 until one row remains. This produces a sequence of nested gene clusters $S_N \supset S_{k_1} \supset S_{k_2} \supset \ldots \supset S_1$ where S_k denotes a cluster of size k.

(4) Estimates the optimal size k using the *Gap statistic* [281].

(5) Orthogonalizes each row of X with respect to $\bar{x}_{S_k}$, the average gene in S_k.

(6) Repeats steps 1-5 with the orthogonalized data to find another optimal cluster, until M clusters are found, with M chosen a *priori*.

Gene shaving applies PCA to find the directions which capture the majority of variance in the data set. However, it does not represent the new coordinates (PCs) using linear combinations of the original features (genes). Instead, it identifies those genes which are most similar to the PCs as coordinates. This generates easily-interpretable results, and the informative genes are readily identifiable. The gene-shaving approach is an unsupervised method, since it does not require that class labels be pre-defined and does not employ training samples. However, as described in [125], this method can be extended into a partly- or fully-supervised approach if class labels and training data are available.

6.3 Class Prediction

6.3.1 *Linear Discriminant Analysis*

Linear discriminant analysis (LDA), sometimes known as *Fisher's linear discriminant* [85, 99], is a classical statistical approach to supervised classification. LDA is designed to find those linear projections of the data which maximize the ratio of between-class variance to within-class variance and effectively separate the k classes.

Suppose there are n observations, each of which is a vector of length m. These observations will be referred to as $X_1 \cdots X_n$. Suppose these n observations are divided into g mutually exclusive classes $G_1 \cdots G_g$, such that G_i is a set of indices of observations. $|G_i|$ is the number of observations in each class.

Class mean μ_i is the mean vector of observations of class G_i. It is defined as

$$\mu_i = \frac{1}{|G_i|} \sum_{j \in G_i} X_j. \tag{6.1}$$

Grand mean μ is defined for all observations as

$$\mu = \frac{1}{n} \sum_{j=1}^{n} X_j. \tag{6.2}$$

In LDA, the within-class and between-class variances are estimated by *within-class scatter* and *between-class scatter*, respectively, and these are then used to formulate the criteria for class separability. There are two ways to define within-class scatter, either using the *covariance matrix* cov_i for each class or using *pooled covariance matrix* S_w.

$$cov_i = \frac{1}{|G_i| - 1} \sum_{j \in G_i} (X_j - \mu_j)(X_j - \mu_j)^T \tag{6.3}$$

$$S_w = \frac{1}{n - g} \sum_{i=1}^{g} (|G_i| - 1)cov_i \tag{6.4}$$

The choice of which definition of within-class scatter to use depends on the data set and the goals of classification problem under investigation. If generalization is more important, the use of S_w is preferred. cov_i is preferred when maximal separability between classes is the goal.

Between-class scatter is computed using the following equation:

$$S_b = \frac{1}{g} \sum_{j=1}^{g} (\mu_j - \mu)(\mu_j - \mu)^T. \tag{6.5}$$

As mentioned, the optimizing criterion of LDA is the ratio of between-class to within-class scatter. For the j-th class, the criterion is represented as follows if cov_j is used as within-class scatter. In this case there is one criterion V_j for each class.

$$V_j = (cov_j)^{-1} \times S_b. \tag{6.6}$$

When S_w is used to measure the within-class scatter, there is only one single criterion:

$$V = (S_w)^{-1} \times S_b. \tag{6.7}$$

To find the projection which most widely separates the different classes, LDA maximizes the criterion using the eigenvector matrix W_j (W) of V_j (V) with the large eigenvalue(s). W_j represent the best projection directions to separate data points in class G_j from the rest. While W are the best overall directions so that the within-class scatter is minimized for every class and the between-class scatter is maximized.

To use LDA for classification, data are first projected onto the best projection directions found earlier. Suppose D is the original data and D_j is the original data of class G_j. Transformed data D' and D'_j are defined as the following, depending on the choice of within-class scatter.

$$D'_j = W_j^T \times D_j, \tag{6.8}$$

$$D' = W^T \times D. \tag{6.9}$$

Once the transformations are completed, a new data point X is first transformed into X' and then the Euclidean distances between X' and each of the group means μ' in the transformed data set are computed. The smallest Euclidean distance among the g distances indicates the predicted class membership of X. More detail regarding LDA is available in [14].

$$dist_j = W_j^T \times x - \mu'. \tag{6.10}$$

$$dist_j = W^T \times x - \mu'. \tag{6.11}$$

Although a simple, classical method, LDA is still one of the most widely-used classification techniques. For example, Hakak et al. [118] and Dudoit et al. [86] applied LDA to microarray data for sample classification. Both LDA and PCA attempt to summarize the variance of data using a set of new coordinates; PCA is unsupervised and emphasizes the reduction of dimensionality, while LDA is supervised and stresses data classification. Additionally, PCA transforms the original data into the PC space, while LDA does not change the original data but only seeks to draw a linear hyperplane between the given classes with the maximal class separability.

Various extensions of Fisher's LDA have been developed, including *quadratic discriminant analysis* (QDA), logistic regression, *diagonal linear discriminant analysis* (DLDA), and *diagonal quadratic discriminant analysis* (DQDA). QDA employs the same input parameters and returns the same results as LDA but uses quadratic rather than linear equations. Logistic regression also takes the same input parameters and returns the same results as LDA and QDA but is more sensitive to data sets containing groupings of unequal size. For example, if the researchers expect only 10% of the cases will be categorized in group 1 and 90% will be categorized in group 2, logistic regression would be a more appropriate choice. Readers may refer to [10] for a detailed discussion of variants of LDA and their applications to microarray data.

6.3.2 *Instance-Based Classification*

In contrast to learning methods that construct a general, explicit classification function (such as LDA), instance-based methods delay the learning phase until a new query instance is probed. Specifically, instance-based methods examine the training samples each time a new query instance is encountered. The relationship between the new query instance and training examples will then be checked in order to assign a class label to the query instance. The following subsections will discuss two representative instance-based methods which have been applied to microarray data: *k*-nearest neighbors (KNN) and weighted voting.

6.3.2.1 *KNN: k-Nearest Neighbor*

The *k-nearest neighbor* method (KNN) is a simple but effective classification technique grounded on the assumption that a test sample x can be best predicted by determining the most common class label among the k training samples to which x is most similar.

A formal description of KNN is given by Amaratunga and Cabrera [10]. Let x_j represent the jth training sample and y_j be the class label for x_j. Let x be the given query instance to be classified, and let N_x be the set of the K nearest neighbors of x in the training set. The KNN method consists of estimating the probability that x belongs to the ith class $p(i|x)$ by the proportion of the K nearest neighbors that belong to the i-th class:

$$\hat{p}(i|x) = \frac{|\{y_j = i | x_j \in N_x\}|}{K}.$$

The ith class which maximizes the probability $\hat{p}(i|x)$ will be assigned as the label of x.

Pomeory et al. [223] applied a modified KNN method to classify the embryonal tumors of the central nervous system (CNS) about which little biological information is known. Irrelevant genes were filtered using the signal-to-noise statistic (see Section 4.4.3) before the KNN classification. The training data consist of 99 samples, and the prediction of the class label was based on the Euclidean distance of the new sample to the k nearest samples ($k = 5$). It was demonstrated that medulloblastomas, the most common malignant brain tumors of childhood, were molecularly distinct from other brain tumors.

Ramaswamy et al. [229] subjected 90 normal tissue samples and 218 tumor samples spanning 14 common tumor types to oligonucleotide microarray gene expression analysis. The expression levels of 16,063 genes and expressed sequence tags (ESTs) were used to evaluate the accuracy of a multiclass classifier based on a KNN and a support vector machine (SVM) algorithm (see Section 6.3.4). The results indicated that the performance of KNN was inferior to SVM when all the genes or ESTs were used as features. However, in the comparison study conducted by Ben-dor et al. [21], KNN demonstrated a high prediction accuracy than SVM after gene filtering. For variants of KNN algorithms and additional information, we refer readers to [197].

6.3.2.2 *Weighted voting*

Golub et al. [114] developed a procedure that applies the expression levels of a fixed set of informative genes (chosen via *neighborhood analysis*, see Section 4.4.3) to predict the class assignments of a new sample. As illustrated in Figure 6.4, each informative gene g_i of the new sample casts a "weighted vote" v_i for one of the classes, AML or ALL, depending on whether its expression level x_i in the sample is closer to μ_{AML} or μ_{ALL} (which denote, respectively, the mean expression levels of AML and ALL in a set of reference samples). The magnitude of the vote is $w_i \cdot v_i$, where w_i is a weighting factor that reflects how well the gene is correlated with the class distinction, which is calculated using Formula 4.7 for selected informative gene g_i based on the training samples of the two classes. The other factor in the weighted vote, v_i reflects the deviation of the expression level in the sample from the average of μ_{AML} and μ_{ALL} and is determined by $v_i = |x_i - (\mu_{AML} + \mu_{ALL})/2|$. The votes from the set of informative genes for each class are then summed to obtain total votes V_{AML} and V_{ALL}. Finally, the sample is assigned to the class with the higher vote total, provided that the *prediction strength* (see Section 6.5.2) exceeds a predetermined threshold.

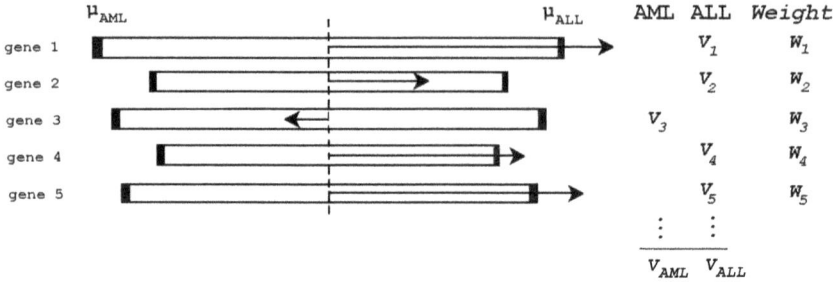

Fig. 6.4   Weighted voting. The prediction of a new sample is based on the "weighted votes" of a set of informative genes. Figure is from [114] with permission from AAAS.

A key advantage of instance-based classification is that instead of estimating the classification function once for the entire instance space, these methods can estimate it locally and differently for each new instance to be classified [197].

6.3.3   *Decision Trees*

Decision tree learning is one of the most widely-used and practical methods in the machine learning field [197]. Each *internal node* of a decision tree specifies a test regarding some attribute of the object, and each branch descending from that node corresponds to one of the possible values of this attribute. Thus, decision trees are only applicable to those objects with *categorical attributes*. For objects with real-valued attributes, such as expression levels of genes, the real values must be *discretized* into several categories. As determined by the test specified by an internal node, the training examples will be divided and distributed along the descending branches. This splitting process continues until all the training examples pertaining to the node share a common label and are considered "pure." Such nodes are called *leaf nodes*.

The construction of a decision tree (the learning process) involves creation of a hierarchy of tests pertaining to attributes. Various decision trees can be built to optimize particular splitting criteria. Some of the most common criteria are [10]:

- *Gini index.* This criterion was used by Breiman et al. [42] in the original version of their *classification and regression tree* (CART) methodology. The objective function is

$$C_g = qq_R p_R + pq_L p_L,$$

where p and q are the proportions of observations going to the left and right buckets, p_L and q_L are the proportions of 0's and 1's in the left-sided bucket, and p_R and q_R are the proportion of 0's and 1's in the right-side bucket, respectively.
- *Entropy.* This criterion is employed in the widely-used C4.5/C5 algorithms introduced by Quinlan [226]:

$$C_e = q(-q_R log q_R - p_R log p_R) + p(-q_L log q_L - p_L log p_L).$$

A decision tree follows a rather simple classification process. The query instance starts from the root node of the tree, tests the attribute specified by this node, and then moves down the tree branch corresponding to the value of the attribute in the query instance. This process is repeated until a leaf node is met. The class label shared by the training objects belonging to this leaf node is then assigned to the query instance.

As noted above, the application of decision-tree classification to microarray data analysis requires that real expression levels be discretized into categorical values. For example, we can assign a boolean value of "true" to a gene if its intensity is above 100; otherwise, the value will be "false." In this way, the decision-tree classifiers may split the samples according to the boolean values of the genes.

Decision tree models are easily built, easily understood, and are able to model fairly complex functions [257]. However, a drawback of decision trees is that they are prone to overfitting of training samples. That is, the trained tree models perform extremely well with the training samples but perform poorly with query samples [10, 197, 257]. This problem may be particularly severe with microarray data sets, which usually contain a large number of genes (features) but very limited training samples. A basic approach to addressing this problem is "pruning" the tree, i.e., restricting the height of the tree. By restricting the number of consecutive branches on the tree, the classification rules become less specific to the training samples and thus can be generalized more successfully to the test samples.

More recent methods to overcome "overfitting" include "bagging" [41] and "boosting" [240], both of which are based on resampling the data. Bagging generates replicate training sets by sampling with replacement from the training set, while boosting retains all the samples but weights each sample differently and generates different trees by adjusting the weights [10]. Each method builds multiple trees and then combines these trees into a single predictive model by voting. Dudoit et al. [86] found that these two techniques improve the performance of decision trees applied to microarray data.

6.3.4   *Support Vector Machines*

Support vector machines (SVM) [37, 292] is a class of linear classification algorithms which tries to maximize the margin of confidence of the classification on training data set and has been considered as well-suited to the noise and high-dimensionality of microarray data [44, 104]. In its simplest form, the decision boundary of SVM is a hyperplane that separates boundary training data of different classes with maximum confidence. Unlike LDA, if no separating hyperplane exists, SVMs will map the samples to a higher-dimensional space where the data become linearly separable.

SVM classifiers are based on the class of hyperplanes $(w \cdot x) + b = 0, w \in$

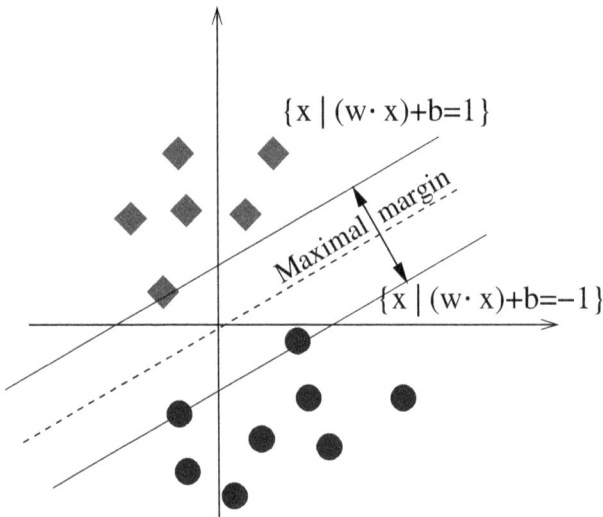

Fig. 6.5   The optimal hyperplane for the separation of balls and diamonds has the maximal margin between the two classes (Figure is adapted from [128]).

$R^N, b \in R$, corresponding to decision functions $f(x) = sign((w \cdot x) + b)$ [128]. This is a convex, quadratic programming problem which is approached using Lagrangian. Figure 6.5 shows that the optimal hyperplane is defined as that with the maximal margin of separation between the two classes. SVMs map the data into some other dot-product space (called the *feature space*) F via a nonlinear map $\Phi : R^N \rightarrow F$, and try to solve the same optimization problem in F. The mapping is not explicitly needed if some kernel function K is used such that

$$\Phi(x_i) \cdot \Phi(x_j) = K(x_i, x_j). \tag{6.12}$$

The mapping is accomplished by raising the dot-product kernel $x \cdot y$ to a positive integer power. For instance, squaring the kernel yields a convex surface in the input space. Raising the kernel to higher powers yields polynomial surfaces of higher degrees.

SVMs are more powerful than linear discriminant methods, since they are able to classify non-linearly separable classes by mapping the samples into higher-dimensional space. Moreover, SVMs are different from other nonlinear classifiers in that they pay special attention to the boundary of separation between the regions corresponding to each class, and this may yield small improvements of the classification prediction rate [10]. To avoid potential overfitting, it is preferable to use simple SVM kernel functions when complex ones are not needed [257].

Fuery et al. [104] applied SVM to an ovarian data set (unpublished data), a leukemia data set [114], and a colon data set [7]. The authors found the results obtained with SVM to be comparable to those of several other methods. Additionally, the applications of SVM are not limited to sample-based analysis. For example, Brown et al. [44] used SVMs in the gene dimension to predict functional roles for uncharacterized yeast ORFs. For a general introduction to SVMs, we refer readers to a collection of four papers in Hearst [128] and the book by Duda et al. [85].

6.4 Class Discovery

Class discovery in sample-based analysis involves the automatic partitioning of samples into several groups, with greater similarities among within-group than between-group samples. Unlike the process of class prediction, class discovery is an "unsupervised learning" problem in the sense that no pre-defined class labels and training samples are available. Distinguishing classes within microarray data is of particular practical interest, since the identification of phenotypes from samples through traditional pathological or clinical methods is usually slow, typically evolving over years of hypothesis-driven research. Moreover, there may be unknown subtypes of tumors which respond differentially to drug treatment (e.g., [114]) or lead to heterogeneous clinical outcomes (e.g., [6]).

To find the class structure hidden in the sample space, the clustering techniques described in Chapter 5 can be used, with samples treated as the data objects and genes as the attributes. Unfortunately, if the entire set of genes (typically several thousand or tens of thousands) in a microarray data set is adopted as the feature vector, the performance of most conventional clustering algorithms will be degraded by the "large p, small n" problem (see Section 6.2).

The class discovery process therefore can be viewed as involving two sub-problems: (1) identifying the hidden class structure embedded in the sample space; and (2) selecting a small subset of informative genes which are strongly correlated to the class structure. In fact, these two issues are closely interrelated. Once the class structure has been correctly identified and the samples have been appropriately assigned to classes, the supervised feature selection methods described in Section 6.2.1 can be used to rank the genes according to their relevance to the classification. Conversely, once the informative genes have been identified, we can apply the clustering techniques discussed in Chapter 5 to partition the samples.

CLIFF [308] and ESPD [273, 274, 275] are two recently-developed strategies for class discovery which exploit this dynamic relationship between genes and samples and combine clustering and gene-selection processes in an iterative manner. Both are based on the intuitive recognition that a valid approximate sample partition can be obtained using the entire set of genes. The approximate partition allows the selection of a moderately-valid gene subset, which will in turn draw the approximate partition closer to the target partition in the next iteration. After several iterations, the sample partition may converge to the true class structure, and the selected genes will be feasible candidates for the set of informative genes.

In the following subsections, we will provide a synthesis of the full presentation of these methods as originally expounded in [308, 273, 274, 275]. Following a formal description of the problem of class discovery, details of the CLIFF and ESPD[1] approaches will be presented.

---

[1] With permission from Oxford University Press.

6.4.1 Problem statement

A *phenotype structure* contains two essential elements: an exclusive and exhaustive partition of the samples such that each group within the partition represents a unique phenotype, and a set of informative genes manifesting the sample partition such that each informative gene displays approximately invariant signals for samples of the same phenotype and highly differential signals for samples from different phenotypes. Class discovery seeks to identify the empirical phenotype structure, the phenotype structure which is controlled by the biological experiment. To be specific, the problem of class discovery is defined as follows.

Problem of class discovery: Given
1. An $n \times m$ data matrix M in which the number of samples is m and the volume of genes is n. Usually, the number of samples and the volume of genes are at different orders of magnitude $(n \gg m)$.
2. $\mathcal{K}$, the number of phenotypes of samples.

The goal of class discovery in microarray data is to identify the phenotype structure, which includes:
1. A $\mathcal{K}$-partition of the samples matching their *empirical phenotype distinction*; and
2. The informative genes that manifests the phenotype distinction.
The quality of the classification result is governed by the degree of similarity of these genes to the empirical phenotype structure.

6.4.2 CLIFF: CLustering via Iterative Feature Filtering

To automatically identify the phenotype structure in a given microarray data set, Xing et al. [308] proposed a sample-based clustering algorithm, *CLIFF* (CLustering via Iterative Feature Filtering), which iterates between the sample-partition process and the gene-filtering process. In the sample-partition process, a graph-based clustering algorithm NCut (Approximate Normalized Cut) [253] is applied to partition the samples on the basis of unfiltered genes. In the gene-filtering process, the current sample partition is used as the *reference partition* to rank the genes. The lower-ranked genes will be removed from the feature set and the remainder form the input to the sample-partition process in the next iteration. Iteration continues until a stable phenotype structure is obtained. The framework of the CLIFF algorithm is shown in Algorithm 6.3.

Algorithm 6.3   : The CLIFF Algorithm

Initialization:

  Apply the two-component Gaussian model to each gene and compute
    the Bayes error for that gene;
  let $G_1 = \{$genes corresponding to the k smallest Bayes errors$\}$;
  Generate partition C_1 based on G_1 using the normalized cut algorithm
    NCut;
  let $C = C_1$

Iterate:

  Use C as a reference partition to compute the information gain for each
    gene;
  let $G_2 = \{$genes corresponding to the k smallest information gain values$\}$;
  Generate partition C_2 based on G_2 using the normalized cut algorithm
    NCut;
  let $C = C_2$

  Use C as a reference partition to compute the Markov blanket score for
    each gene;
  let $G_3 = \{$genes corresponding to the k smallest Markov blanket scores$\}$;
  Generate partition C_3 based on G_3 using the normalized cut algorithm
    NCut;
  let $C = C_3$

  if C and G_3 converge, **end** iteration.

6.4.2.1 *The sample-partition process*

To partition the samples, CLIFF first transforms the gene expression data
set into a weighted graph G where each vertex v_i in the graph corresponds
to one sample x_i in the data set, and each pair of vertices (v_i, v_j) are
connected by an edge $e(v_i, v_j)$. In CLIFF, Pearson's correlation coefficient
(see Section 5.2.2) is used to measure the similarity between two samples
x_i and x_j, and the weight of the corresponding edge $e(v_i, v_j)$ is determined
by

$$w(v_i, v_j) = exp\{-\frac{1 - \rho(x_i, x_j)}{\sigma}\},$$

where $\rho(x_i, x_j)$ is the similarity between x_i and x_j, and σ is a parameter
relating the sensitivity of clustering to the "strength" of the similarity.

  Given a weighted graph $G(V, E, W)$ where V, E, and W are the sets of vertices, edges, and edge weights, respectively, the sum of weights $w(A, B)$ between two subsets $A, B \subseteq V$ is defined as $w(A, B) = \sum_{u \in A} \sum_{v \in B} w(u, v)$. For a *cut* of G which splits V into two subsets A and $\bar{A}$, the *normalized weight* of the cut is defined as

$$\text{Ncut}(A, \bar{A}) = \frac{w(A, \bar{A})}{w(A, V)} + \frac{w(A, \bar{A})}{w(\bar{A}, V)}.$$

In the simple situation where the samples are partitioned into two subsets, an optimal partition can be obtained from an *optimal normalized cut* of G which has a minimum normalized weight. However, computing an optimal normalized cut is NP-hard. Therefore, CLIFF adopts an approximate approach, the *approximate normalized cut algorithm* [253], which pursues a second optimal result with a higher efficiency. To generalize the partition from two-way to multi-way, one can perform recursive two-way cuts until each resulting subset of the vertices is a singleton. An alternative method is to extend the approximate normalized cut algorithm so that it arrives at a simultaneous K-way cut [253].

6.4.2.2 *The gene-filtering process*

In [308], non-informative genes were divided into three categories: 1) *non-discriminative genes* (genes in the "off" state); 2) *irrelevant genes* (genes do not respond to the physiological event); and 3) *redundant genes* (genes that are redundant or secondary responses to the biological or experimental conditions distinguishing different samples).

To filter non-discriminant genes, CLIFF assumes that a gene will be in either the "on" or "off" state in each sample. Following this assumption, CLIFF applies a *two-component Gaussian model* to the expression levels of each gene:

$$P(x_i | \theta_i) = \prod_{d=1}^{m} \prod_{k=0}^{1} (\pi_{i,k} [\frac{1}{\sqrt{2\pi}\sigma_{i,k}} exp\{-\frac{(x_{di} - \mu_{i,k})^2}{2(\sigma_{i,k})^2}\}])^{z_d^k},$$

where x_i is the vector of measurements $\{x_{i1}, \ldots, x_{im}\}$ of gene i and $k = (0, 1)$ denotes the state of the gene (e.g., 1 denotes the state "on," and 0 stands for "off"). Moreover, $\theta_i = (\theta_{i,0}, \theta_{i,1})$, where $\theta_{i,k} = (\pi_{i,k}, \mu_{i,k}, \sigma_{i,k})$ is the set of the parameters of the Gaussian model and z_d^k is 1 if sample d has state k, and 0 otherwise.

The parameters of the Gaussian model $\theta_{i,k} = (\pi_{i,k}, \mu_{i,k}, \sigma_{i,k})$ can be determined by the EM algorithm [69] (see Section 5.3.4). Gene i is considered to be "on" if $\pi_{i,1}P(x_i|\theta_{i,1}) \geq \pi_{i,0}P(x_i|\theta_{i,0})$ and "off" otherwise. Using these parameters, the next step is the calculation of the *Bayes error*, which is the probability of misclassifying a sample drawn from the Gaussian model. Genes with a high Bayes error are considered to be non-discriminative and are filtered from the feature set.

Note that the Gaussian model is an "unsupervised" approach since the reference partition of samples is not used to rank the genes. CLIFF also incorporates two additional supervised scoring methods. These two scores filter "irrelevant" and "redundant" genes on the basis of the *information gain* and the *Markov blanket*, respectively. The genes which survive these two filters will be used in sample-partition process of the next iteration.

Xing and Karp applied CLIFF to a public leukemia data set [114] with 72 leukemia samples and 7130 genes. The experimental results showed that CLIFF outperformed the k-means algorithm and the NCut algorithm without gene selection and produced a result that was very close to the original expert labeling of the sample set. However, CLIFF is sensitive to outliers and noise, since the gene-filtering process is largely governed by the NCut algorithm, which is not robust to noise and outliers.

6.4.3   ESPD: Empirical Sample Pattern Detection

A crucial observation of the phenotype structure in microarray data is that the informative genes display approximately invariant signals for samples within the same phenotype and highly differential signals for samples with different phenotypes. Based on this observation, Tang et al. [273, 274, 275] proposed a series of measures to evaluate the degree of consistency within the given phenotypes, the extent of divergence across various phenotypes, and the quality of phenotype structures. A heuristic algorithm and a mutual-reinforcing adjustment algorithm were designed to detect the phenotype structure.

6.4.3.1   *Measurements for phenotype structure detection*

Each properly-constructed phenotype structure should simultaneously satisfy two requirements. First, the expression levels of each informative gene should be similar in all samples of the same phenotype. Second, the expression levels of each informative gene should be clearly dissimilar in each pair of phenotypes within the phenotype structure. Two statistical metrics are useful in determining and quantifying these factors: *intra-consistency* measures the similarity of the gene expressions within each sample group of a candidate phenotype structure, and *inter-divergency* measures the dissimilarity of the candidate gene set between different sample groups. These metrics are combined to provide the *phenotype quality* metric, which is used to qualify a phenotype structure.

Let $S = \{\vec{s_1}, \ldots, \vec{s_m}\}$ be a set of samples and $G = \{\vec{g_1}, \ldots, \vec{g_n}\}$ be a set of genes. A phenotype structure includes a $\mathcal{K}-$partition of samples $\{S_1, \ldots, S_\mathcal{K}\}$, where $S = \bigcup_{i=1}^{\mathcal{K}} S_i$ and $S_i \cap S_j = \emptyset$ for $(1 \leq i < j \leq \mathcal{K})$, and a set of genes $G' \subseteq G$.

Assume S' is one of the sample groups within a sample partition $\{S_1, \ldots, S_\mathcal{K}\}$ and $M_{G',S'} = \{w_{i,j} | \; \vec{g_i} \in G', \vec{s_j} \in S'\}$ is the corresponding sub-matrix with respect to S' and G'. The *variance* of each row in the sub-matrix is defined as:

$$Var(i, S') = \frac{\sum_{\vec{s_j} \in S'}(w_{i,j} - \overline{w}_{i,S'})^2}{|S'| - 1}, \tag{6.13}$$

where $\overline{w}_{i,S'} = \frac{1}{|S'|}\sum_{\vec{s_j} \in S'} w_{i,j}$. The variance of each row measures the variability of a given gene over all samples within the sub-matrix. A small variance value indicates that the gene has consistent values.

Intra-consistency, or the degree of consistency of a given gene's expression levels over a group of samples, is indicated by the average of variances in the subset of genes. Formally, intra-consistency is defined as:

$$Con(G', S') = \frac{1}{|G'|} \sum_{\vec{g_i} \in G'} Var(i, S'), \tag{6.14}$$

$$= \frac{1}{|G'| \cdot (|S'| - 1)} \sum_{\vec{g_i} \in G'} \sum_{s_j \in S'} (w_{i,j} - \overline{w}_{i,S'})^2.$$

Low intra-consistency values indicate that the expressions of the genes across the samples are relatively invariant and that the corresponding sub-matrix therefore has a greater likelihood of being a unique phenotype.

The concept of intra-consistency is intuitive, and it is better suited to effective phenotype- and informative-gene-mining than are local similarity measurements. In Table 6.1, we compare the intra-consistency with two typical local pattern similarity measurements, residue (Equation 6.15) and mean squared residue (Equation 6.16) on the two examples shown in Figure 6.6.

Let $G' \subset G$ and $S' \subset S$ be subsets of genes and samples. The mean

Table 6.1   Several different local pattern measurements for the two data sets in Figure 6.6.

Measurement	Data (A)	Data (B)
residue	0.1975	0.4506
mean squared residue	0.0494	0.4012
intra-consistency	339.0667	5.3000

squared residue [52] of the sub-matrix $M_{G',S'}$, $MSR(G',S')$, is:

$$\frac{1}{|G'||S'|} \sum_{i\in G',j\in S'} (w_{i,j} - w_{i,S'} - w_{G',j} + w_{G',S'})^2, \qquad (6.15)$$

where

$$w_{i,S'} = \frac{1}{|S'|}\sum_{j\in S'} w_{i,j}, \; w_{G',j} = \frac{1}{|G'|}\sum_{i\in G'} w_{i,j}, \; and \; w_{G',S'} = \frac{1}{|G'||S'|}\sum_{i\in G',j\in S'} w_{i,j}$$

The residue score [316] of sub-matrix $M_{G',S'}$, $Residue(G',S')$, is:

$$\frac{1}{|G'||S'|} \sum_{i\in G',j\in S'} |(w_{i,j} - w_{i,S'} - w_{G',j} + w_{G',S'})|. \qquad (6.16)$$

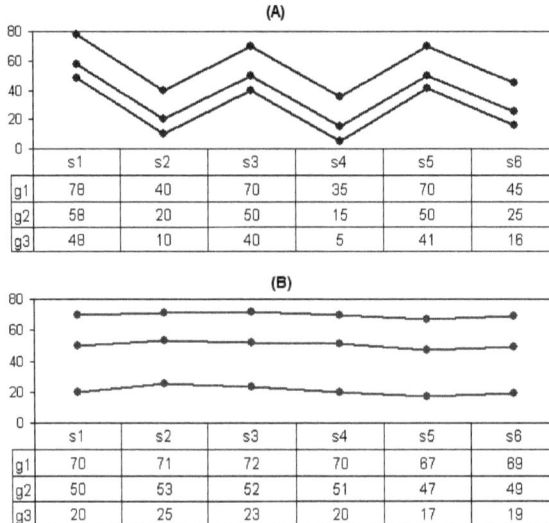

(A)

	s1	s2	s3	s4	s5	s6
g1	78	40	70	35	70	45
g2	58	20	50	15	50	25
g3	48	10	40	5	41	16

(B)

	s1	s2	s3	s4	s5	s6
g1	70	71	72	70	67	69
g2	50	53	52	51	47	49
g3	20	25	23	20	17	19

Fig. 6.6   An example of the performance of the intra-consistency metric. The horizontal
axis shows different samples.

Residue and mean squared residue measure the similarity of a group of genes while intra-consistency measures the variance among genes. A smaller residue or mean squared residue indicates greater similarity in gene patterns. These two metrics indicate a greater degree of similarity on the genes in Figure 6.6(A) than those in Figure 6.6(B). However, as can be seen from the figures, the genes in Figure 6.6(B) are much more consistent with the samples, and only the intra-consistency metric reflects that fact.

Inter-divergency
We introduce the *inter-divergency* to quantize how a subset of genes can distinguish two phenotypes of samples.

The inter-divergency of a set of genes G' in two groups of samples (denoted by S_1 and S_2) is defined as

$$Div(G', S_1, S_2) = \frac{\sum_{\vec{g}_i \in G'} \left| \overline{w}_{i,S_1} - \overline{w}_{i,S_2} \right|}{|G'|}, \qquad (6.17)$$

where $\overline{w}_{i,S_1}$ and $\overline{w}_{i,S_2}$ are defined as in Equation 6.13. The measure is normalized by $|G'|$ to avoid the possible bias arising from the large volume of genes. A higher inter-divergency indicates that the genes more effectively distinguish the samples in different groups.

Phenotype quality of a candidate phenotype structure
The goal of the above metrics is to identify a phenotype distinction and informative space characterized by a high level (or low value) of intra-consistency within each phenotype and a large inter-divergency value between each pair of phenotypes. Greater consistency and divergency produces a clearer and more distinct phenotype structure.

Suppose a set of samples S is partitioned into $\mathcal{K}$ exclusive groups, $S_1, \dots, S_{\mathcal{K}}$, where $S_i \cap S_j = \emptyset$ for $(1 \leq i < j \leq \mathcal{K})$. Given a set of genes G', we can define the following quality measure Ω to quantize the quality of the candidate phenotype structure:

$$\Omega = \frac{1}{\sum_{S_i, S_j (1 \leq i, j \leq \mathcal{K}; i \neq j)} \frac{\sqrt{Con(G', S_i) + Con(G', S_j)}}{Div(G', S_i, S_j)}}. \qquad (6.18)$$

A high *phenotype quality* value is required to qualify a phenotype structure.

Figure 6.7 illustrates three data sets which manifest the same two phenotypes, with samples shown on the horizontal axis. Table 6.2 provides the the *intra-consistency*, *inter-divergency*, and *phenotype quality* measurements of the data sets shown in Figure 6.7.

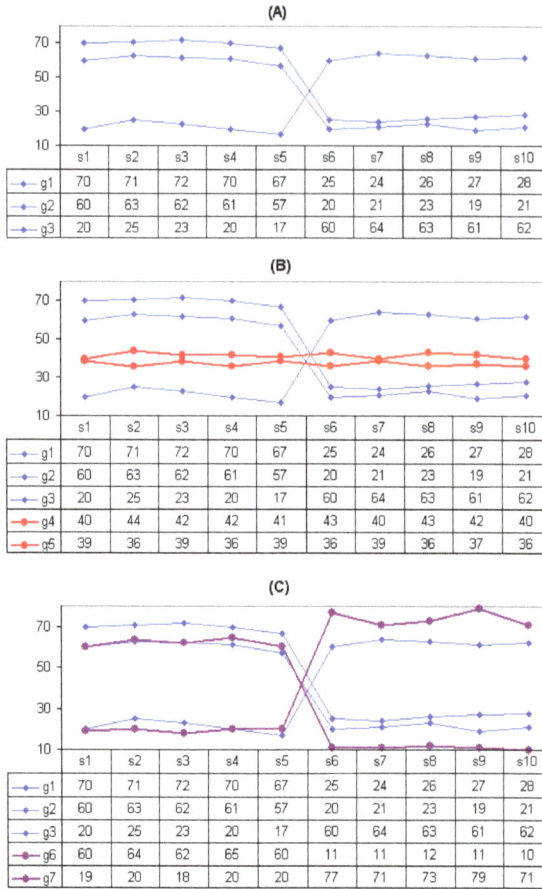

(A)

	s1	s2	s3	s4	s5	s6	s7	s8	s9	s10
g1	70	71	72	70	67	25	24	26	27	28
g2	60	63	62	61	57	20	21	23	19	21
g3	20	25	23	20	17	60	64	63	61	62

(B)

	s1	s2	s3	s4	s5	s6	s7	s8	s9	s10
g1	70	71	72	70	67	25	24	26	27	28
g2	60	63	62	61	57	20	21	23	19	21
g3	20	25	23	20	17	60	64	63	61	62
g4	40	44	42	42	41	43	40	43	42	40
g5	39	36	39	36	39	36	39	36	37	36

(C)

	s1	s2	s3	s4	s5	s6	s7	s8	s9	s10
g1	70	71	72	70	67	25	24	26	27	28
g2	60	63	62	61	57	20	21	23	19	21
g3	20	25	23	20	17	60	64	63	61	62
g6	60	64	62	65	60	11	11	12	11	10
g7	19	20	18	20	20	77	71	73	79	71

Fig. 6.7   An example showing the effect of the phenotype quality metric. Samples $s1 \sim s5$ belong to one phenotype and $s6 \sim s10$ are in the other phenotype.

The data set in Figure 6.7(A) includes three genes (g_1, g_2, g_3) which manifest two phenotypes. The data set in Figure 6.7(B) includes all genes within Data(A) and two additional genes (g_4 and g_5) which exhibit low variance within each phenotype. The first row of Table 6.2 provides the average intra-consistency values of two phenotypes without and with the inclusion of these two genes. It is evident the value for intra-consistency $R(x, y)$ diminished with the addition of these genes, increasing the quality of the result. However, informative genes should display not only low intra-consistency but also high inter-divergency. For example, in Figure 6.7(B),

Table 6.2 The intra-consistency, inter-divergency and phenotype quality of the data sets shown in Figure 6.7.

Measurement	Data(A)	Data(B)	Data(C)
intra-consistency	4.2500	3.4400	4.5200
inter-divergency	41.6000	25.2000	46.1600
phenotype quality	14.2687	9.6074	15.3526

genes (g_4 and g_5) are invariant over all samples. They cannot distinguish different phenotypes and thus are not informative. As a result, the inter-divergency value also diminishes, reducing the overall phenotype quality. In summary, the phenotype quality metric indicates that the two additional genes should not be included in the informative space, even though their inclusion did reduce the intra-consistency value.

The data set shown in Figure 6.7(C) is also based on Data(A), with the addition of genes g_6 and g_7. These added genes exhibit both low variance within each phenotype and high dissimilarity between the two phenotypes. Table 6.2 indicates that, although the average intra-consistency of two phenotype groups is slightly increased by their addition (from 4.2500 to 4.5200), they also produced a significant increase in inter-divergency (from 41.60 to 46.16). The overall phenotype quality therefore rose (from 14.2687 to 15.3526), indicating that the phenotype quality was improved by the inclusion of g_6 and g_7 in the informative space. The phenotype quality of the data set in Figure 6.7(C) therefore is the highest among the three analyzed in that figure.

6.4.3.2 *Algorithms*

As mentioned previously, phenotype-structure mining involves two linked activities: discovering the empirical phenotype distinctions among samples and searching for informative genes. Once informative genes have been identified, they can be used as features in the discovery of phenotypes via existing partitioning methods. The process can also be reversed, with identified phenotypes serving as the basis for the selection of informative genes via the methods described in Section 6.2. Therefore, the mining of phenotypes and identification of informative genes should be dually reinforced.

Clearly, a naïve method to obtain a globally-optimal solution would test each possible $\mathcal{K}$-partitions of the samples, identify corresponding informative genes for each partition, and then select the best by comparing the phenotype quality of each partition with the introduced subset of genes.

However, this method yields a very inefficient algorithm.

In the following discussion, we will present two methods [274, 275] which offer a more practical approach to a solution. The first is a heuristic searching algorithm. The algorithm starts with a candidate informative space containing a random partition of samples and a subset of genes and then iteratively adjusts the partition and the gene set. It employs the *stimulated annealing technique* [169] to arrive at a locally-optimal solution. The second method is a novel mutual-reinforcing adjustment algorithm to approximate the optimal solution. This approach incorporates deterministic, robust techniques to dynamically manipulate the relationship between samples and genes and conducts an iterative adjustment of genes and samples to approximate the empirical phenotype structure.

Both algorithms to be discussed here maintain a candidate phenotype structure which includes a set of genes as candidates for the informative genes and a partition of samples as the candidate for the phenotype distinction. The best phenotype quality will be approached by iteratively adjusting the candidate structure.

Formally, the *candidate phenotype structure* consists of the following items:

- A partition of samples $\{S_1, S_2, \ldots, S_{\mathcal{K}}\}$ such that $S_i \cap S_j = \emptyset$ for $(1 \leq i < j \leq \mathcal{K})$ and $S = \bigcup_{i=1}^{\mathcal{K}} S_i$.
- A set of genes $G' \subseteq G$ which is *a candidate set of informative genes*.
- The *phenotype quality* (Ω) of the candidate structure, calculated on the basis of the partition $\{S_1, S_2, \ldots, S_{\mathcal{K}}\}$ on G'.

An *adjustment* is an indivisible operation upon a sample or a gene which can change the current candidate structure of the algorithm. An adjustment may be any of the following operations:

- For a gene $\vec{g_i} \notin G'$, *insert* $\vec{g_i}$ into G'. That is, $G' = G' \cup \vec{g_i}$ (Figure 6.8(a));
- For a gene $\vec{g_i} \in G'$, *remove* $\vec{g_i}$ from G'. That is, $G' = G' - \{\vec{g_i}\}$ (Figure 6.8(b)); or
- For a sample $\vec{s_i}$ in partition S', *move* $\vec{s_i}$ to group S'' where $S' \neq S''$ (Figure 6.8(c)).

To measure the effect of an adjustment to a candidate structure, the *quality gain* of the adjustment indicates the change of the phenotype quality, i.e., $\Delta\Omega = \Omega' - \Omega$, where Ω and Ω' are the phenotype quality values of the

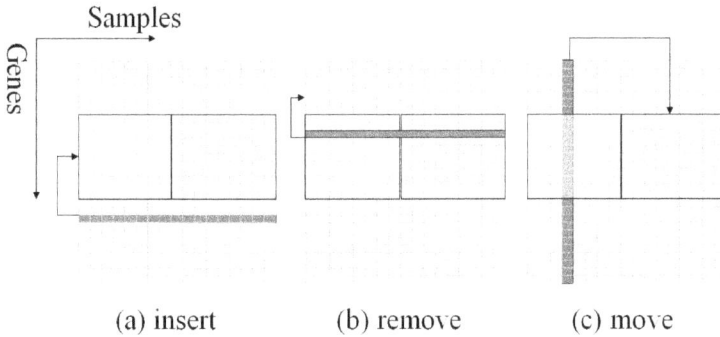

Fig. 6.8   A simplified illustration of the adjustment operations.

structure before and after the adjustment, respectively.

The goal of this process is to move from a starting candidate structure via a series of adjustments to reach a structure which maximizes the accumulated quality gain. Both algorithms record the best structure which achieves the highest quality to date.

A heuristic algorithm

An immediate solution to the problem is to design a heuristic algorithm based on the *simulated annealing* technique. Simulated Annealing is a stochastic simulation method originally proposed by Kirkpatrick et al. [169]. The inspiration for simulated annealing comes from the physical process of cooling molten materials down to the solid state. It has proven to be a simple and robust algorithm and is useful in addressing a wide range of complex combinatorial optimization problems that are difficult for conventional optimization problems. The method iteratively deploys several tools to change the state E of the system at temperature T and either accepts or rejects each change. Proposed changes are evaluated using the "Metropolis" criterion: if the state improves with the change ($\Delta E > 0$), the change is accepted unconditionally; otherwise, it is accepted with probability $P = \exp^{\Delta E/T}$. When the "temperature" T of the system is zero, changes are accepted only if E increases. Algorithms of this sort are characterized as "hill-climbing" or "greedy." At the other end of the spectrum, if the "temperature" of the system is very large, all changes are accepted, and we simply move at random. It can be shown that, if the temperature decreases sufficiently slowly [169], the probability of arriving at a global optimum tends to certainty.

A heuristic algorithm can follow this same model, starting from a random candidate phenotype structure and heuristically moving to a better status by conducting iterative adjustments, eventually approaching the empirical phenotype structure. This heuristic method takes an approach similar to the algorithms employed by δ-cluster [316] and CLARANS [205].

Algorithm 6.4   : The Heuristic Algorithm of ESPD

Initialization phase:
  a) Create a $\mathcal{K}$-partition of samples and select a set of genes G' randomly,
  b) Calculate *phenotype quality (Ω_0)* for the initial structure.
Iterative adjusting phase:
  Repeat:
    List a sequence of genes and samples randomly;
    For each gene or sample along the sequence, do:
    1.1) if the entity is a gene, compute $\Delta\Omega$ for the possible insert/remove;
      else if the entity is a sample, compute $\Delta\Omega$ for the best movement;
      end if
    1.2) if $\Delta\Omega \geq 0$, perform the adjustment;
      else if $\Delta\Omega < 0$, perform the adjustment with probability
      $p = \exp(\frac{\Delta\Omega}{\Omega \times T(i)})$;
      end if
  Until no positive adjustment can be performed.

The algorithm consists of two phases: *initialization phase* and *iterative adjusting phase* (see Algorithm 6.4). In the initialization phase, an initial structure is generated randomly and the corresponding quality value, Ω_0, is computed. In the iterative adjusting phase, genes and samples are examined individually during each iteration. Each gene can be either inserted or removed from the current structure, and the corresponding quality gain ($\Delta\Omega$) is then calculated. There are ($\mathcal{K} - 1$) possible adjustments for each sample, each involving relocation of the sample to one of the other ($\mathcal{K} - 1$) groups. The quality gain is also calculated for each sample adjustment, and the adjustment that produces the largest quality gain is selected. The adjustment of a gene or sample will be definitely performed if $\Delta\Omega$ is positive. If $\Delta\Omega$ is negative, the adjustment will be performed with a probability $p = e^{\frac{\Delta\Omega}{\Omega \times T(i)}}$, where $T(i)$ is a decreasing stimulated annealing function [169], and i is the iteration number.

The probability function p employed here has two components. The first component, $\frac{\Delta\Omega}{\Omega}$, considers the quality gain from a proportional standpoint, with a greater reduction in Ω correlating with a lower probability that the adjustment will be performed. The second component, $T(i)$, is a decreasing stimulated annealing function, where i is the iteration number. When $T(i)$ is large, p will be close to 1, and there will be a high probability that the adjustment will be performed. As the iteration proceeds, $T(i)$ becomes smaller, and the probability p also diminishes. In this algorithm, $T(0) = 1$, and $T(i) = \frac{1}{1+i}$, producing a slow annealing function. A slow annealing more effectively approaches a globally optimal solution, but more iterations will be required.

As indicated in [169], a simulated annealing search can reach the globally optimal solution if the simulated annealing function is sufficiently slow and there is an adequate number of iterations. The upper bound of the number of iterations is the total number of possible solutions. However, a laborious stimulated annealing search may be practically infeasible, and real applications must consider the trade-off between running time and finding the optimal solution. Thus, this method sets the termination criterion as *whenever no positive adjustment is performed in an iteration.* When the iteration stops, the partition of samples and the candidate gene set in the best structure will be output.

The time complexity of the heuristic algorithm is dominated by the iteration phase. The time to compute Ω initially is in $O(m \cdot |G'|)$, where $|G'|$ is the number of genes selected in the first candidate structure. The effect introduced by $\mathcal{K}$, which is typically very small, can be ignored. In each iteration, the time complexity depends on the calculation of Ω' for the possible adjustments. Since Equations 6.14, 6.17, and 6.18 are all accumulative, we can simplify the formula by computing only the changed portion of the measurements. For example, if we add one gene $\vec{g_1}$ into gene set G' ($G' \Rightarrow G''$) , the updated *inter-divergency* will be:

$$
\begin{aligned}
Div(G'', S_1, S_2) &= \frac{\sum_{\vec{g_i} \in G''} \left| \overline{w}_{i,S_1} - \overline{w}_{i,S_2} \right|}{|G''|} \\
&= \frac{|G'|}{|G'| + 1} \left(Div(G', S_1, S_2) + \frac{\left| \overline{w}_{1,S_1} - \overline{w}_{1,S_2'} \right|}{|G'|} \right) \text{(6.19)}
\end{aligned}
$$

The extra time required to compute $Div(G'', S_1, S_2)$ is therefore $O(m)$. It can be proved that the time cost of computing Ω' is $O(m)$ for each gene, and $O(m \cdot n)$ for each sample. There are n genes and m samples involved in each iteration. Therefore, the algorithm's time complexity is $O(n \cdot m^2 \cdot l)$, where l is the number of iterations.

The mutual-reinforcing adjustment algorithm

The heuristic algorithm is highly sensitive to the order of genes and sample adjustments considered in each iteration. To give every gene or sample an equal opportunity, all possible adjustments are randomized at the beginning of each iteration. However, since the number of samples is far less than the number of genes, samples have fewer opportunities for adjustment since they are treated equally with all genes during the adjustment process. In addition, because the number of samples is quite small, even one or two noisy or outlier samples may highly impact the quality of results and the adjustment decisions. The heuristic approach cannot detect or eliminate the influence of noise or outliers in the samples. Furthermore, it is difficult to mine phenotypes and informative genes directly from high-dimensional noisy data.

An effective strategy to address these challenges is to divide the genes and samples into smaller groups. It is intuitively obvious that the patterns within one small group of perhaps a few hundred genes will be more consistent and easily detected and that outlier samples can be more readily separated from other samples.

The implementation of this approach starts with the division of the gene expression matrix into a set of smaller sub-matrices with clearly-detectable data distributions and patterns. A sub-structure with high phenotype quality can then be identified from these sub-matrices. The globally-optimal phenotype structure can be gradually approached by analyzing and merging the levels of consistency and divergency thus discovered from the sub-structure.

Figure 6.9 illustrates an example gene expression matrix formed by 18 samples and 25 genes. There are two empirical phenotypes, one comprising the first eight samples and the other containing the next ten samples. Nine informative genes distinguish these two phenotypes. Two colored rectangles in Figure 6.9(A) represent the empirical phenotype structure. In Figure 6.9(B), the samples are divided into four groups while the genes are divided into five groups. The same two colors are used to show the distribution of the samples and genes within the empirical phenotype structure. According to definition presented in Equation 6.18, this empirical phenotype structure has a high phenotype quality value. In Figure 6.9(B), each one of the sub-matrix A, B, and C is formed by samples from a single phenotype and a subset of informative genes. Sub-matrices "A versus B" and "A versus C" can be regarded as the sub-structures of the empirical phenotype structure. Based on the definition of phenotype quality, the sub-structure of the empirical phenotype structure should also have a high phenotype value.

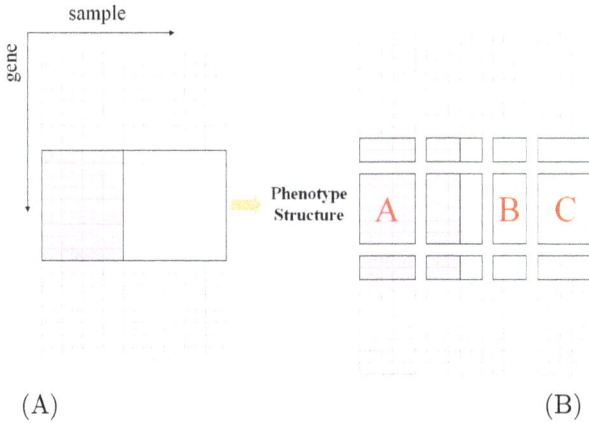

(A)   (B)

Fig. 6.9   Distribution of the empirical phenotype structure before and after clustering the matrix. (A) The empirical phenotype structure before clustering. (B) The empirical phenotype structure after clustering.

Algorithm 6.5   : The Mutual-Reinforcing Adjustment Algorithm of ESPD

Initialization phase:
   Let $G' = G$
Iteration phase:
   Repeat:
      1. partitioning the matrix
         1.1) group samples S into m_s ($m_s > \mathcal{K}$) groups;
         1.2) group genes in G' into n_g groups.
      2. identifying the reference partition
         2.1) compute *reference degree* for each sample group;
         2.2) select $\mathcal{K}$ groups from m_s groups of samples;
         2.3) do partition adjustment.
      3. adjusting the genes
         3.1) compute Ω for *reference partition* on G';
         3.2) perform possible adjustment of each gene
         3.3) update G', go to step 1.
   Until no any positive adjustment can be performed.
Refinement phase:
   find the highest *pattern quality candidate structure*.
   do *structure* refinement.

The mutual-reinforcing adjustment algorithm (see Algorithm 6.5) implements this concept, building upon the heuristic framework but adopting a more robust, deterministic, and noise-insensitive approach to sample adjustment. The algorithm proceeds via an iteration phase and a refinement phase. In each iteration, the entire set of samples and the current set of candidate informative genes are divided into small groups. These subsets are analyzed and the relationships thus discovered are used to post a partial or approximate phenotype distinction called a *reference partition* of samples. This reference partition is then used to direct the adjustment of genes. In turn, the adjusted informative genes guide the detection of a more refined phenotype structure in the next iteration. After each iteration, the candidate set of informative genes is adjusted and changed. The next iteration will start with this informative gene set and perform further adjustment until the optimal informative gene set is secured.

In the refinement phase, all the samples and genes are inspected once to ensure that none were overlooked. The final result is then output. The alterations in sample and gene status in each iteration are shown in Figure 6.10. Details of the algorithm will be discussed below.

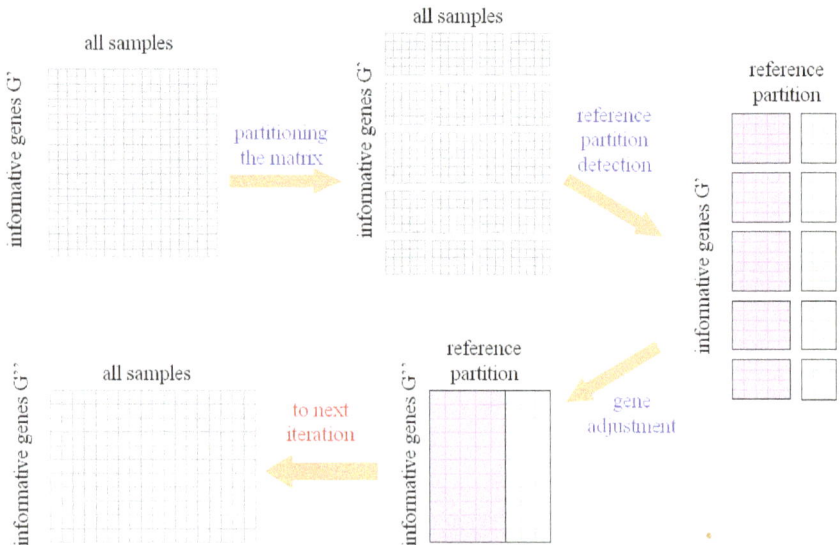

Fig. 6.10   Changes of current gene and sample set during an iteration phase.

Step 1: Partitioning the matrix

This step uses clustering techniques to divide the complete set of samples and the set of candidate informative genes into smaller groups. At the beginning of the first iteration, the set of candidate informative genes contains all the genes in the given data set, while for other iterations, the current candidate informative gene set and all samples will be the input.

To partition the data matrix, we apply a clustering technique to both the gene and sample dimensions. When clustering samples, the current informative genes serve as the features. The clustering of genes uses all the samples as features. After the clustering procedure, the matrix is divided into several sub-matrices, as shown in Figure 6.10.

The CAST (Cluster Affinity Search Technique) algorithm [22] (see Section 5.3.3.2) is applied to group genes and samples, and Pearson's correlation coefficient is chosen to calculate the similarity matrix. In CAST, a threshold must be specified to control the coherence of the cluster. In biological applications, experience shows that a given biological function typically involves between several dozen and several hundred genes [114]. Therefore, the threshold is set so that a majority of groups will be of this size. For samples, the threshold can be chosen such that the number of groups ranges within $\mathcal{K} \sim 2 \cdot \mathcal{K}$ to maintain a good compromise between the number of clusters and their degree of separation.

Let us consider a case in which, after partitioning the matrix using CAST, we have m_s ($m_s > \mathcal{K}$) exclusive groups of samples, $\{S_1, S_2, \ldots, S_{m_s}\}$ and n_g groups of genes in set G' of candidate informative genes, i.e., $G' = \{G_1, G_2, \ldots, G_{n_g}\}$. Samples or genes in each group will therefore share similar patterns with the others in the same group. □

Step 2: Reference partition detection

In this phase, $\mathcal{K}$ groups of samples (among the m_s sample groups) will be selected to "represent" the *phenotype distinction of the samples*. This $\mathcal{K}$-sample-group is called a *reference partition*. In the reference-partition detection phase, the informative gene set will not be changed.

The *reference partition* is selected among the sample groups which have small intra-consistency value (i.e., being highly consistent) and which exhibit high inter-divergency among groups. The purpose of selecting such a reference partition is to approximate the phenotype distinction as closely as possible and to eliminate noise-related interference caused by outlier samples.

A *reference degree* is defined for each sample group S_j by accumulating the intra-consistency values over the n_g gene groups generated from the previous step. This reference degree measures the likelihood that a given sample group will be included into the reference partition. That is,

$$Ref(S_j) = \log |S_j| \sum_{G_i \in G'} \frac{1}{Con(G_i, S_j)}. \tag{6.20}$$

The reference degree is proportional to the logarithm of the number of samples in a group, so very small groups will be assigned a low reference degree. In practice, a very small group of samples is unlikely to be a phenotype. A high *reference degree* indicates that the group of the samples are consistent on most of the genes. Such groups should have a high likelihood to represent a phenotype of samples.

This step starts with the selection of a sample group S_{p_0} which has the highest reference degree. We then identify a second sample group S_{p_1} with the lowest intra-consistency value and the highest inter-divergency with respect to S_{p_0} among the remaining groups. A third group is then selected by considering the inter-divergency with respect to both S_{p_0} and S_{p_1}. The identification of subsequent groups is governed by a *selection criterion* which selects the xth sample group S_{p_x} by combining its intra-consistency and inter-divergency with respect to the sample groups already selected:

$$Ran(S_{p_x}) = \log |S_{p_x}| \sum_{G_i \in G'} \frac{\sum_{t=0}^{x-1} Div(G_i, S_{p_x}, S_{p_t})}{Con(G_i, S_{p_x})}. \tag{6.21}$$

The algorithm then calculates Ran values for the groups which have not been selected. The group having the highest Ran value will be selected as the xth sample group. A tie is broken by choosing the group containing the largest number of samples. In total, $(\mathcal{K}-1)$ sample groups, along with the fist group S_{p_0}, are selected to form the reference partition.

The reference partition selected following this method is based on the results of the matrix-partitioning phase. Therefore, samples from other groups that may also be appropriate candidates to include in the representative partition may be overlooked. A *partition adjustment step* addresses this issue by adding other samples that may improve the quality of the reference partition.

The probability that a sample $\vec{s}$ not yet included in the reference partition should be added is determined as follows. First, the quality gain achieved by inserting $\vec{s}$ into each group in the reference partition is calculated. The group with the highest quality gain is called the *matching group* of $\vec{s}$. If the quality gain is positive, then $\vec{s}$ will be inserted into its matching group; otherwise, $\vec{s}$ will be inserted into its matching group with a probability $p = e^{\left(\frac{\Delta\Omega}{n \times T(i)}\right)}$. □

Step 3: Gene adjustment

In this step, the reference partition derived from the last step is used to guide gene adjustment. In Figure 6.10, the colored area indicates the reference partition of the samples. As a result of this step, the current informative gene set will be changed. The measure of quality used here is not computed on the basis of the full partition of samples but rather using only the reference partition and the current candidate gene set G'. Thus, the current best structure maintained by the algorithm is also based on the reference partition.

The gene-adjustment process is similar to that in the heuristic searching algorithm. All the genes will be examined individually in a random order. The possible adjustment is to remove genes from G' or to insert genes into G' for genes not in the candidate set. The adjustment of each gene will be conducted if the quality gain is positive and will occur with a probability $p = e^{\left(\frac{\Delta\Omega}{n \times T(i)}\right)}$ if the quality gain is negative.

As shown in Figure 6.10, after each iteration, the candidate gene set G' is changed to G''. The next iteration starts with the new candidate gene set G' and the complete set of samples. In each iteration, we use the candidate gene set to improve the reference partition and use the reference partition to improve the candidate gene set. Therefore, it is a mutually-reinforcing process. □

The iteration phase terminates when no positive adjustment is conducted in the last iteration. The results are then generated on the basis of the best recorded structure. Since the reference partition corresponding to the best structure may not encompass all the samples, a refinement phase is conducted.

Refinement is accomplished by adding all samples not included in the reference partition into their matching groups. Thus, the refined reference partition becomes a full partition, which will be output as the phenotypes of the samples. A gene adjustment process is then conducted, accepting all adjustments which yield a positive quality gain. The genes in the candidate set of genes G' are output as informative genes.

It can be shown that the mutual-reinforcing adjustment algorithm has the same time complexity as the heuristic-search method; i.e., $O(n \cdot m^2 \cdot l)$. However, as compared to the random adjustment of samples in the latter approach, the mutual-reinforcing method pursues a more deterministic, robust, and noise-insensitive adjustment sampling strategy by improving the reference partition.

6.4.3.3   *Experimental results*

The heuristic algorithm and the mutual-reinforcing adjustment algorithm have been tested for class discovery with several real data sets. The results of these trials will be detailed below.

Multiple sclerosis data sets

The multiple sclerosis (MS) data set [206] contains two pair-wise group comparisons of interest. The first data subset, "MS vs. Controls," is comprised of microarray data from 15 MS samples and 15 age and sex-matched controls, while the second subset, "MS-IFN," is made up of microarray data from 14 MS samples taken prior to and 24 hours after interferon-β (IFN) treatment. Each sample is measured over $4,132$ genes.

Leukemia data sets

The leukemia data set was taken from a collection of leukemia patient samples reported by Golub et al. in [114]. This well-known data set often serves as benchmark [255] for microarray analysis methods. It contains measurements corresponding to acute lymphoblastic leukemia (ALL) and acute myeloid leukemia (AML) samples from bone marrow and peripheral blood. Two matrices are involved: one includes 38 samples (27 ALL vs. 11 AML, denoted as G1), and the other contains 34 samples (20 ALL vs. 14 AML, denoted as G2). Each sample is measured over $7,129$ genes.

Colon cancer data set

The samples comprising this data set were taken from colon adenocarcinoma specimens snap-frozen in liquid nitrogen within 20 minutes of removal from patients [7]. From some of these patients, paired normal colon tissue also was obtained. The cell lines used (EB and EB-1) and the process of RNA extraction and hybridization to the array are described in [7]. The microarray data set consists of 22 normal and 40 tumor colon tissue samples. In this data set, each sample contains 2,000 genes.

Hereditary breast cancer data set

Testing was also performed using a hereditary breast cancer data set [129]. Hedenfalk et al. reported on a microarray experiment concerning the genetic basis of breast cancer. Tumors from 22 women were analyzed, and three sample types were included in a single data matrix. These samples represented seven women known to have the BRCA1 mutation, eight known to have BRCA2, and seven "sporadics" who were free of cancer. Each sample was measured over 3,226 genes.

Empirical Phenotype Detection
To evaluate the performance of the algorithms discussed above, their effectiveness in detecting empirical phenotypes of samples was compared with that of several other state-of-the-art tools, including CNIO [81], J-Express [232], CIT [90], and CLUSFAVOR [221]. From many subspace clustering algorithms (see Section 7.3), δ-cluster [316] was chosen for comparison. Table 6.3 illustrates the results obtained by the various approaches using rand index (discussed in Section 5.7.2). These results indicated that the heuristic search and mutual-reinforcing adjustment methods consistently achieved clearly better phenotype structure detection results than the other methods.

In Figure 6.11, an interactive visualization tool [326] (see Section 8.4) which maps the n-dimensional data set onto two-dimensional space is used to view the changes in object distribution throughout the mutual-reinforcing adjustment process. As indicated by this figure, prior to the application of the mutual-reinforcing adjustment approach, the samples are uniformly scattered, with no distinct classes. As the iterations proceed, sample clusters progressively emerge, until, at the termination of the iterations, the samples are clearly separated into two groups. The green and red dots indicate the actual partition of the samples, while the straight line shows the partition of the mutual-reinforcing adjustment approach, with a small circle highlighting the incorrectly-classified samples. This visualization provides a clear illustration of the iterative process.

Informative Gene Selection
The informative genes identified via the heuristic and mutual-reinforcing approaches were also evaluated. Evaluation was complicated by the lack of a commonly-accepted ground truth for informative genes. Even for the Leukemia-G1 data set, which often serves as the benchmark for microarray analysis methods [255], various researchers have identified different informative genes. In [114], a supervised method, termed *neighborhood analysis* (see Section 4.4.3), was applied to select the top 50 genes as the distinguishing criteria between ALL and AML classes. In [280], a *statistical regression modeling* approach was used to identify another set of 50 informative genes within the same data set. Among these two 50-gene sets, 29 genes are overlapping.

Figure 6.12 displays the informative genes from the Leukemia-G1 data

Table 6.3   Rand Index value reached by applying different methods.

Data Set	MS vs. IFN	MS vs. Control	Leukemia -G1	Leukemia -G2	Colon	Breast
number of genes	4132	4132	7129	7129	2000	3226
number of samples	28	30	38	34	62	22
number of phenotypes	2	2	2	2	2	3
J-Express	0.4815	0.4851	0.5092	0.4965	0.4939	0.4112
CLUTO	0.4815	0.4828	0.5775	0.4866	0.4966	0.6364
CIT	0.4841	0.4851	0.6586	0.4920	0.4966	0.5844
CNIO	0.4815	0.4920	0.6017	0.4920	0.4939	0.4112
CLUSFAVOR	0.5238	0.5402	0.5092	0.4920	0.4939	0.5844
δ-cluster	0.4894	0.4851	0.5007	0.4538	0.4796	0.4719
ESPD Heuristic Searching	0.8052	0.6230	0.9761	0.7086	0.6293	0.8638
ESPD Mutual-Reinforcing	0.8387	0.6513	0.9778	0.7558	0.6827	0.8749

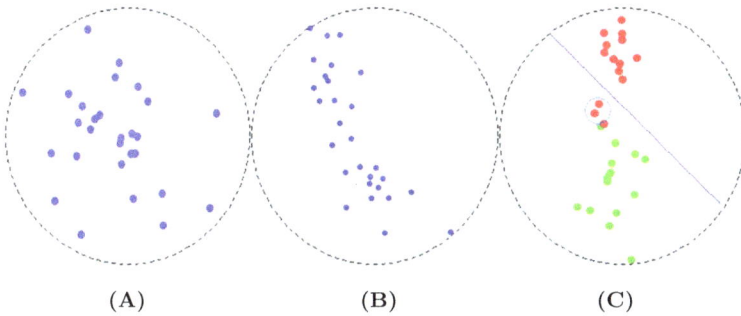

Fig. 6.11 The mutual-reinforcing adjustment approach as applied to the MS-IFN group. (A) shows the distribution of the original 28 samples. Each point represents a sample with 4132 genes mapped to two-dimensional space. (B) shows the distribution in the middle of the adjustment. (C) shows the distribution of the same 28 samples after the iterations. 76 genes were selected as informative genes, therefore each point represents a two-dimensional mapping of a 76-dimensional vector.

set as detected by the mutual-reinforcing adjustment approach. In this figure, each column represents a sample, while each row corresponds to an informative gene. The first 27 samples belong to ALL group, while the rest 11 samples belong to AML group; this agrees with the ground truth of the sample partition. Forty-nine genes are output as the informative genes. The description and probe for each gene are also provided in the figure. Different colors in the matrix indicate the various expression levels. It is evident from Figure 6.12 that the top 22 genes distinguish ALL-AML phenotypes according to an "on-off" pattern while the rest 27 genes follow an "off-on" pattern.

This result was also compared with the neighborhood analysis and statistical regression modeling, both widely-accepted supervised methods [114, 280]. The number shown in the "match" column in Figure 6.12 indicates whether the corresponding gene was in the top 50 gene sets generated by these two supervised methods. A "2" in this column indicates that the gene was among the top 50 selected by both supervised methods, while a "1" indicates that it was selected by just one of the two methods. Interestingly, as shown in Figure 6.12, 41 of the 49 informative genes identified by the mutual-reinforcing adjustment approach were selected by one or the other of those methods. One may conclude that, even without supervision, the mutual-reinforcing adjustment method learns well from the real-world data set.

Figure 6.13 presents the phenotype structure detection results for the

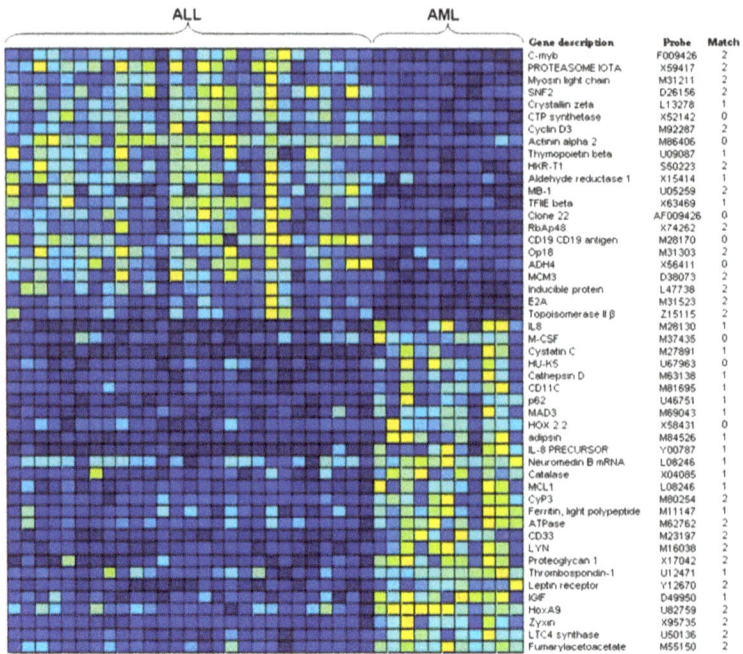

Fig. 6.12 A phenotype structure detection result of Leukemia-G1 dataset. The mutual reinforcing adjustment approach selected 49 informative genes. Each row in the heat plot represents a gene and each column is a sample. The two columns to the right of the heat plot provide the gene description and probe accession numbers. The column labeled "match" indicates whether a given gene in the informative gene space from the mutual reinforcing adjustment approach was also identified in the neighborhood analysis and statistical regression modeling methods.

Leukemia-G1 data set of 50 experiments as generated by the heuristic and mutual-reinforcing adjustment methods. Figure 6.13 (A) and (B) provide the number and percentage of informative genes identified by the heuristic-searching method that were also selected via the neighborhood analysis or statistical regression modeling approaches. Figure 6.13 (C) and (D) provide the same information in comparison with the mutual-reinforcing adjustment approach. The upper polylines of (A) and (C) indicate the number of informative genes output by each experiment. The polylines marked "matching NA" indicate the number and the percentage of informative genes output by the heuristic/mutual-reinforcing algorithm that were also contained in the 50-gene set output by the neighborhood analysis method. The polylines labeled "matching SRM" provide comparable information to that generated

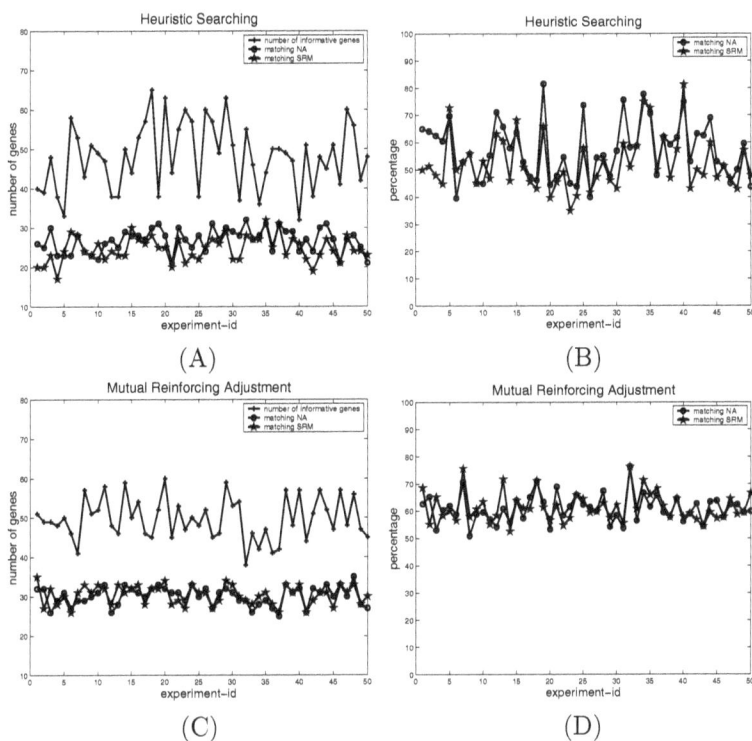

(A)

(B)

(C)

(D)

Fig. 6.13   The phenotype structure detection results on Leukemia-G1 data set of 50 executions of the heuristic-searching and mutual-reinforcing adjustment approaches. (A) and (C) show the numbers of informative genes that match the *neighborhood analysis* (NA) method or the *statistical regression modeling* (SRM) approach by the two methods, respectively. (B) and (D) show the percentages of informative genes that match the *neighborhood analysis* method or the *statistical regression modeling* approach by the two methods, respectively.

by the statistical regression modeling approach.

Figure 6.13 shows that the heuristic-searching method generated between 32 and 65 informative genes from the various experiments. On average, about 57% of these genes matched those generated by the neighborhood analysis method, and 52% matched those output by statistical regression modeling. The mutual-reinforcing adjustment approach produced between 38 and 60 informative genes from the various experiments. Approximately 61% of these genes matched those generated by the neighborhood analysis method, and 62% matched the output produced by statistical regression modeling. Overall, the two unsupervised methods matched 58% of the genes output by the supervised methods. In general, unsupervised approaches are much more complex than supervised methods, since no prior knowledge (such as class labels and training examples) is provided. The fact that these two unsupervised methods produced a percentage of informative genes comparable to the two supervised methods indicates the effectiveness of these unsupervised approaches in mining informative genes.

6.5 Classification Validation

In general, sample-based analysis results in a classification among samples. As previously noted, this may be approached either as "supervised" (class prediction) or "unsupervised" classification (class discovery). Each approach requires a different technique for the validation of results. Class discovery is essentially a clustering method, and the results can be validated using the methods described in Section 5.7. The validation of class-prediction results must examine both the accuracy and the reliability of the prediction. In the following subsection, we will focus on the validation of class prediction. Details regarding the validation of class discovery are provided in Section 5.7.

6.5.1 *Prediction Accuracy*

The *prediction accuracy* of a classification algorithm is the percentage of test samples for which class labels are correctly assigned by the algorithm. Accuracy measurement starts by partitioning the data set into a *training set* and a *test set* and constructing a classifier on the training set. Prediction accuracy is then calculated using the test set. Typically, one-third of the available examples are allocated to the test set, and the other two-thirds are used for training [197]. This method (termed *training-test approach*) is straightforward and effective. However, when the data set contains only a limited number of data objects, it is desirable to build the classifier with as many training examples as possible. In that instance, the test set is usually reserved for use with the learning process, and a *cross-validation strategy* is employed to measure the prediction accuracy.

The *k-fold cross-validation* approach performs the validation procedure k times. A common deployment of k-fold cross-validation is as follows:

(1) Equally partition the full set of examples into k subsets $D_1, ..., D_k$;
(2) Perform the validation k times; each run uses a different D_i ($1 \leq i \leq k$)

as the test set and a combination of the remaining $k - 1$ subsets as the training set. Calculate the prediction accuracy p_i;

(3) Report the average prediction accuracy $\frac{\sum p_i}{k}$ as the cross-validation result.

A special case of k-fold cross-validation, termed *leave-one-out cross validation*, sets k as the total number of examples. In other words, each validation run uses only one example for validation, while all the other examples are for training.

Cross-validation may be less effective than the training-test approach, since the examples are used for both training and testing. However, it is particularly well-suited to the analysis of microarray data, which commonly features a small number of samples. Moreover, the cross-validation method is often used in the training process. For example, in Section 6.2.1.4, we have seen that the cross-validation result of a KNN algorithm was used to select the informative genes in a GA algorithm.

6.5.2 *Prediction Reliability*

The *reliability* of a prediction made by a classifier implies the confidence of this prediction. In practice, this measure provides cautions to users when the prediction is critical (e.g., in diagnosing cancer in a patient).

Typically, the prediction reliability is provided by a classifier and closely bundled with the classification algorithm. For example, Golub et al. [114] measured the reliability using the metric of *prediction strength* based on their *weighted voting* algorithm (see Section 6.3.2.2). To be specific, suppose the samples consist of two groups C_1 and C_2. Given the test sample s, each informative gene g_i votes for either C_1 or C_2. The votes for C_1 and C_2 are summed as V_1 and V_2, respectively, and the *prediction strength* for s is defined as $|\frac{V_1 - V_2}{V_1 + V_2}|$. Clearly, if most of the informative genes uniformly vote s for C_1 (or for C_2), the value of the predication strength will be high. High values of prediction strength indicate a biologically-significant prediction.

In [114], the output of the classifier was one of the three values: "AML", "ALL", and "UNKNOWN". The classifier outputs "UNKNOWN" if the prediction strength is below a threshold. The classifier was applied to a leukemia data set and 29 of the 34 samples were predicted with 100% accuracy to be either "AML" or "ALL", while the remaining 5 samples were classified as "UNKNOWN". In essence, this method constructs a predictor based on derived clusters and converts the reliability assessment of sample clusters to a "supervised" classification problem. In [321], Yeung et al. extended this idea and proposed an approach to cluster validation for gene clusters (see Section 5.7.3.2).

6.6   Summary

This chapter has reviewed the most prevalent approaches to sample-based analysis, with a particular focus on sample clustering and classification. We have seen that the major challenge of sample-based clustering is the proper selection of informative genes. Typically, this selection is made via supervised techniques employing phenotype information taken from samples as training sets. These techniques generally provide highly accurate identification of clusters and informative genes. Unsupervised sample-based clustering and informative gene selection, necessary in situations where no prior knowledge is available, is more complex and has been addressed through two strategies. The first strategy reduces the number of genes prior to clustering samples. Approaches following this strategy rely on a statistical model [9, 79], and their effectiveness is strongly shaped by the nature of the data distribution [319].

An alternative strategy utilizes the relationship between genes and samples to perform gene selection and sample clustering simultaneously in an iterative paradigm. Section 6.4 presented a detailed discussion of CLIFF [308] and ESPD [275], two novel approaches which pursue this strategy. With both of these methods, the iteration converges to an accurate partition of the samples and a set of informative genes. In particular, the ESPD technique proposed by Tang et al. [275] redefines the process of mining phenotype structures as a coupled effort to optimize the quality of both candidate sample partitions and informative genes. A series of statistical measurements evaluates the quality of the mining results and uses these results to coordinate between sample phenotype discovery and informative space detection. These measurements delineate local pattern qualities based on a partition of samples derived from a subset of genes. A phenotype quality function which serves as the objective function of the optimization process is defined based on these statistical measurements.

The ESPD approach employs two novel unsupervised learning algorithms: a heuristic search and a mutual-reinforcing adjustment algorithm. These have been discussed here in some detail, including a presentation of an iterative pattern adjustment to approach the optimal solution at which the pattern quality is maximized. These methods dynamically manipulate the relationship between samples and genes and conduct an iterative adjustment of both components to approximate both the empirical phenotype structure and set of informative genes. Experimental results summarized in this chapter indicate that the ESPD approach is effective, scalable, and yields accurate mining results suitable for real-world applications.

Chapter 7

Pattern-Based Analysis

Contributor: Daxin Jiang

7.1 Introduction

As explored in Chapters 5 and 6, gene expression data can be analyzed through either the gene dimension or the sample dimension. In gene-based approaches, the genes are considered to be the data objects, and the entire set of experimental conditions (either samples or time points) is used as the set of attributes. Conversely, for sample-based methods, the samples are treated as data objects, while the genes serve in the role of attributes. Since the number of genes in a microarray data experiment is often by far larger than the number of samples and many genes may be irrelevant to the selected biological problem, a small subset of "informative" genes usually serves in the role of attributes. Once the informative genes have been determined, sample-based approaches will seek out sample patterns with respect to the entire set of informative genes.

The various gene- and sample-based approaches described in previous chapters search for patterns of objects using the full set of attributes. This assumes that all gene or sample patterns within a specific microarray data set will share exactly the same set of attributes. However, current thinking in molecular biology holds that only a small subset of genes participates in any particular cellular process, and that any cellular process takes place only in a subset of the samples. In many cases, it would be interesting to find a subset of genes which exhibit similar expression profiles for a subset of samples, or, conversely, to find a subset of samples which share a coherent expression pattern with respect to a subset of genes. This calls for a new analytical theme which involves identifying patterns that consist of subsets of genes and samples. This approach is termed *pattern-based analysis*.

In this chapter, we will discuss two approaches to pattern-based analysis in the context of microarray data. The first approach, termed *association rule mining*, arises from the analysis of business transaction data, where the goal is to identify such rules as "a customer who buys products A and B will also buy product C with probability $x\%$." Association rule mining is a classical topic in the field of data mining, and numerous mining algorithms are discussed in the literature of the field. Recently, these techniques have been applied to microarray data (e.g., see [290]) and have revealed various interesting rules. For example, this technique has shown that the up-regulation of certain genes may infer the down-regulation of other genes, with a specified confidence level.

Another approach to pattern-based analysis is *subspace clustering*, which was first proposed to extract clusters embedded in the subspaces of high dimensional data sets [5]. Liu and Wang [185] have classified subspace clustering models into two categories based on the measure of similarity. The first category is distance-based (e.g. [2, 3, 5, 51, 148]), using a *pair-wise* proximity measure to evaluate the distance or similarity between data objects. The second category, pattern-based approaches (e.g., [52, 185, 218, 296, 316]), moves beyond this pair-wise similarity analysis to evaluate the overall similarity or coherence of the whole cluster. We will adopt this same categorization, drawing a distinction between distance-based subspace clustering and pattern-based subspace clustering, or, briefly, pattern-based clustering.

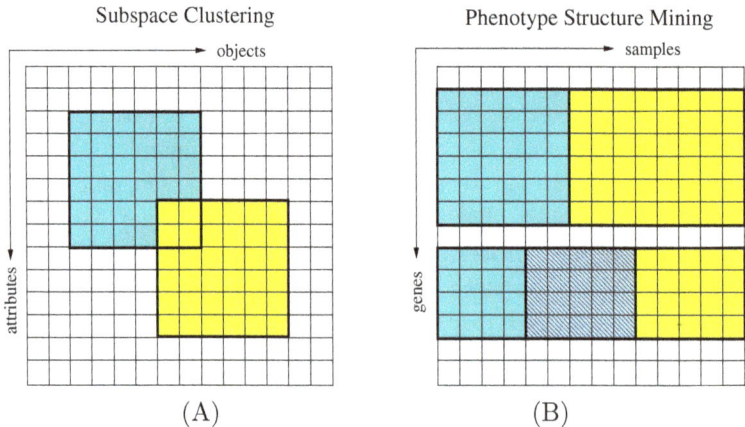

Fig. 7.1 Differences of subspace clustering and phenotype structure mining. (A) shows a data matrix containing two subspace clusters. (B) shows a data matrix containing two phenotype structures. The upper structure contains two phenotypes while the lower one contains three phenotypes.

Upon a cursory examination, the problem of subspace clustering may appear similar to the detection of phenotype structures discussed in Section 6.4. However, there are significant inherent differences between these two approaches, as follows (see Figure 7.1):

- In subspace clustering, the attribute subsets associated with various clusters may be either shared or dissimilar. In contrast, phenotype structure mining requires a unique and exclusive set of genes to characterize each phenotype partition.
- Subspace clusters may overlap and share common objects, and some objects may not belong to any subspace cluster. Phenotype structure detection requires that all samples belong to a phenotype and that the phenotypes be mutually exclusive.
- The pattern similarity measurements presented for subspace clustering (e.g., [52, 316]) can only detect whether genes rise and fall simultaneously within a subspace cluster. In phenotype structure mining, the values of informative genes are required to be intra-consistent, so new metrics are required to detect the informative genes.
- Subspace-clustering algorithms detect only locally-correlated attributes and objects without considering the degree of dissimilarity between clusters. To effectively mine both phenotypes and informative genes, it is necessary to identify those genes which can differentiate all phenotypes for each phenotype structure.

Section 7.2 of this chapter will discuss association rule mining with applications to microarray data. Subspace clustering techniques will be presented in 7.3, with a particular focus on pattern-based clustering since biologists are often more interested in the overall shape of the patterns instead of the absolute magnitude of distance [316]. The methods to be discussed in Sections 7.2 and 7.3 are applicable to data with two types of variables. Recent developments in microarray technology have generated novel data which contain three types of variables (genes, samples, and time points). The extension of pattern-based analysis to such novel three-dimensional expression data will be discussed in Section 7.4.

7.2 Mining Association Rules

As noted above, association-rule mining was originally developed to find interesting associations or correlation relationships among data in a large database, such as business transaction records. The discovery of interesting association rules is valuable in many business decision-making processes, such as catalog design, cross-marketing, and loss-leader analysis [120]. During the past several years, many mining algorithms have been developed to suit a variety of applications. In general, these algorithms can be categorized into breadth-first search (BFS) and depth-first search (DFS) methods, based on how the association-rule space is visited [135].

To find the association rules in microarray data, each data set is represented by a $M \times N$ matrix $\{x_{ij} | 1 \leq i \leq M, 1 \leq j \leq N\}$, where x_{ij} represents the expression level of gene j under experiment i. The data set is then discretized according to whether the expression levels x_{ij} exceed certain user-specified thresholds. This process results in a discretized matrix $\{r_{ij} | 1 \leq i \leq M, 1 \leq j \leq N\}$, where the values r_{ij} are of three different levels: *unchanged* (denoted as #), *up-regulated* (denoted as *uparrow*), and *down-regulated* (denoted as *downarrow*) [290]. Using the discretized matrix $\{r_{ij}\}$, an experiment i can be represented as a *transaction* consisting of N *items* (genes), with each item r_{ij} taking one of the three values (#, ↑, ↓).

In this section, we will first introduce the basic concepts underlying association rule mining. We will then present two classical mining algorithms, Apriori [4] (BFS) and FP-growth [121] (DFS), along with a common pattern-mining algorithm, CARPENTER [212], which is specifically designed for mining gene expression data. Finally, we will discuss the management of the very large number of association rules which arise in the analysis of microarray data.

7.2.1   Concepts of Association-Rule Mining

Following the definitions provided in [4], let $\mathcal{I} = \{i_1, \ldots, i_n\}$ be a set of distinct literals, called *items*. A set $X \subseteq \mathcal{I}$ with $|X| = k$ is called a *k-itemset* or simply *an itemset*. Let $\mathcal{D}$ be a set of transactions, where each transaction T is an itemset. Associated with each transaction is a unique identifier, its *TID*. A transaction T *contains* or *supports* an itemset X, if $X \subseteq T$. An *association rule* is an expression $X \Rightarrow Y$, where $X \subset \mathcal{I}, Y \subset \mathcal{I}$, and $X \cap Y = \emptyset$. The itemset X has *support* x in the transaction database $\mathcal{D}$ if $x\%$ of transactions T in $\mathcal{D}$ contain X, i.e.,

$$supp(X) = \frac{|\{T \mid X \subseteq T, T \in \mathcal{D}\}|}{|\mathcal{D}|}.$$

The rule $X \Rightarrow Y$ has *support* s in the transaction database $\mathcal{D}$ if $s\%$ of transactions T in $\mathcal{D}$ contain $X \cup Y$, i.e.,

$$supp(X \Rightarrow Y) = \frac{|\{T \mid (X \cup Y) \subseteq T, T \in \mathcal{D}\}|}{|\mathcal{D}|}.$$

The rule $X \Rightarrow Y$ has *confidence* c in the transaction database $\mathcal{D}$ if $c\%$ of transactions T in $\mathcal{D}$ that contain X also contain Y, i.e.,

$$conf(X \Rightarrow Y) = \frac{supp(X \cup Y)}{supp(X)}.$$

Support should not be confused with confidence; while the support of an itemset or a rule indicates the statistical significance of the itemset or the rule, the confidence is a measure of the rules' strength. Usually, we are interested in the rules with support and confidence above certain thresholds *minsupport* and *minconf*, respectively.

In this formulation, the problem of rule mining can be decomposed into two steps: *frequent itemsets identification* and *rule generation*. The first step entails the identification of all *frequent itemsets* $\mathcal{F} = \{X \mid supp(X) \geq minsupp\}$. Once the set of all frequent itemsets, as well as their supports, are known, the second step involves derivation of the desired association rules from $\mathcal{F}$. This step is very simple; for each $X \in \mathcal{F}$, the confidence of all possible rules $X - Y \Rightarrow Y$ are checked, where $Y \subset X$ and $Y \neq \emptyset$, and those rules which fall below $minconf$ are excluded.

The main challenge of mining association rules lies in the first step, identification of frequent itemsets. It is intuitively obvious that a linear increase in $|\mathcal{I}|$ will result in an exponential growth of the number of itemsets to be considered. For example, Figure 7.2 gives the search space of $\mathcal{I} = \{a, b, c, d\}$ [135], a lattice containing $2^{|\mathcal{I}|} = 2^4 = 16$ nodes. A naïve search in this lattice cannot be finished in practical time when $\mathcal{I}$ is large. In fact, the problem of mining frequent itemsets has been proved to be NP-hard [314].

Fortunately, the itemset support has the *downward closure property*: all subsets of a frequent itemset must also be frequent [4]. As a result, there is a border in the lattice structure separating the frequent and infrequent itemsets [135], with the frequent itemsets located above the border and the infrequent itemsets below. The basic principle of mining frequent itemsets is to employ this border to prune the search space efficiently. Once the border is determined, itemsets below the border can be pruned according to the downward closure property. In general, a search of the lattice can follow one of two strategies: a breadth-first search (BFS) or a depth-first

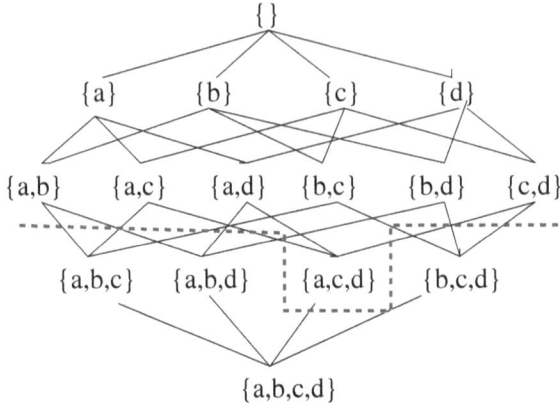

Fig. 7.2   Lattice of $\mathcal{I} = \{a, b, c, c\}$. The dashed border in the lattice structure separates the frequent and infrequent itemsets.

search (DFS). The following discussion will include a representative and widely-used algorithm for each strategy.

7.2.2   The Apriori Algorithm

The Apriori algorithm [4] adopts a breath-first-search approach to the itemset lattice and uses k-itemsets to explore $(k + 1)$-itemsets. The algorithm scans the database in the first round to simply count the occurrences of each item. It then finds the set of frequent 1-itemset (denoted as L_1) with respect to a given support threshold *minsupp*. Subsequent rounds of the algorithm (for example, round k) consists of two phases. First, the frequent $(k - 1)$-itemsets L_{k-1} found in the $(k - 1)$th round are used to generate the candidate itemsets C_k. Second, the database is scanned and the k-itemsets in C_k are checked. If a k-itemset X in C_k is not frequent, it will be removed from C_k. The remaining k-itemsets in C_k constitutes L_k and will be used for the $(k + 1)$th round. These two phases iterate until the set of frequent k-itemsets L_k is empty. In the following, we will discuss the two phases in each round in more detail.

Candidate Generation: The candidate set C_k is generated by self-joining L_{k-1}. Let l_1 and l_2 be itemsets in L_{k-1} and $l_i.item_j$ be the jth item of l_i, where $i = 1, 2$ and $1 \leq j \leq k - 1$. l_1 and l_2 are called "joinable" if 1) $l_1.item_p = l_2.item_p$ for $1 \leq p \leq k - 2$; and 2) $l_1.item_{k-1} < l_2.item_{k-1}$ (this condition is applied to prevent duplication). If l_1 and l_2 are "joinable", the join operation will return a k-itemset l_3 such that $l_3.item_p = l_1.item_p$ for $1 \le p \le k-1$ and $l_3.item_k = l_2.item_{k-1}$. The self-joining of L_{k-1} refers to joining every pair of "joinable" itemsets in L_{k-1} and placing the result in C_k.

Candidate Pruning: According to the downward closure property and the process of candidate generation, it is obvious that all the frequent k-itemsets must appear in C_k. However, each member of C_k may or may not be frequent. To determine the k-itemsets in C_k, the database is scanned once to count the support for each itemset in C_k. However, when the database is very large, counting the support of an itemset could be expensive. The Apriori algorithm uses the downward closure property again to avoid counting unpromising candidates. Specifically, for each itemset l_i in C_k, Apriori will check each $(k-1)$-subset x of l_i. If x does not appear in L_{k-1}, l_i is removed from C_k, since such an x is not frequent and thus l_i cannot be a frequent itemset.

7.2.3 The FP-Growth Algorithm

Although the Apriori algorithm fully exploits the downward closure property and significantly reduces the size of candidate sets, it may suffer from two nontrivial costs [121]:

- It may need to generate a huge number of candidate sets. For example, if there are 10^4 frequent 1-itemsets, Apriori will need to generate more than 10^7 candidate 2-itemsets. Moreover, to discover a frequent pattern of size 100, it must generate in total more than $2^{100} \approx 10^{30}$ candidates.
- Since Apriori must scan the database in each round, it may need to repeatedly scan the database to ascertain the support for a large set of candidates by pattern-matching. For example, to check a long frequent pattern containing 100 items, the algorithms would have to perform 100 scans of the database.

To address these problems, Han et al. [121] proposed a novel approach to mining frequent itemsets without candidate generation, called *frequent-pattern growth* (FP-growth). The FP-growth method conducts a depth-first search on the itemset lattice and adopts a divide-and-conquer strategy. The algorithm consists of two phases. In the first phase, the algorithm compresses the database by constructing a *frequent-pattern tree* (FP-tree) to represent frequent items while retaining the itemset association information. In the second phase, the algorithm partitions the frequent patterns and projects the compressed database into a set of *conditional databases*. The conditional databases are then mined individually in a recursive manner until a designed stop criterion is met. In the following, we will describe these two phases in greater detail.

FP-tree construction: This phase involves scanning the database twice. The first scan derives the set of frequent items (1-itemsets) and counts their supports. The set of frequent items is then sorted in the order of descending support count, termed the *L order*. The root of the FP-tree is created and labeled with "null." In the second scan of the database, the items in each transaction are processed in the L order and a branch is created for each transaction. If two transactions share a *prefix*, they will share a (sub)path on the tree. Additionally, to facilitate tree traversal, a *header table* is created and the occurrences of each item are linked with the *node-links*.

Frequent pattern growth: This phase starts by taking each frequent item as an *initial suffix pattern*. It then projects the original database on this suffix pattern by collecting the set of *prefix paths* in the original FP-tree which co-occur with the suffix pattern. This process results in a *conditional pattern base* with respect to the suffix pattern. A *conditional FP-tree* is constructed from the conditional pattern base, and mining continues recursively on such a tree until the conditional FP-tree becomes empty or contains a single path. The pattern growth is achieved by the concatenation of the suffix pattern with the frequent patterns generation from a conditional FP-tree.

7.2.4   *The CARPENTER Algorithm*

Pan et al. [212] proposed an algorithm, CARPENTER, which was specifically designed for mining *closed frequent patterns* in long biological data sets such as microarray data sets. An frequent itemset X is considered *closed* if there exists a proper superset $Y \supset X$ such that $supp(Y) = supp(X)$ [323]. Pan et al. [212] noted that a particular challenge in mining (closed) frequent itemsets from microarray data is that the data sets typically contain thousands or tens of thousands of genes (items) but only tens or hundreds of samples (transactions). Most previous algorithms (including Apriori [4], FP-growth [121], and CHARM [323]) for mining (closed) frequent patterns may not be able to process such high-dimensional data in practical time.

Most previous (closed) frequent pattern mining methods are effective when applied to data sets with small average row lengths, where a row length refers to the number of items in a transaction. If i is the maximum row length, there could be 2^i potential frequent itemsets; usually $i < 100$. However, row lengths for microarray data sets can range up to tens of thousands of items, resulting in impractically onerous searches over the itemset space. To address this challenge, CARPENTER adopts a strategy of searching the row-set space rather than the itemset space, since the number of rows (m) in a typical microarray data set is usually in the more manageable range of a hundred to a thousand (i.e., $m \ll i$).

Table 7.1   An example data table.

i	r_i
1	a,b,c,g,k,m
2	a,d,e,h,i,j
3	a,c,f,i,k,m
4	b,d,f,h,k,m

Table 7.2   The transposed table of Table 7.1.

f_j	$\mathcal{R}(f_i)$
a	1,2,3
b	1,4
c	1,3
d	2,4
e	2
f	3,4
g	1
h	2,4
i	2,3
j	2
k	1,3,4
m	1,3,4

Let $F = \{f_1, f_2, \ldots, f_m\}$ be a set of items; a gene expression data set can be represented by a table $R = \{r_1, \ldots, r_n\}$ where each row r_i in the table is a set of items, i.e., $r_i \subseteq F$. Table 7.1 shows an example data set in which the items are represented using letters a through m. The first row r_1 contains the itemset $\{a, b, c, g, k, m\}$.

As mentioned above, unlike previous algorithms which search by enumeration of items, CARPENTER enumerates the rows of tables. To facilitate such row-wise enumeration, CARPENTER first transposes the original data table such that each tuple t_i in the transposed table (called TT) corresponds to one item f_i. Moreover, each tuple t_i in TT also records the id of rows in which f_i appears. Table 7.2 provides the transposed table generated from Table 7.1.

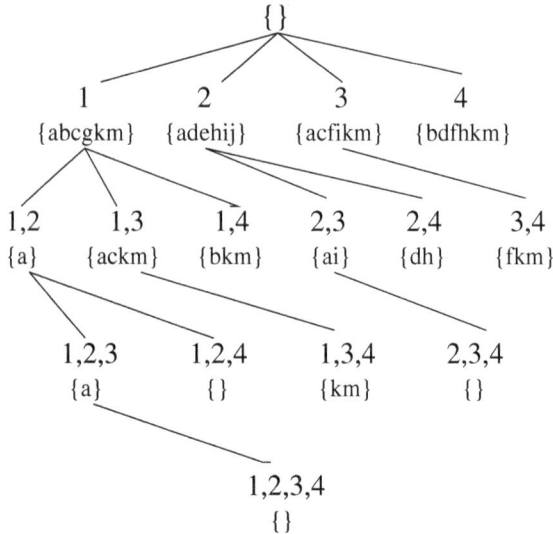

```
                              {}
                  ┌───────┬──┴──┬──────────┐
                  1       2     3          4
               {abcgkm} {adehij} {acfikm} {bdfhkm}
              ┌──┴──┬──────┬─────────┐
             1,2   1,3    1,4   2,3   2,4    3,4
             {a}  {ackm}  {bkm} {ai}  {dh}  {fkm}
                 └────┬──────┬──────┐
                1,2,3   1,2,4   1,3,4   2,3,4
                {a}     {}      {km}    {}
                    └──────┬──────┘
                        1,2,3,4
                          {}
```

Fig. 7.3　The row enumeration tree of the example data set in Table 7.1.

To find frequent closed patterns, the CARPENTER algorithm then performs a depth first search (DFS) of the row-set enumeration tree. By imposing a total order (such as lexicographic order) on the row sets, the algorithm can perform a systematic search of the tree and find the complete set of closed frequent patterns. Figure 7.3 shows the row enumeration tree of the example data set in Table 7.1. For the details of the CARPENTER algorithm, please refer to [212].

7.2.5　*Generating Association Rules in Microarray Data*

As mentioned above, association-rule mining can be decomposed into two steps: *frequent itemsets identification* and *rule generation*. In Section 7.2, we discussed several general frequent-itemset identification algorithms as well as one specifically designed for microarray data. Once the set of all frequent itemsets are known, the generation of association rules is very simple; for each frequent itemset X, the confidence of all possible rules

$X - Y \Rightarrow Y$ are checked, where $Y \subset X$ and $Y \neq \emptyset$, and those rules which fall below threshold *minconf* are excluded.

Microarrays usually contain thousands of genes, and classical mining algorithms thus often generate millions of association rules, far more than can be efficiently interpreted by biologists. In [290], a post-mining process is proposed to assist biologists in evaluating the meaningfulness of these results. Rule selection is achieved by providing several rule-evaluation operators, including rule-grouping, filtering, browsing, and data-inspection operators, that allow biologists to simultaneously validate several individual gene regulation patterns. By applying these operators iteratively, biologists can explore a significant proportion of the initially-generated rules within an acceptable timeframe and identify those which address biological questions of particular interest.

7.2.5.1 *Rule filtering*

This operator allows biologists to impose various constraints on the syntactic structure of the discovered biological rules by using *templates* to reduce the number of rules to be explored. In general, these templates can be specified using the following notations:

RulePart **HAS** quantifier **OF** $C_1, C_2, \ldots, C_N$ [**ONLY**].

Here *RulePart* can be "BODY", "HEAD", or "RULE", and this term specifies the part of the rule (antecedent, consequent, or the whole rule, respectively) on which the constraint is placed. $C_1, C_2, \ldots, C_N$ is a *comparison set*. Each element C_i specify one gene or a group of genes (possibly accompanied by expression levels). *Quantifier* is either one of the keywords "ALL," "ANY," or "NONE" or an expression (such as a numeric value or a range of numeric values) that specifies the number of genes in the comparison set that have to be contained in *RulePart* in order to accept this template. Finally, an optional key word "ONLY" can be used to indicate that *RulePart* include only the genes that are present in the $C_1, C_2, \ldots, C_N$ list.

The syntax defined above can be used to specify a variety of biological rules with constraints. For example:

(1) All rules that contain at least one of the following genes: G1, G5, and G7: **RULE HAS (ANY) OF** G1, G5, G7.
(2) All rules that contain some of the genes G1, G5, G7, but no other genes: **RULE HAS (ANY) OF** G1, G5, G7 **ONLY**.
(3) All rules that contain *exactly one* of the following genes: G1, G5, G7.

Moreover, only rules with up-regulated G1, down-regulated G5, and up-regulated or down-regulated (but not unchanged) G7 are acceptable: **RULE HAS (1) OF** G1↑, G5↓, G7={↑,↓}.

(4) All rules that contain a DNA repair gene (i.e., a gene that belongs to a pre-defined group of DNA repair genes) in the head (consequent): **HEAD HAS (ANY) OF** [DNA_Repair].

The template-based rule filtering operator can be further combined using boolean operators "AND," "OR," and "NOT." For example, "all rules that have up to 3 DNA repair genes and no other genes in the body, as well as gene G7 in the head of the rule" would be denoted by **BODY HAS (1-3) OF** [DNA_Repair] **ONLY AND HEAD HAS (ANY) OF** G7.

7.2.5.2   *Rule grouping*

The association rules discovered through mining algorithms may include many which differ only slightly, and it would be helpful to enable biologists to explore and analyze these similar rules as a group rather than individually. This would greatly diminish the sheer volume of rules needing examination and would also provide biologists with a high-level overview and the capability to engage in exploratory top-down rule analysis.

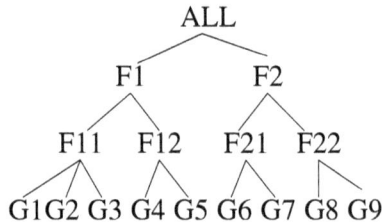

Fig. 7.4   An example of a gene hierarchy (Figure is adapted from [290] with permission from ACM).

Tuzhilin et al. [290] have suggested a hierarchical structure for rule-grouping in which genes (leaf nodes) are grouped by function according to levels of internal nodes. The hierarchical structure can be specified by the domain expert. For example, Figure 7.4 presents an example of such a hierarchy, where genes G1, G2, and G3 are combined into functional category F11, and genes G4 and G5 are combined into functional category F12. Functional categories F11 and F12 are further combined into category F1, and, finally, functional categories F1 and F2 are combined into a category

ALL that represents all genes and corresponds to the root node of the tree.

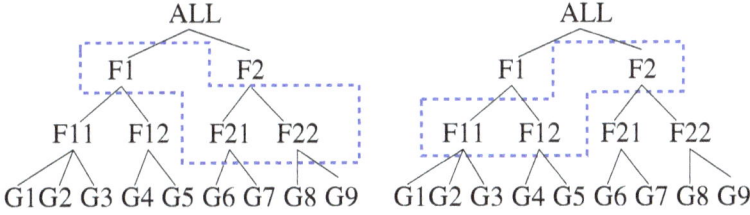

```
        ALL                        ALL
      /     \                    /     \
   F1        F2              F1          F2
  /  \      /   \          /   \       /    \
F11  F12  F21  F22       F11   F12   F21   F22
/\    /\   /\    \       /\     /\    /\     \
G1G2 G3 G4 G5 G6 G7 G8 G9   G1G2 G3 G4 G5 G6 G7 G8 G9
```

Fig. 7.5 Examples of different gene aggregation levels (Figure is adapted from [290] with permission from ACM).

The identified rules are grouped by specifying the *gene aggregation level* in a given gene hierarchy. A gene aggregation level is defined as the subset of all nodes (either leaf or non-leaf) of the gene hierarchy which occur along each path from a leaf node to the root. Figure 7.5 presents two examples of gene aggregation levels (marked with dotted lines). The gene aggregation level is usually selected by a biologist with specialized domain knowledge.

After the rule aggregation level has been specified, the discovered rules are aggregated by performing a syntactic transformation. Individual genes and their values (expression levels) in the rules are replaced with the corresponding node in the specified aggregation level. For example, the rule $G1\uparrow$ & $G2\downarrow \Rightarrow G5\downarrow$ indicates that, whenever gene 1 is up-regulated and gene 2 is down-regulated, gene 5 is likely down-regulated. By applying the aggregation level from Figure 7.5, both $G1\uparrow$ and $G2\downarrow$ will be replaced with F1 in the body of the rule, and $G5\downarrow$ will be replaced with F2 in the head of the rule. As a result, the rule $G1\uparrow$ & $G2\downarrow \Rightarrow G5\downarrow$ will be grouped into *aggregated rule* $F1 \Rightarrow F2$.

In addition to rule-filtering and grouping operators, [290] proposed several other operators to allow biologists to view either individual rules or groups of rules and inspect the transactions (experiments) which support specific rules.

7.3 Mining Pattern-Based Clusters in Microarray Data

In general, the problem of mining pattern-based clusters is NP-hard [52]. Two categories of approaches have been proposed to tackle this problem in practical time. The first category encompasses *heuristic approaches*. In essence, these approaches adopt a heuristic search in the gene-sample space to find approximately-optimal coherent clusters. However, due to the inherent characteristics of a heuristic search, the quality of clusters cannot be guaranteed. The second category of methods consists of a series of *deterministic approaches* which report the complete set of pattern-based clusters and thus guarantee the quality of clusters. These approaches exploit several interesting observations noted in mining pattern-based clusters from microarray data (e.g., the data are often sparse and the pattern-based clusters have the *downward closure property*, see Section 7.2.1) and typically rely on efficient pruning techniques to improve search efficiency. In the following, we will present several representative algorithms for each category. For detailed surveys of pattern-based clustering methods, please refer to [189].

7.3.1 Heuristic Approaches

7.3.1.1 Coupled two-way clustering (CTWC)

Getz et al. [109] model the pattern-based cluster as a "stable" combination of features $(\mathcal{F}_i)$ and objects $(\mathcal{O}_j)$, where both $\mathcal{F}_i$ and $\mathcal{O}_j$ can be either genes or samples. The cluster is "stable" in the sense that, when only the features in $\mathcal{F}_i$ are used to cluster the corresponding $\mathcal{O}_j$, $\mathcal{O}_j$ does not split below some threshold. CTWC provides a heuristic alternative to the brute-force enumeration of all possible combinations. Only subsets of genes or samples that are identified as stable clusters in previous iterations are candidates for the next iteration.

CTWC begins with only one pair of gene set and sample set (G_0, S_0), where G_0 is the set containing all genes and S_0 is the set that contains all samples. A hierarchical clustering method, called the *super-paramagnetic clustering algorithm* (SPC) [31] (see Section 5.4.2.2), is applied to each set, and the stable clusters of genes and samples yielded by this first iteration are G_1^i and S_1^j. CTWC dynamically maintains two lists of stable clusters (*gene list GL* and *sample list SL*) and a *pair list* of pairs of gene and sample subsets (G_n^i, S_m^j). For each iteration, one gene subset from GL and one sample subset from SL that have not been previously combined are coupled and clustered mutually as objects and features. Newly-generated stable clusters are added to GL and SL, and a pointer that identifies the parent pair is recorded in the pair list to indicate the origin of the clusters. The iteration continues until no new clusters are found which satisfy some criterion, such as stability or critical size.

CTWC was applied to a leukemia data set [114] and a colon cancer data set [7]. For the leukemia data set, CTWC converges to 49 stable gene clusters and 35 stable sample clusters in two iterations. For the colon cancer data set, 76 stable sample clusters and 97 stable gene clusters were reported by CTWC in two iterations. The experiments demonstrated the capability of CTWC to identify sub-structures of gene expression data which cannot be clearly identified when all genes or samples are used as objects or features.

However, CTWC searches for clusters in a heuristic manner, and the clustering results are therefore sensitive to initial clustering settings. For example, suppose (G,S) is a pair of stable clusters. If, during the previous iterations, G was separately assigned to several clusters according to features S', or S was separated in several clusters according to features G', then (G,S) can never be found by CTWC in the following iterations. Another drawback of CTWC is that clustering results are sometimes redundant and hard to interpret. For example, for the colon cancer data, a total of 76 sample clusters and 97 gene clusters were identified. Among these, four different gene clusters partitioned the samples in a normal/cancer classification and were therefore redundant, while many of the clusters were not of interest, i.e., hard to interpret. More satisfactory results would be produced if the framework can provide a systematic mechanism to minimize redundancy and rank the resulting clusters according to significance.

7.3.1.2 *Plaid model*

The *plaid model* [176] regards gene expression data as a sum of multiple "*layers*," where each layer may represent the presence of a particular biological process with only a subset of genes and a subset of samples involved. The generalized plaid model is formalized as $Y_{ij} = \sum_{k=0}^{K} \theta_{ijk} \rho_{ik} \kappa_{jk}$, where the expression level Y_{ij} of gene i under sample j is considered coming from multiple sources. To be specific, θ_{ij0} is the background expression level for the whole data set, and θ_{ijk} describes the contribution from layer k. The parameter ρ_{ik} (or κ_{jk}) equals 1 when gene i (or sample j) belongs to layer k, and equals 0 otherwise.

The clustering process searches the layers in the data set individually, using the *EM algorithm* [69] (see Section 5.3.4) to estimate the model parameters. After the extraction of the first $K - 1$ layers, the Kth layer is identified by minimizing the sum of squared errors $Q = \frac{1}{2} \sum_{i=1}^{n} \sum_{j=1}^{m} (Z_{ij} - \theta_{ijK} \rho_{iK} \kappa_{jK})^2$, where $Z_{ij} = Y_{ij} - \theta_{ij0} - \sum_{k=1}^{K-1} \theta_{ijk} \rho_{ij} \kappa_{jk}$ is the residual from the first $K - 1$ layers. The clustering process stops when the variance of expression levels within the current layer is below a specified threshold.

The plaid model was applied to a yeast gene expression data set combined from several time-series under different cellular processes [176]. Some interesting clusters were found among the 34 layers that were extracted from the data set. For example, the second layer was recognized as being dominated by genes that produce ribosomal proteins involved in protein-synthesis process in which mRNA is translated. However, the plaid model is based on the questionable assumption that, if a gene participates in several cellular processes, then its expression level is the sum of the terms involved in the individual processes. Thus, the effectiveness and interpreting ability of the discovered layers need further investigation.

7.3.1.3   *Biclustering and δ-Clusters*

Cheng et al. [52] introduced the *bicluster* concept to model a pattern-based cluster. Let G' and S' be subsets of genes and samples. The pair (G', S') specifies a sub-matrix with the *mean-squared residue score* $H(G', S') = \frac{1}{|G'||S'|} \sum_{i \in G', j \in S'} (w_{ij} - \eta_{iS'} - \eta_{G'j} + \eta_{G'S'})^2$, where $\eta_{iS'} = \frac{1}{|S'|} \sum_{j \in J} w_{ij}$, $\eta_{G'j} = \frac{1}{|G'|} \sum_{i \in I} w_{ij}$, $\eta_{G'S'} = \frac{1}{|G'||S'|} \sum_{i \in G', j \in S'} w_{ij}$ are the row and column means and the means in the submatrix (also see 6.4.3.1). A submatrix is called a δ-*bicluster* if $H(G', S') \leq \delta$ for some $\delta > 0$.

The lowest score $H(I, J) = 0$ indicates that the gene expression levels within the submatrix are invariant. This includes the trivial or constant biclusters where there is no fluctuation. These trivial biclusters may not be very interesting but need to be discovered and masked so more interesting ones can be found. The row variance $V(I, J) = \frac{1}{|J|} \sum_{j \in J} (a_{ij} - a_{Ij})^2$ may be an accompanying score to reject trivial biclusters. That is, a low mean-squared residue score together with a large row variance suggest a good criterion for identifying a pattern-based cluster.

However, the problem of finding a minimum set of biclusters to cover all the elements in a data matrix has been shown to be NP-hard. A greedy method which provides an approximation of the optimal solution and reduces the complexity to polynomial-time was introduced in [52]. To find a bicluster, the score H is computed for each possible addition or deletion of a row or column, and the action that decreases H the most is applied. If no action will decrease H or if $H \leq \delta$, a bicluster is returned. This algorithm relies on the brute-force deletion and addition of rows or columns and requires computational time $O((n + m) \cdot mn)$, where n and m are the number of genes and samples, respectively. Clearly, use of this algorithm is time-consuming when applied to large gene expression data sets. A more efficient *biclustering algorithm* based on addition or deletion of multiple row or column with time-complexity $O(mn)$ was also proposed in [52]. After one bicluster is identified, the elements in the corresponding sub-matrix are replaced (masked) by random numbers. The biclusters are successively extracted from the raw data matrix until a pre-specified number of clusters have been identified. However, this biclustering algorithm does have several drawbacks. First, the algorithm stops when a pre-specified number of clusters have been identified. To encompass the majority of elements in the data matrix, the number specified is usually large. The biclustering algorithm cannot guarantee that the biclusters identified earlier in the process will be of superior quality to those identified later, which adds to the difficulty of the interpretation of the resulting clusters. Second, biclustering "masks" the identified biclusters with random numbers, preventing the identification of overlapping biclusters.

Yang at al. [316] presented a subspace clustering method named "δ-clusters" to capture K embedded subspace clusters simultaneously. They used the average *residue* across every entry in the submatrix to measure the coherence within a submatrix. A heuristic move-based method called FLOC (FLexible Overlapped Clustering) was applied to search K embedded subspace clusters. FLOC starts with K randomly selected submatrices as the subspace clusters, then iteratively tries to add or remove each row or column into or out of the subspace clusters to lower the residue value to a local minimum. The time complexity of the δ-clusters algorithm is $O((n+m)*n*m*k*l)$, where k is the number of clusters and l is the number of iterations. δ-clusters algorithm also requires pre-specification of the number of clusters. The advantage of the "δ-clusters" approach is that it is robust to missing values, since the residue of a submatrix is computed using only existing values. This approach can also detect overlapping embedded subspace clusters.

7.3.2 *Deterministic Approaches*

7.3.2.1 δ-pCluster

In [296], Wang et al. propose the model of δ-*pCluster*. To constrain the coherence of patterns in a cluster C, δ-pCluster requires the change of differences of any two data objects x, y in C with respect to two attributes

a, b in C should be below a threshold δ. To be formal, let $\mathcal{O}$ be a subset of objects in the database ($\mathcal{O} \subseteq \mathcal{D}$), and let $\mathcal{T}$ be a subset of attributes ($\mathcal{T} \subseteq \mathcal{A}$). The pair $(\mathcal{O}, \mathcal{T})$ specifies a submatrix. Given x, $y \in \mathcal{O}$, and a, $b \in \mathcal{T}$, the *pScore* of the 2×2 matrix is defined as

$$pScore(\begin{bmatrix} d_{xa} & d_{xb} \\ d_{ya} & d_{yb} \end{bmatrix}) = |(d_{xa} - d_{xb}) - (d_{ya} - d_{yb})|.$$

The pair $(\mathcal{O}, \mathcal{T})$ forms a δ-pCluster if, for any 2×2 submatrix X in $(\mathcal{O}, \mathcal{T})$, we have $pScore(X) \leq \delta$ for some $\delta > 0$. In particular, a δ-pCluster containing at least min_o objects and min_a attributes (i.e., $||\mathcal{O}|| \geq min_o$ and $||\mathcal{T}|| \geq min_a||$) is considered as a *significant δ-pCluster*, where min_o and min_a are user-specified thresholds. The goal of mining δ-pClusters is to find the complete set of significant δ-pClusters.

By definition, δ-pClusters have the *downward closure property* (see Section 7.2.1). That is, if $C = (\mathcal{O}, \mathcal{T})$ is a δ-pCluster, every sub-matrix of C is also a δ-pCluster. Frequent itemsets possess this property but contain only one type of variable (the items), while δ-pClusters contain two types of variables (objects and attributes). Therefore, although the conceptual aspects of basic pruning techniques applied to mining frequent itemsets by methods such as Apriori can be extended to the identification of δ-pClusters, the technical details are very different.

The δ-pCluster mining algorithm [296] consists of the following three major steps:

- Step 1. Find the attribute-pair and object-pair *maximal dimension sets* (MDS) using sliding windows. An object/attribute-pair MDS is a maximal δ-pCluster containing only two objects or attributes. These MDS are the building blocks of the δ-pClusters, since a non-trivial δ-pCluster must have at least two objects and two attributes.

- Step 2. Prune unpromising attribute-pair and object-pair MDS iteratively. An object-pair MDS $(\{o_1, o_2\}, \mathcal{T})$ can be pruned if the number of attributes in $\mathcal{T}$ is less than min_a, since, in this case, o_1 and o_2 cannot appear together in any significant δ-pCluster. Similarly, an attribute-pair MDS $(\mathcal{O}, \{a_1, a_2\})$ can be pruned if the number of objects in $\mathcal{O}$ is less than min_o since, in this case, a_1 and a_2 cannot appear together in any significant δ-pCluster. This process of pruning attribute-pair MDS and object-pair MDS is conducted iteratively until no more MDS can be pruned.

- Step 3. Generate significant δ-pClusters using a prefix tree. After the pruning in Step 2, the algorithm inserts the surviving object-pair MDSs into a prefix tree. Given an ordering of attributes R, two object-pair MDSs sharing the same prefix with respect to R will share the corresponding path from the root in the tree. Finally, a post-order traversal of the prefix tree is conducted to examine the nodes whose depths are no less than min_a, and report the significant δ-pClusters.

In a more recent study [218], Pei et al. proposed the mining of *maximal* δ-clusters. This approach completely avoids the identification of redundant clusters by mining only the maximal pattern-based clusters, which are considered to be skylines of all pattern-based clusters. An efficient algorithm, *MaPle*, was developed to mine the complete set of non-redundant pattern-based clusters.

7.3.2.2 *OP-Cluster*

In [185], Liu and Wang presented another model, termed the *OP-Cluster*. In this model, coherence within a cluster is constrained by the relative order of the attribute values of the objects within the cluster. To be specific, an object O exhibits the *"UP"* pattern in a permutation $A' = \{a_{i1}, \ldots, a_{in}\}$ of attributes $A = \{a_1, \ldots, a_n\}$ if the attribute values $\langle d_{i1}, \ldots, d_{in} \rangle$ of O on A' is in a *non-decreasing* order. Let $\mathcal{O}$ be a subset of objects in the database $(\mathcal{O} \subseteq \mathcal{D})$, and let $\mathcal{T}$ be a subset of attributes $(\mathcal{T} \subseteq \mathcal{A})$. The pair $(\mathcal{O}, \mathcal{T})$ forms an *Order Preserving Cluster* (*OP-Cluster*) if there exists a permutation of attributes in $\mathcal{T}$, in which every object in $\mathcal{O}$ shows the *"UP"* pattern. In practice, the requirement of *"non-decreasing order"* is often relaxed by a *group function*; i.e., if the difference between the values of two attributes is not significant, they are considered as a "group" and not ordered.

Based on the concepts of the "UP" pattern and the group function, a gene expression data matrix can be converted into a set of *sequences*, allowing the application of techniques used to identify *frequent sequential patterns*. For example, suppose the gene expression levels are measured for samples (A,B,C,D,E). The expressions of genes g_1 and g_2 are (1, 4, 4.5, 8, 10) and (2, 5, 7, 4.5, 9), respectively. Assume that the group function regards two values with difference smaller than 1 to be "equivalent." The corresponding sequence of groups for g_1 is then A(BC)DE, and for g_2 is A(DB)CE. Since ABCE is a common subsequence of them, g_1 and g_2 form an OP-Cluster on the attribute set of ABCE.

An efficient algorithm for the identification of OP-Clusters was proposed in [185]. The algorithm conducts a depth-first traversal of an *OPC-tree*, a structure similar to an FP-tree (see Section 7.2.3), to generate frequent subsequences by recursively concatenating legible suffixes with the existing frequent prefixes.

The OP-Cluster [185] and δ-pCluster [296, 218] approaches have several features in common. First, both models control the coherence of pattern-based clusters by restricting the divergence among objects within a threshold δ. Second, both models have the downward closure property. Third, both approaches seek to find *significant* and *maximal* pattern-based clusters. Finally, both approaches are deterministic, in that they guarantee to find the *complete* set of the pattern-based clusters in the data set. In fact, Liu and Wang [185] have shown that the OP-Cluster is more flexible than the δ-pCluster and treat the latter as a special case of the former.

7.4 Mining Gene-Sample-Time Microarray Data

7.4.1 *Three-dimensional Microarray Data*

The previous discussion has focused on *two-dimensional* (2D) microarray data, which can be divided into two categories:

- *Gene-time data sets* which record the expression levels of various genes during important biological processes over a series of time points, and
- *Gene-sample data sets* which compile the expression levels of various genes across related samples.

Both types of data sets can be represented by a two-dimensional gene expression matrix, where each row is a gene and each column is either a sample (in a gene-sample data set) or a time point (in a gene-time data set). Each cell in the matrix represents the expression level of a certain gene for a certain sample or at a certain time point.

With the latest advances in the microarray technology, the expression levels of a set of genes with respect to a set of samples can be monitored synchronically over a series of time points [300]. Unlike the previous gene-time or gene-sample microarray data sets, these new data sets have three types of variables: *genes*, *samples* and *time points*. Such three-dimensional (3D) microarray data is termed *gene-sample-time microarray data*, or, briefly, *GST data*. Figure 7.6 represents the structure of a GST microarray data set.

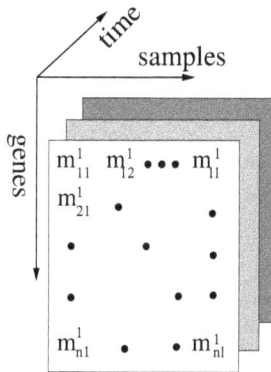

Fig. 7.6 The structure of *GST* microarray data.

As discussed throughout this book, many previous studies have investigated the mining of interesting patterns from microarray data. For example, various clustering algorithms have been proposed to identify co-expressed genes which exhibit coherent patterns over a time-series (see Chapter 5). Both supervised and unsupervised approaches are available to partition samples into homogeneous groups (see Chapter 6). In addition, pattern-based clustering approaches (see Section 7.3) have recently been developed to identify subsets of objects which exhibit similar patterns for subsets of attributes. However, all these studies and methods have been intended to address conventional gene-time or gene-sample microarray data sets. These clustering models cannot be used to describe the inherent correlation among the three types of variables in GST data, and these algorithms cannot be directly extended to explore the coherent patterns in the 3D data. The following section will present two approaches, *coherent gene clusters* [155, 156][1] and *tri-clusters* [328], which have been developed specifically for analysis of 3D data.

7.4.2 *Coherent Gene Clusters*

Previous studies on gene-sample microarray data (e.g., [6, 7, 114, 271]) have indicated that strong correlations may exist between gene expression patterns and certain diseases. It is natural to extend the similar analysis to GST microarray data. Interestingly, a 3D GST data set can also be pro-

---

[1] With kind permission of Springer Science and Business Media.

jected into an 2D matrix so that each cell contains the time series pertinent
to a gene with respect to a sample (see Figure 7.7). Thus, it is interesting
to identify a subset of genes G and a subset of samples S in a GST microarray data set for which each gene $g \in G$ has coherent patterns across
the samples in S during the time series. For example, in Figure 7.7, gene
g_{i1}, g_{i2}, and g_{i3} show coherent patterns across samples s_{j1}, s_{j2}, and s_{j3},
respectively. Such subsets of genes and samples are termed a *coherent gene*
cluster.

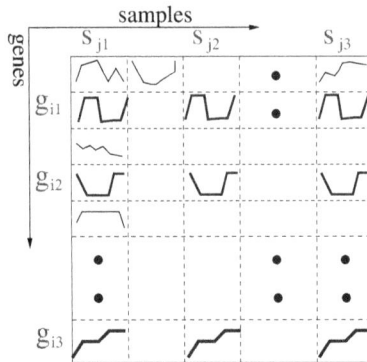

Fig. 7.7   The projected GST microarray data and the coherent gene cluster.

The coherent gene cluster model addresses a significant problem in the
clinical use of a variety of drug therapies. For example, IFN-β is the most
widely prescribed immunomodulatory therapy for multiple sclerosis (an autoimmune disease of the brain and spinal cord). The therapy is known to
exercise all its biological effects via gene transcription, but there are no
validated markers for its long-term efficacy in multiple sclerosis. Although
double-blind, randomized, placebo-controlled clinical trials have established
that IFN-β treatment reduces the progression of disability in multiple sclerosis, only thirty to forty percent of patients respond well to the therapy.
To define the mechanism of IFN-β and investigate the partial responsiveness of various patients, the expression levels of large numbers of genes
were monitored for thirteen multiple sclerosis patients during a ten-point
time-series [300].

There is considerable inter-individual heterogeneity in responsiveness
to IFN-β. In other words, patients can differ both in the magnitude and
rate of their gene expression profiles. However, the underlying mechanisms
are not fully characterized. The coherent gene cluster model is directly

relevant to this characterizing, since it is capable of identifying patients whose responses are similar and defining the time courses of genes that distinguish these patient subsets. Moreover, the genes within the coherent clusters may suggest candidate genes which are correlated to the physical response.

As this example indicates, coherent gene clusters may be a valuable source of biological hypotheses. The sample sets within clusters may correspond to certain phenotypes, such as diseased or healthy patients or patients manifesting different treatment responses, while the corresponding set of genes may be a source of candidate genes which are correlated to the phenotypes.

The functions of genes in an organism are complex, and there are typically multiple coherent clusters in a data set. Different clusters may correlate to different phenotypes, such as age and gender. Therefore, to ensure that all potentially-useful hypotheses are identified, it is necessary to mine *all* the coherent clusters in a data set.

7.4.2.1 *Problem description*

Given a 3D microarray data set, let $G\text{-}Set = \{g_1, \ldots, g_n\}$ be the set of genes, let $S\text{-}Set = \{s_1, \ldots, s_l\}$ be a set of samples, and let $T\text{-}Set = \{t_1, \ldots, t_T\}$ be the set of time points. A three dimensional microarray data set is a real-valued $n \times l \times T$ matrix $M = G\text{-}Set \times S\text{-}Set \times T\text{-}Set = \{m_{ij}^k\}$ where $i \in [1, n], j \in [1, l], k \in [1, T]$.

As previously noted, a 3D GST data set can also be projected into an 2D matrix, such that each cell $\vec{m}_{i,j}$ contains the time series $\langle m_{i,j}^1, \ldots, m_{i,j}^T \rangle$ with respect to gene g_i under sample s_j (see Figure 7.7). In this subsection, we will not make a strict distinction between the two notations $m_{i,j}$ and $\vec{m}_{i,j}$, so that $m_{i,j}$ will be used to refer to the vector $\langle m_{i,j}^1, \ldots, m_{i,j}^T \rangle$.

The methods discussed in this section are intended to identify those genes that are coherent within a subset of samples throughout the entire time series. Various methods are available to measure the correlation between two time series. Users of gene expression data are often interested in the overall trends exhibited by the expression levels rather than the absolute magnitudes. To satisfy these objectives, Pearson's correlation coefficient ρ (see Section 5.2) is used as the coherence measure, since it is robust to shifting and scaling patterns [316]. The value of the correlation coefficient ranges between -1 and 1, with a larger value indicating greater coherence between the two vectors (see Section 5.2.2).

Definition 7.1 Coherent gene submatrix. Given a GST data matrix M, a gene $g_i \in$ *G-Set* is *coherent* across a subset of samples $S \subseteq$ *S-Set*, if, for any given pair of samples $s_{j_1}, s_{j_2} \in S$, $\rho(m_{i,j_1}, m_{i,j_2}) \geq \delta$, where δ is a *minimum coherence threshold* specified by the user. A subset of genes $G \subseteq$ *G-Set* is *coherent* across a subset of samples $S \subseteq$ *S-Set*, if every gene $g_i \in G$ is coherent across samples in S. We call a submatrix $(G \times S)$ a *coherent gene submatrix* if G is coherent across S. A coherent gene submatrix having u genes and v samples is said a (u, v)-coherent gene submatrix.

Proposition 7.1 Trivial coherent gene submatrices. *For any gene g_i and any sample s_j, $(\{g_i\} \times \{s_j\})$ is a $(1,1)$-coherent gene submatrix, and $(G\text{-}Set \times \{s_j\})$ is $(|G\text{-}Set|, 1)$-coherent gene submatrix.*
Proof. The proposition follows directly from the definition of the coherent gene submatrix. ∎

To avoid the triviality indicated in Proposition 7.1, we stipulate that a coherent gene submatrix must consist of at least two genes and two samples.

Proposition 7.2 Anti-monotonicity. *Let $(G \times S)$ be a coherent gene submatrix. Then, for any subsets $G' \subseteq G$ and $S' \subseteq S$, $(G' \times S')$ is also a coherent gene submatrix.*
Proof. Any gene pair in G' and any sample pair in S' must satisfy the coherence requirement, since both are in the cluster $(G \times S)$. ∎

The anti-monotonicity of coherent submatrices results in significant redundancy, which may be reduced by seeking only the maximal submatrices. A coherent gene submatrix $(G \times S)$ is *maximal* if there exists no other coherent gene submatrix $(G' \times S')$ such that $G \subseteq G'$, $S \subseteq S'$. A user may also wish to eliminate very small clusters, which are often formed by chance. To accomplish this, a user can specify the minimum numbers of genes and samples in a submatrix. Generally, given min_g and min_s as user-defined thresholds for minimum gene size and minimum sample size, a submatrix $(G \times S)$ is considered *significant* if $|G| \geq min_g$ and $|S| \geq min_s$.

Definition 7.2 Coherent gene cluster. Given a GST microarray matrix M, a minimum coherence threshold δ, a minimum gene size threshold min_g, and a minimum sample size threshold min_s, a submatrix $(G \times S)$ of M is a *coherent gene cluster* if it satisfies the following constraints: (1) $(G \times S)$ is a coherent gene submatrix; (2) $(G \times S)$ is maximal; and (3) $|G| \geq min_g$ and $|S| \geq min_s$.

The problem of *mining coherent gene clusters* is to find the complete set of coherent gene clusters in a given data set with respect to certain user-specified parameters.

A naïve method for the identification of maximal coherent gene clusters would test every possible combination of genes and samples thoroughly and report the maximal clusters. Such a method would be costly and impractical for real data sets. For example, a data set with $1,000$ genes and 20 samples would generate up to $(2^{1000}-1)\times(2^{20}-1-20)\approx1.12\times10^{307}$ combinations, each of which would need to be tested.

7.4.2.2 *Maximal coherent sample sets*

Two algorithms have been proposed to compute maximal coherent gene clusters [155, 156]. Both algorithms include a step to test the coherence of a subset of genes with regard to a subset of samples. To facilitate the tests, for each gene g_k, we compute the sets of samples S such that (1) $|S| \geq min_s$; (2) g_k is coherent on S; and (3) there exists no proper superset $S' \supset S$ such that g_k is also coherent on S'. S is called a *maximal coherent sample set* of g_k. Please note that, in general, a gene may have more than one maximal coherent sample set.

All the maximal coherent sample sets for a gene g_k can be computed efficiently using the following two-step process. In the first step, gene g_k is tested for coherency with each pair of samples (s_i, s_j). A binary triangle matrix $\{c_{i,j}\}$ is populated, where $1 \leq i < j \leq |S\text{-}Set|$. $c_{i,j}$ is set to 1 if gene g_k is coherent on samples s_i and s_j; i.e., $\rho(m_{k,i}, m_{k,j}) \geq \delta$, otherwise, $c_{i,j} = 0$.

Once the matrix $\{c_{i,j}\}$ is populated, the second step involves identifying maximal coherent sample sets of g_k which can be reduced to the problem of finding all maximal cliques of size at least min_s in graph $\mathcal{G}_k = (S\text{-}Set, E)$, where (s_i, s_j) is an edge in the graph if and only if $c_{i,j} = 1$. Here, we follow the terminology that a clique is a set of vertices such that the induced subgraph is a complete graph. A clique S is considered *maximal* if there exists no other clique S' such that $S \subset S'$. More than one maximal clique may occur in a graph. Unlike the conventional clique problem where the clique of the maximal size is found, here, we need to find the complete set of maximal cliques in the graph.

Theorem 7.1   Complexity of preprocessing. *The problem of computing the complete set of maximal cliques that have at least min_s vertices is NP-hard.*

Proof. It is well known that the conventional clique problem is NP-complete. Therefore, the counting problem of finding the complete set of cliques of size min_s or larger is in #P. Since a #P problem corresponding to any NP-Complete problem must be NP-Hard, the problem of computing the complete set of maximal cliques in a graph is NP-Hard. ∎

Fortunately, real *GST* microarray data sets are often sparse, and the number of samples is typically below one hundred. The number of maximal cliques for each gene is therefore quite small, and the samples can often be partitioned into small exclusive subsets. Experimental results show that, by employing the efficient search and pruning techniques to be presented below, it is practically feasible to find the complete set of maximal cliques. In the following, we will show how to find the maximal cliques of a sample set by a depth-first search in a sample-set enumeration tree.

Given a set of samples $S = \{s_1, \ldots, s_l\}$, the set 2^S (i.e., all combinations of samples) can be enumerated systematically. For example, consider a set of samples $S = \{a, b, c, d\}$. The complete set of non-empty combinations of samples can be divided into 4 exclusive subsets: (1) those containing sample a; (2) those containing sample b but no a; (3) those containing sample c but no a or b; and (4) $\{d\}$. They are shown as the immediate children of the root in Figure 7.8.

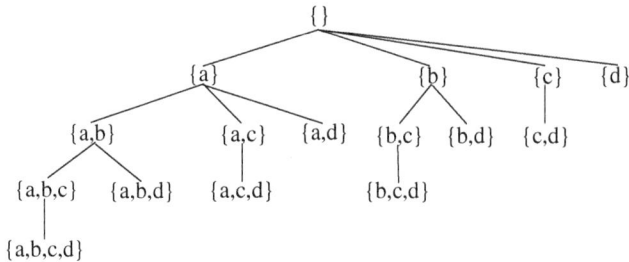

Fig. 7.8   Enumeration of sample combinations.

These subsets can be further partitioned. For example, the first subset can be further divided into three exclusive sub-subsets: (1) those containing samples a and b; (2) those containing samples a and c but not b; and (3) a, d.

The tree shown in Figure 7.8 is called a *set enumeration tree* [236] with respect to $\{a, b, c, d\}$. It provides a conceptual tool for systematic enumeration of the complete set of combinations.

We can conduct a *recursive, depth-first search* of the sample-set enumeration tree to detect the maximal cliques of the samples. Given a set of samples S, the set enumeration tree has $2^{|S|}$ nodes. In practice, we never need to materialize such a tree. Instead, we only need to maintain a path from the root of the tree to the current working node as a working set. Such a path contains at most $(|S| + 1)$ nodes.

It is obvious that each node in the set enumeration tree contains a unique subset of samples, and the sample subset can therefore be used to refer to the node. At node $\{s_{i_1}, \dots, s_{i_k}\}$ $(1 \leq i_1 < \cdots < i_k \leq l)$, we also establish a list $Tail$ which contains samples that can be used to extend the node to larger subsets of samples in the subtree. We have the following result.

Lemma 7.1 *At node $v = \{s_{i_1}, \dots, s_{i_k}\}$ of the sample set enumeration tree, where $(1 \leq i_1 < \cdots < i_k \leq l)$, a sample $s_j \notin Tail$ if (1) $j \leq i_k$; or (2) there exists some $1 \leq r \leq k$ such that $c_{i_r,j} = 0$. Moreover, for v's parent node $v' = \{s_{i_1}, \dots, s_{i_{k-1}}\}$, v's $Tail$ is a subset of that of v'.*
Proof. The first claim of the lemma follows from the definition of a set enumeration tree. If $j \leq i_k$, s_j cannot appear in any node of the subtree of v. Furthermore, if $c_{i_r,j} = 0$, then the gene is not coherent with respect to c_j and $c_{i_r,j}$.

To demonstrate the second claim, we need only note that v contains a superset of samples in v'. If the gene is coherent on every sample in v and a sample s that is not in the tail of v, it must also be coherent on every sample in v' and s. ∎

Clearly, any subtree that cannot lead to a coherent sample set of at least min_s samples can readily be pruned.

Pruning Rule 7.4.1 **Pruning small sample sets.**
At a node $v = \{s_{i_1}, \dots, s_{i_k}\}$, the subtree of v can be pruned if $(k + |Tail|) < min_s$.

For example, for a set of l samples, the complete set of sample combinations can be divided into l exclusive subsets as shown above. We only need to search the first $(l - min_s + 1)$ subsets, since each of the last $(min_s - 1)$ subsets contains less than min_s samples.

Moreover, if the samples at the current node and its $Tail$ are subsumed by some maximal coherent sample set already identified, the recursive search can also be pruned, since it cannot lead to any new maximal coherent sample set.

Pruning Rule 7.4.2   **Pruning subsumed sets.**
At a node $v = \{s_{i_1}, \ldots, s_{i_k}\}$, if $\{s_{i_1}, \ldots, s_{i_k}\} \cup Tail$ is a subset of some maximal coherent sample set, then the subtree of the node can be pruned.

Algorithm 7.1 presents a preprocessing approach which is based on Lemma 7.1 and the above pruning rules. Readers familiar with the techniques of depth-first mining of maximal/closed frequent patterns will recognize the similarity of these pruning concepts to those employed in frequent closed itemset mining (e.g., [216]). These approaches differ, however, in that frequent pattern mining enumerates complete databases, and this algorithm does not need to scan the database once the triangle matrix $\{c_{i,j}\}$ has been constituted.

Algorithm 7.1   : **Computing Maximal Coherent Sample Sets**

Input: the GST data set, the coherence threshold δ, and the minimum
      sample size threshold min_s;
Output: the maximal coherent sample sets for each gene;
Method:
    for each gene g_k do
        generate matrix $c_{i,j}$ for g_k;
        for $i = 1$ to $(l - min_s + 1)$ call *search-clique*($\{s_i\}, \{s_{i+1}, \ldots, s_l\}$);
    end-for

Procedure: *search-clique(head, tail)*
    // *head* records the samples in the current node
    suppose s_i is the last sample in *head*
        remove samples s_j from *tail* such that $c_{i,j} = 0$; // Lemma 7.1
    if $(|head \cup tail| < min_s)$ // Pruning 7.4.1
        or $(head \cup tail \subset S)$ s.t. S is a maximal clique // Pruning 7.4.2
    then return;
    if $tail = \emptyset$ then output a maximal clique;
    else do
        let $j = \min\{k | s_k \in tail\}$, $tail = tail - \{s_j\}$;
        call *search-clique*($head \cup \{s_j\}, tail$);
    until $tail = \emptyset$;
    return;

7.4.2.3   *The mining algorithms*

The approach taken in Algorithm 7.1 can now be used as a guide to identify maximal coherent gene clusters. There, the computation of maximal coherent sample sets proceeds via a systematic enumeration of combinations of samples in a recursive depth-first search, using several techniques to aggressively prune unpromising subspaces. Similarly, the identification of maximal coherent gene clusters also involves a systematic enumeration of the gene and sample combinations, pruning the unfruitful combinations.

This process can give priority to either samples or genes. A sample-based approach, termed the *Sample-Gene Search*, starts with the systematic enumeration of all combinations of samples. The maximal subsets of genes that form coherent gene clusters with respect to each subset of samples are then identified, and these are checked to ensure that the clusters are maximal. Conversely, a *Gene-Sample Search* starts with gene enumeration, and the maximal subsets of samples that form coherent gene clusters with each subset of genes are identified and checked for maximality. In both cases, proper pruning techniques should be applied to excise unpromising combinations and search branches at an early stage.

The framework that underlies these two search methods is set forth in Algorithm 7.2. Proper pruning techniques should be developed to prune the unpromising combinations and search branches as early as possible.

Algorithm 7.2   : The Framework of Two Search Methods

The *Sample-Gene Search*
    depth-first enumerate subsets of samples
    for each subset of samples S **do**
        find the maximal subsets of genes G such that
            $G \times S$ is a coherent gene cluster;
        test whether $(G \times S)$ is a maximal coherent gene cluster;
    end-for

The *Gene-Sample Search*
    depth-first enumerate subsets of genes
    for each subset of genes G **do**
        find the maximal subsets of samples S such that
            $G \times S$ is a coherent gene cluster;
        test whether $(G \times S)$ is a maximal coherent gene cluster;
    end-for

At first glance, the algorithm may suggest that the two methods are fully symmetric. However, since the roles of genes and samples in a data set are not symmetric, the technical details of the two approaches are in fact substantially different. In this section, we will focus on the Sample-Gene Search to illustrate the basic concepts common to both approaches. For details of the Gene-Sample Search, please refer to [156].

Table 7.3   The maximal coherent sample
sets for genes.

Gene	Maximal coherent sample sets
g_1	$\{s_1, s_2, s_3, s_4, s_5\}$
g_2	$\{s_1, s_2, s_4\}, \{s_1, s_5\}$
g_3	$\{s_1, s_2, s_3, s_4, s_5\}$
g_4	$\{s_1, s_2, s_3\}, \{s_5, s_6\}$
g_5	$\{s_1, s_5, s_6\}$

An effective Sample-Gene Search must address the following issues, each of which will be discussed in more detail below:

- Identification of the maximal sets of genes which are coherent with respect to each subset of samples. This should occur during the systematic enumeration of sample combinations.
- Identification of sample sets that can be pruned during the enumeration process.
- Identification and elimination of searches that cannot lead to any potential maximal coherent clusters.
- Determination of those coherent gene clusters which are subsumed by other clusters.

Maximal Coherent Gene Sets for Sample Sets

For each combination of samples S, the *maximal coherent gene set* G_S is computed. The genes in G_S must be coherent with respect to S, and no proper superset $G' \supset G_S$ also has this property.

Given a gene g, if there exists a maximal coherent sample set S_g such that $S \subseteq S_g$, then $g \in G_S$. In other words, G_S can be derived by a single scan of the maximal coherent sample sets of all genes. If a maximal coherent sample set is a superset of S, then the corresponding gene g is inserted into G_S. However, repeating a complete scan of the maximal coherent sample sets of all genes for each combination of samples is expensive and inefficient. A more efficient solution is to use an *inverted list*.

This concept can be illustrated via an example involving five genes and six samples. The maximal coherent sample sets for each gene are listed in Table 7.3. Let us label each maximal coherent sample set by the gene g_k and the set-id b_j. For example, gene g_2 has two maximal coherent sample sets, $g_2.b_1 = \{s_1, s_2, s_4\}$ and $g_2.b_2 = \{s_1, s_5\}$. An *inverted list* L_s can then be compiled for each sample s. This comprises a list of all maximal coherent sample sets containing s, as shown in Table 7.4.

Table 7.4   The inverted lists for Table 7.3.

Sample	The inverted list
s_1	$\{g_1.b_1, g_2.b_1, g_2.b_2, g_3.b_1, g_4.b_1, g_5.b_1\}$
s_2	$\{g_1.b_1, g_2.b_1, g_3.b_1, g_4.b_1\}$
s_3	$\{g_1.b_1, g_3.b_1, g_4.b_1\}$
s_4	$\{g_1.b_1, g_2.b_1, g_3.b_1\}$
s_5	$\{g_1.b_1, g_2.b_2, g_3.b_1, g_4.b_2, g_5.b_1\}$
s_6	$\{g_4.b_2, g_5.b_1\}$

Subsequent computation of the maximal coherent gene sets for a subset of samples such as $\{s_1, s_2, s_3\}$ needs not search the complete list provided in Table 7.3. Instead, only the intersection of the inverted lists for each of the samples needs to be examined; in this case, this is $\{g_1.b_1, g_3.b_1, g_4.b_1\}$. The membership of these intersecting lists indicates that $\{g_1, g_3, g_4\}$ is the maximal coherent gene set for this example.

Pruning Irrelevant Samples

For a combination of samples $S = \{s_{i_1}, \ldots, s_{i_k}\}$, where $i_1 < \cdots < i_k$, let S_{tail} be the set of samples that can be used to extend S to a larger set $S' \subset S \cup S_{tail}$ such that there are at least min_g genes coherent on S'. It is obvious that a sample $s_j \notin S_{tail}$ if $j \leq i_k$. Moreover, in consonance with Lemma 7.1, if the maximal coherent gene set of $S \cup \{s_j\}$ contains less than min_g genes, then $s_j \notin S_{tail}$. For example, in the case shown in Tables 7.3 and 7.4, sample s_6 cannot be used to extend sample set $S = \{s_2\}$, since there is no gene coherent on both s_2 and s_6.

Moreover, similar to Pruning rule 7.4.1, if $|S| + |S_{tail}| < min_s$, then S cannot lead to any coherent gene cluster having min_s or more samples, and thus can be pruned.

Pruning Unpromising Coherent Gene Clusters

In a situation parallel to that addressed by pruning rule 7.4.2, unpromising combinations that cannot lead to any new maximal coherent gene cluster can be pruned. Two pruning techniques are available for this purpose.

Using the example introduced above (Tables 7.3 and 7.4), let us find the maximal coherent gene cluster $(\{g_1, g_3\} \times \{s_1, s_2, s_3, s_5\})$ before searching sample set $S = \{s_1, s_3\}$. For $S = \{s_1, s_3\}$, $S_{tail} = \{s_5\}$ and $G_{S \cup S_{tail}} = \{g_1, g_3\}$. That is, both $S \cup S_{tail}$ and $G_{S \cup S_{tail}}$ are subsumed by the maximal coherent gene cluster. The recursive search of S cannot lead to any maximal coherent gene cluster and thus can be pruned.

This can be generalized to any search for a sample combination S'. If any maximal coherent gene cluster $(G \times S)$ is encountered such that

$S' \cup S'_{tail} \subseteq S$ and $G_{S' \cup S'_{tail}} \subseteq G$, then any recursive search from S' results in a coherent gene cluster subsumed by $(G \times S)$, and thus can be pruned.

Additionally, if a maximal coherent gene cluster $(G \times S)$ has been identified such that $S' \subseteq S$ and every maximal coherent sample set containing S' also contains S, then the recursive search of S' cannot lead to any maximal coherent gene cluster either, and thus S' can be pruned. Using the example in Tables 7.3 and 7.4, suppose we search the sample set $S' = \{s_2\}$ after the maximal coherent gene cluster $(\{g_1, g_2, g_3, g_4\} \times \{s_1, s_2\})$ has been identified. Table 7.3 indicates that every maximal coherent sample set containing s_2 also contains s_1. In other words, there exists no maximal coherent gene cluster containing s_2 but no s_1. Thus, the search of S' can be pruned.

Algorithm 7.3   : **The Sample-Gene Search**

Input: the maximal coherent samples sets for genes // from Algorithm 7.1
Output: the complete set of coherent gene clusters
Method:
    generate the inverted list for samples;
    for $i = 1$ to $(|S\text{-}Set| - min_s)$ do
        let $S = \{s_i\}$ and $S_{tail} = \{s_{i+1}, \ldots, s_{|S\text{-}Set|}\}$;
        call *recursive-search*(S, S_{tail});
    end-for

Procedure: *recursive-search*(S, S_{tail})
    remove irrelevant samples from S_{tail};
    if $(|S| + |S_{tail}| < min_s)$ then return;
    derive the intersection of inverted lists for samples in S;
    if S meets any pruning criteria then return;
    while $S_{tail} \neq \emptyset$ do
        let $i = \min\{j|s_j \in S_{tail}\}$;
        let $tail = tail - \{s_i\}$;
        call *recursive-search*$(S \cup \{s_i\}, S_{tail})$;
    end-while
    derive the maximal coherent gene set G_S;
    output $(G_S \times S)$ as a maximal coherent gene cluster
        if it is not subsumed by any maximal
    coherent gene cluster found before
End

Determining maximal coherent gene clusters A search of a combination of samples S should ascertain whether $G_S \times S$ is a maximal coherent gene cluster. This is accomplished by examining the maximal coherent gene clusters $(G' \times S')$ such that $S' \supset S$. Since a depth-first search has been conducted in the set enumeration tree, such maximal coherent gene clusters should be reported either before S is searched or in the subtree rooted at S.

The *Sample-Gene Search* algorithm (Algorithm 7.3) reflects the issues and approaches discussed above.

7.4.2.4 *Experimental results*

The algorithms presented in Section 7.4.2 were tested on both a real GST microarray data set and on synthetic data sets. The GST microarray data set reported in [300] consists of microarray measurements of 4,324 genes in thirteen multiple sclerosis (MS) patients prior to treatment and at various intervals ranging from one hour to three months after IFN-β treatment. MS patients exhibited heterogeneous responses to these treatments. For example, those patients with recurring MS responded better to IFN-β treatments than did patients with a progressive form of the disease. However, even the former group exhibited considerable inter-individual heterogeneity in clinical responses to IFN-β therapies. At this point, the effects of IFN-β treatment at the genomic level in humans are poorly understood. Researchers are interested in distinguishing the heterogeneous clinical responses to IFN-β therapy among the patients. Characterized gene-expression dynamics correlated to these heterogeneous responses may assist in exploring the causal mechanisms at the molecular level.

Data Preprocessing
The original MS microarray data contain outliers, missing values, and experimental bias. Data preprocessing began by applying a global normalization strategy to discard outliers [300] and to estimate missing values using KNN imputation [284] (see Section 3.5.2). This process also standardized the data by setting the gene expression levels of each patient at each time point to a mean of zero and a standard deviation of one. Principle component analysis (PCA) [143] (see Section 6.2.2.1) was then applied to remove the systematic variation caused by experimental bias. Genes which exhibited "flat patterns" across the entire set of samples were also filtered. Any gene which did not exhibit a significant (minimum ten percent) change in expression level for any sample during the entire time-series was eliminated in this step. It was assumed that these genes were likely to be irrelevant to the IFN-β response under investigation.

Clustering Results

The Sample-Gene Search algorithm was applied to the processed data with $min_s = 3$, $min_g = 50$ and $\delta = 0.8$. 25 highest-quality coherent gene clusters were selected from the mining results. For each cluster $C = (G \times S)$, the genes showing "flat patterns" across S are removed. The clusters are reported at http://www.cse.buffalo.edu/DBGROUP/bioinformatics/GST.

Statistical Significance of Clusters

To estimate the statistical significance of the selected clusters, 100 permutated data sets were generated from the original data. Since clusters containing only a single gene or sample are considered to be trivial (Proposition 7.1), size limits were set at $min_s = 2$, $min_g = 2$, and $\delta = 0.8$, and the algorithm was applied to each permutated data set. The clusters generated from the permutated data sets were recorded in $\mathcal{C}$, and the histograms of the number of samples and genes in these clusters were calculated.

Figure 7.9 illustrates the distribution of the percentage of clusters in $\mathcal{C}$ with respect to the number of samples in the clusters. For example, 18.54% of the clusters in $\mathcal{C}$ contained two samples, while no clusters included more than five samples. Figure 7.10 shows the distribution of the number of clusters in $\mathcal{C}$ with respect to the number of genes in those clusters, given a specific number of samples. For example, in Figure 7.10(a), all clusters containing only two samples have between 160 and 260 genes, and the distribution is approximately Gaussian. However, when the number of samples in a cluster increases, the number of genes per cluster decreases dramatically. For example, the number of genes per cluster ranges between two and 24 when $|S| = 3$ (Figure 7.10(b)), and the range drops to between two and six genes per cluster when $|S| = 4$ (Figure 7.10(c)) and between two and three genes when $|S| = 5$ (Figure 7.10(d)). This observation indicates that, simply by chance, fewer genes will form coherent patterns across a larger subset of samples.

Let c denote the random variable of cluster, and g and s be the number of genes and samples per cluster, respectively. The probability of c is a joint distribution $P(c) = P(g, s) = P(s) \cdot P(g|s)$. The p-value of a given cluster $C = (G \times S)$ is defined as

$$p\text{-}value(C) = P(s \geq |S|) \cdot P(g \geq |G| \mid s = |S|). \tag{7.1}$$

It is evident that the p-$value(C)$ indicates the likelihood that a cluster C with at least $|G|$ genes and $|S|$ samples will have been formed by chance. Thus, a smaller p-value indicates a greater statistical significance. In this

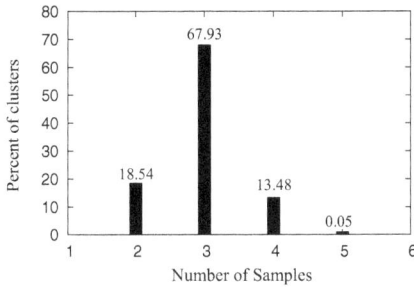

Fig. 7.9   The percentage of clusters with respect to the number of samples per cluster.

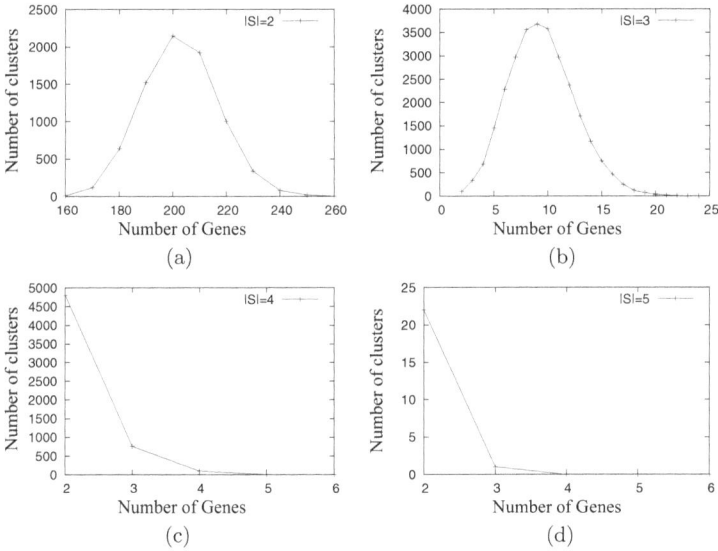

Fig. 7.10   The number of clusters with respect to the number of genes per cluster when (a) $|S| = 2$, (b) $|S| = 3$, (c) $|S| = 4$, and (d) $|S| = 5$.

test case, size limits were set at $min_s = 3$ and $min_g = 50$. Given these conditions, Figure 7.10(b) shows that the p-value of a cluster with at least three samples and 50 genes is approximately zero. Therefore, the clusters reported by the Sample-Gene Search algorithm are statistically significant.

Gene Annotation of Clusters

The Gene Ontology (GO) [30] was used to annotate the genes in the clusters. GO is organized as a hierarchical direct acyclic graph (DAG). GO is comprised of three independent ontologies which describe the attributes of molecular function, biological process, and cellular components for a gene product. The GO vocabulary is expanding rapidly and currently includes over 11,000 terms. A significant portion of the genes (3,501 from a total of 4,324) in the MS data set used in this study have been annotated by GO terms.

Suppose the total number of genes in the data set associated with a GO term T is M. If we randomly draw p genes from the complete gene set $G\text{-}Set$, the probability that q of the selected p genes are associated with T can be approximated by the hypergeometric distribution [277]:

$$P(q \mid |G\text{-}Set|, M, p) = \frac{\binom{M}{q}\binom{|G\text{-}Set|-M}{p-q}}{\binom{|G\text{-}Set|}{p}},$$

and the *p-value* of a cluster $C = (G \times S)$ with q genes in term T is

$$p\text{-}value(C, T) = \sum_{i=q}^{Min(M,|G|)} \frac{\binom{M}{i}\binom{|G\text{-}Set|-M}{|G|-i}}{\binom{|G\text{-}Set|}{|G|}}. \qquad (7.2)$$

Among the 25 clusters reported by the Sample-Gene Search algorithm, fifteen were significantly enriched in one of the second-level GO terms pertaining to biological processes. Among these, six clusters were strongly correlated with "response to external stimuli," and nine clusters were associated with "cell communication." The testing process also examined the remaining ten clusters which were not enriched in any second-level GO terms. Several of these clusters contained a large percentage of novel but unannotated genes. For example, only 19 of 107 genes in Cluster 2 were annotated. However, the 107 genes in Cluster 2 exhibited coherent patterns across a wide range of eight samples (Figure 7.11). This cluster may provide valuable information toward the prediction of the function of these unknown genes. Moreover, some of the ten clusters which were not enriched in second-level GO terms were found to be enriched in third-level GO terms. For example, Cluster 11 was enriched in "transport" (p-value = 0.0184), and Cluster 24 was enriched in "oxygen and reactive oxygen species metabolism" (p-value = 0.00648). The expression patterns and the detailed gene annotations for all clusters are available at http://www.cse.buffalo.edu/DBGROUP/bioinformatics/GST. Since IFN-β has anti-viral, anti-proliferative, and immunomodulatory effects, the clustering results are biologically feasible.

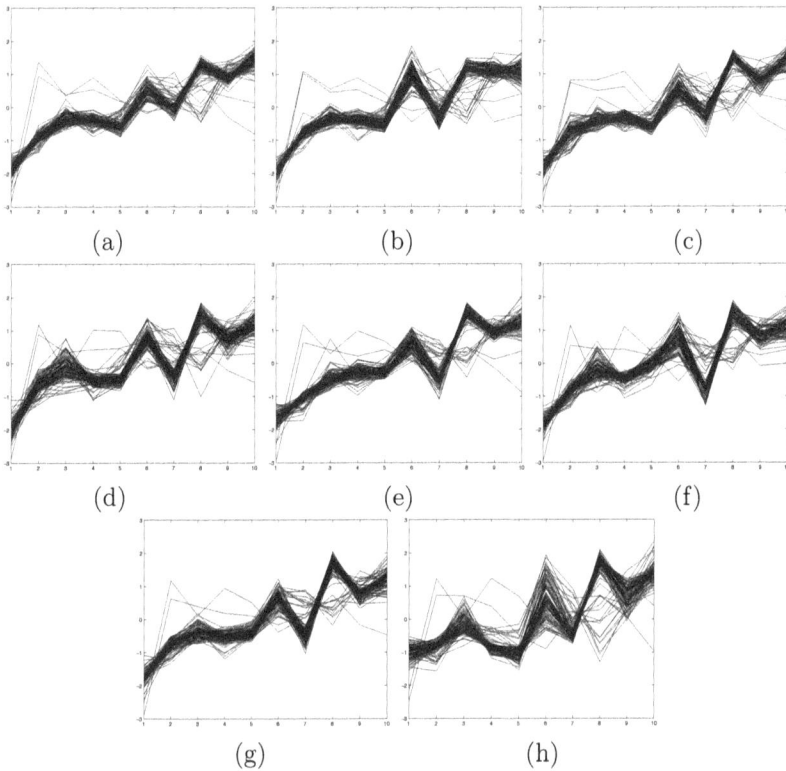

Fig. 7.11   The temporal expression patterns of genes with respect to individual samples in Cluster 2. Each figure is the parallel coordinates of the genes belonging to one sample, where the X-axis represents the time-series and the Y-axis represents the gene expression level.

Interesting Observations on Clusters

The usefulness of the coherent gene cluster model was assessed in reference to certain "benchmark" mRNA sequences that are known to be IFN-β-responsive. For example, 80 of the 659 genes in Cluster 1 were associated with the response to external stimuli (p-value 0.0262). This cluster contained several genes known to be induced by IFN-β treatment [72, 265]. These potentially significant genes include a signal transducer and activator of transcription 1 (STAT-1, Hs.479043), which is associated with the IFN-β receptor and forms part of the transcription-factor complex that binds the interferon-responsive promoter sequence; guanylate binding protein-1 (Hs.62661), myxovirus resistance protein-2 (Hs.926) and the double-stranded RNA-dependent protein kinase (PKR, Hs.131431), all of which are involved in the anti-viral response.

As another example, 46 of 174 genes in cluster 12 are involved in cell communication (p-value 0.00749), and 13 are known to associate with cell-cell signaling (p-value 0.00487). Those genes, together with other unknown or poorly-understood genes in the same cluster, may serve as switches in the genetic network and hence play an essential role in biological processes. Thus, studying the time-series of the genes in coherent gene clusters may assist in understanding the regulatory mechanisms underlying the response to IFN-β treatment.

It is noteworthy that only a few classes of temporal patterns were distinguished within each coherent gene cluster (e.g., only one major pattern was found in Cluster 2 (Figure 7.11)). This held true even though the computational model permitted genes with diverse temporal profiles to be present in the same cluster if they maintained consistent profiles across the subset of patients. This suggests that there are groups of genes with similar temporal profiles that may be activated in patients. The emergence of significant p-values in most clusters suggests that the gene groups are functionally coordinated. Biologically, this finding is potentially very valuable because the promoter sequences of these gene groups can be analyzed to determine whether they share common regulatory pathways. Further analysis incorporating clinical information could potentially reveal whether the subsets of patients differ in their clinical phenotypes.

7.4.3 Tri-Clusters

7.4.3.1 The tri-cluster model

Section 7.4.2 addressed the process of finding the *coherent gene clusters* in three-dimensional microarray data. In a coherent gene cluster, the genes exhibit coherent expression patterns across the samples throughout an entire time series. In [328], Zhao and Zaki explored this data from another viewpoint and extended the two-dimensional δ-pCluster model to a three-dimensional *tri-cluster* model.

Given a three-dimensional microarray data set, let $G = \{g_1, \ldots, g_n\}$ be the set of genes, let $S = \{s_1, \ldots, s_m\}$ be a set of samples, and let $T = \{t_1, \ldots, t_l\}$ be the set of time points. A three dimensional microarray data set is a real-valued $n \times m \times l$ matrix $D = G \times S \times T = \{d_{ijk}\}$ whose three dimensions correspond to genes, samples, and times, respectively. A *tri-cluster C* is a submatrix of the data set D, where $C = X \times Y \times Z = \{c_{ijk}\}$, with $X \subseteq G, Y \subseteq S$ and $Z \subseteq T$, which satisfies certain conditions of coherence. For example, a simple condition might be that all values $\{c_{ijk}\}$ are identical or approximately equal.

When exploring the coherent patterns in 3D microarray data, two particular types of tri-clusters, *scaling clusters* and *shifting cluster*, are often of interest. Let $C_{2,2} = \begin{bmatrix} c_{ia} & c_{ib} \\ c_{ja} & c_{jb} \end{bmatrix}$ be any arbitrary 2×2 submatrix of C, i.e., $C_{2,2} \subseteq X \times Y$ (for some $z \in Z$), or $C_{2,2} \in X \times Z$ (for some $y \in Y$) or $C_{2,2} \in Y \times Z$ (for some $x \in X$). If a cluster C satisfies $c_{ib} = \alpha_i c_{ia}$ and $c_{jb} = \alpha_j c_{ja}$, and, in addition, $|\alpha_i - \alpha_j| \leq \epsilon$, C is called a *scaling cluster*, indicating that the expression values differ by an approximately-constant (within ϵ) multiplicative factor α. If a cluster C has $c_{ib} = \beta_i + c_{ia}$ and $c_{jb} = \beta_j + c_{ja}$, and, in addition, $|\beta_i - \beta_j| \leq \epsilon$, C is called a *shifting cluster*, since the expression values differ by an approximately-constant additive factor β.

In particular, a cluster $C = X \times Y \times Z$ is a *valid cluster* if it satisfies the following constraints:

- Let $r_i = |\frac{c_{ib}}{c_{ia}}|$ and $r_j = |\frac{c_{jb}}{c_{ja}}|$ be the ratios of two column values for given rows i and j. $\frac{max(r_i, r_j)}{min(r_i, r_j)}$ is required to be smaller than or equal to $1 + \epsilon$, where ϵ is a maximum ratio value.
- If $c_{ia} \times c_{ib} < 0$ then $sign(c_{ia}) = sign(c_{ja})$ and $sign(c_{ib}) = sign(c_{jb})$, where $sign(x)$ returns -1/1 if x is negative or nonnegative (in the preprocessing step, expression values of zero are replaced with a small random positive correction value). This facilitates mining data sets having negative expression values and avoids nonsensical results such as expression ratio -5/5 is equal to 5/-5.
- For any $c_{i_1 j_1 k_1} \in C$ and $c_{i_2 j_2 k_2} \in C$, let $\delta = |c_{i_1 j_1 k_1} - c_{i_2 j_2 k_2}|$. The following maximum range thresholds along each dimension should be satisfied: a) If $j_1 = j_2$ and $k_1 = k_2$, then $\delta \leq \delta^x$, b) if $i_1 = i_2$ and $k_1 = k_2$, then $\delta \leq \delta^y$, and c) if $i_1 = i_2$ and $j_1 = j_2$, then $\delta \leq \delta^z$, where δ^x, δ^y and δ^z represent the maximum range of expression values allowed along the gene, sample, and time dimensions, respectively.
- The cluster should be non-trivial; i.e., $|X| \geq mx$, $|Y| \geq my$ and $|Z| \geq mz$, where mx, my, and mz denote minimum cardinality thresholds for each dimension.
- The cluster should be maximal. That is, there should not exist another cluster C' in the data set such that C' satisfies the above four constraints and C' subsumes C.

7.4.3.2   *Properties of tri-clusters*

Two properties can be derived from the definition of tri-clusters:

- *Symmetry Property.* Let $C = X \times Y \times Z$ be a tricluster, and let $C_{2,2} = \begin{bmatrix} c_{ia} & c_{ib} \\ c_{ja} & c_{jb} \end{bmatrix}$ be any arbitrary 2×2 submatrix of C. Let $r_i = |\frac{c_{ib}}{c_{ia}}|$, $r_j = |\frac{c_{jb}}{c_{ja}}|$, $r_a = |\frac{c_{ja}}{c_{ia}}|$, and $r_b = |\frac{c_{jb}}{c_{ib}}|$, then $\frac{max(r_i,r_j)}{min(r_i,r_j)} - 1 \leq \epsilon \Leftrightarrow \frac{max(r_a,r_b)}{min(r_a,r_b)} - 1 \leq \epsilon$.

- *Shifting Cluster.* Let $C = X \times Y \times Z = \{c_{xyz}\}$ be a maximal tri-cluster. Let $e^C = \{e^{c_{xyz}}\}$ be the tri-cluster obtained by the exponential function to value in C. If e^C is a scaling cluster, then C is a shifting cluster.

The symmetry property of tri-clusters allows clusters to be mined by searching over the dimensions with the least cardinality. This may substantially improve the mining efficiency by managing the large variations in the dimensionality of genes, samples, and time points. For example, the gene dimension usually contains thousands of or tens of thousands of genes, while the dimensionality of samples and time points is usually small, typically below one hundred. The second property of tri-clusters allows the mining of shifting clusters, although the definition of a valid tri-cluster represents a scaling cluster. For proof of these two properties, please refer to [328].

The definition of valid tri-clusters is general. Different choices of dimensional range thresholds (δ^x, δ^y, and δ^z) produce different kinds of clusters:

- If $\delta^x = \delta^y = \delta^z = 0$, then the cluster has identical values along all dimensions.
- If $\delta^x = \delta^y = \delta^z \approx 0$, then the cluster has approximately identical values along all dimensions.
- If $\delta^x \approx 0$, $\delta^y \neq 0$, and $\delta^z \neq 0$, then each gene $g_i \in X$ of the cluster $C = (X \times Y \times Z)$ has similar expression values across the different samples Y and the different time points Z, and different genes' expression values cannot differ by more than the threshold δ^x. Similarly, other cases can be obtained by setting i) $\delta^x \neq 0$, $\delta^y \approx 0$, and $\delta^z \neq 0$ or ii) $\delta^x \neq 0$, $\delta^y \neq 0$, and $\delta^z \approx 0$.
- If $\delta^x \approx 0$, $\delta^y \approx 0$, and $\delta^z \neq 0$, then the cluster has similar values for genes and samples, but the time-series are allowed to differ by some arbitrary scaling factor. Similar cases are obtained by setting i) $\delta^x \approx 0$, $\delta^y \neq 0$, and $\delta^z \approx 0$, and ii) $\delta^x \neq 0$, $\delta^y \approx 0$, and $\delta^z \approx 0$.

- If $\delta^x \neq 0$, $\delta^y \neq 0$ and $\delta^z \neq 0$, then the cluster exhibits scaling behavior on genes, samples and time points, and the expression values are bounded by δ^x, δ^y and δ^z, respectively.

7.4.3.3 *Mining tri-clusters*

To find the complete set of valid clusters in 3D microarray data, Zhao and Zaki [328] proposed the TRICLUSTER algorithm which consists of the following main steps. First, the algorithm constructs for each time point t_k a *range multigraph* which is a compact representation of all similar value ranges in the data set between any two sample columns. The algorithm then searches for constrained maximal cliques in the multigraph to yield the set of "bi-clusters"[2] for the time point t_k. TRICLUSTER then constructs a higher-order multigraph using the bi-clusters as vertices from each time point. The clique mining of the high-order multigraph will generate the final set of tri-clusters. Optionally, the TRICLUSTER algorithm merges/deletes some clusters having large overlaps.

Step 1: Construct range multigraph. Given a data set D, the minimum size thresholds, mx, my and mz, and the maximum ratio threshold ϵ, let s_a and s_b be any two sample columns in some time t of D and let $r_x^{ab} = \frac{d_{xa}}{d_{xb}}$ be the ratio of the expression values of gene g_x in columns s_a and s_b. A *ratio range* is defined as an interval of ratio values, $[r_l, r_u]$, with $r_l \leq r_u$. Let $\mathcal{G}_{ab}([r_l, r_u]) = \{g_x : r_x^{ab} \in [r_l, r_u]\}$ be the set of genes, termed the *gene-set*, whose ratios w.r.t. columns s_a and s_b lie in the given ratio range, and, if $r_{ab}^x < 0$, all the values in the same column have same signs (negative/non-negative).

In the first step, TRICLUSTER seeks to summarize the promising ratio ranges that may contribute to a particular bi-cluster using a sliding window similar to that employed by the δ-pCluster (see Section 7.3.2.1). Once the set of promising ratio ranges and the corresponding gene-sets are identified, a weighted, directed, range multigraph $M = (V, E)$ is constructed, where $V = S$ (the set of all samples), and, for each ratio range $[r_l, r_u]$ associated with samples s_a and s_b ($a < b$), there exists a weighted, directed edge (from s_a to s_b) with weight $w = \frac{r_u}{r_l}$. In addition, each edge in the range multigraph has associated with it the gene-set corresponding to the range on that edge.

---

[2]The term "bi-cluster" as used in this section differs from its use in Section 7.3.1.3. Here, bi-clusters can be regarded as the projection of tri-clusters to the gene and sample dimensions.

Step 2: Mine bi-clusters from range multigraph. The range multigraph represents in a compact way all the valid ranges that can be used to mine potential bi-clusters corresponding to each time slice and thus filters out most of the unrelated data. In concept, this is similar to the pre-processing step of mining the maximal coherent sample set prior to the mining of the coherent gene clusters (see Section 7.4.2.2). In this step, the TRICLUSTER algorithm also deploys a depth-first search (DFS) on the range multigraph to mine all the bi-clusters. It first checks whether the current cluster meets the requirements of a bi-cluster; if so, the cluster will be reported and any cluster C' in the cluster set $\mathcal{C}$ which is subsumed by C will be removed. The algorithm then generates a new candidate cluster by expanding the current candidate by one additional sample and constructs the appropriate gene-set for the new candidate. This process continues recursively until all combinations of genes and samples have been checked.

Step 3: Get Tri-clusters from bi-cluster graph. After the maximal bi-cluster set $\mathcal{C}^t$ for each time point t has been computed through the step described above, the bi-clusters are used to mine the maximal tri-clusters. This is accomplished by enumerating the subsets of the time points. Technically, this step is very similar to Step 2; the algorithm first ascertains whether the current cluster meets the requirements of a valid tri-cluster; if so, the cluster will be exported and the subsumed tri-clusters will be deleted. The algorithm then generates a new candidate cluster by expanding the current candidate by one additional time point and constructs the appropriate sample-gene-set for the new candidate by an *intersect* operation. For example, suppose $C^{t_1} = \{g_1, g_4, g_8\} \times \{s_1, s_2, s_4, s_6\}$, $C^{t_2} = \{g_1, g_2, g_8\} \times \{s_1, s_3, s_4, s_6\}$, and $C^{t_3} = \{g_1, g_3, g_8\} \times \{s_1, s_4, s_6, s_7\}$ are biclusters at time points t_1, t_2, and t_3, respectively. The result of intersection $C^{t_1} \cap C^{t_2} \cap C^{t_3}$ is $\{g_1, g_8\} \times \{s_1, s_4, s_6\}$, and this yields a candidate cluster $\{g_1, g_8\} \times \{s_1, s_4, s_6\} \times \{t_1, t_2, t_3\}$. This recursive process continues until all combinations of bi-clusters and time points have been evaluated, when all the valid tri-clusters will have been reported.

Step 4: Merge and prune clusters. Microarray data can contain significant noise, and users may not be able to provide correct values for various parameters. Moreover, many clusters obtained from the previous steps may have large overlaps, increasing the difficulty of distinguishing important clusters. To address this, the final step of the TRICLUSTER algorithm merges or deletes certain clusters with large overlaps.

    Given a cluster $C = X \times Y \times Z$, the *span* of C is defined to be the set of gene-sample-time tuples that belong to C; i.e., $L_C = \{(g_i, s_j, t_k)|g_i \in X, s_j \in Y, t_k \in Z\}$. For two clusters $A = X_A \times Y_A \times Z_A$ and $B = X_B \times Y_B \times Z_B$, the following derived spans are defined:

- $L_{A \cup B} = L_A \cup L_B$,
- $L_{A-B} = L_A - L_B$, and
- $L_{A+B} = L_{(X_A \cup X_B) \times (Y_A \cup Y_B) \times (Z_A \cup Z_B)}$.

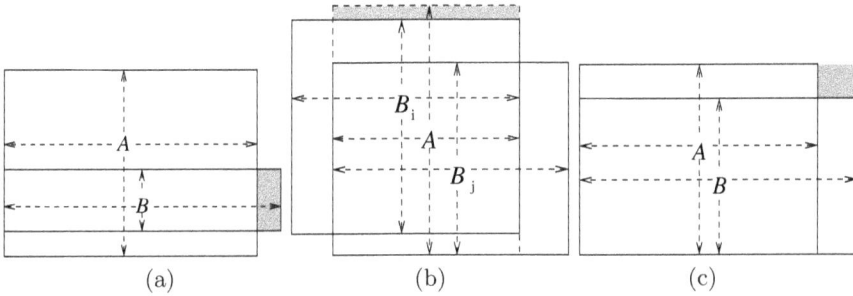

(a)   (b)   (c)

Fig. 7.12   Three pruning or merging cases (Figures is from [328] with permission from ACM).

TRICLUSTER will either delete or merge the clusters when any of the following three overlap conditions are met:

(1) For any two clusters A and B, if $L_A > L_B$, and if $\frac{|L_{B-A}|}{|L_B|}$ is smaller than a parameter η, then delete B. As illustrated in Figure 7.12(a), this will result in the deletion of the cluster with the smaller span (B) if that cluster contains only a few extra elements.
(2) This is a generalization of case 1. For a cluster A, if there exists a set of clusters $\{B_i\}$ such that $\frac{|L_A - L_{\cup_i B_i}|}{|L_A|} < \eta$, then delete cluster A. As illustrated in Figure 7.12(b), A is largely covered by the union of $\{B_i\}$ and can be deleted.
(3) For two clusters A and B, if $\frac{|L_{(A+B)-A-B}|}{|L_{A+B}|}$ is smaller than a parameter γ, merge A and B into one cluster $(X_A \cup X_B) \times (Y_A \cup Y_B) \times (Z_A \cup Z_B)$. This case is illustrated in Figure 7.12(c).

Zhao and Zaki [328] tested the TRICLUSTER algorithm on a yeast-cell-cycle data set containing 7,679 genes, fourteen time points, and thirteen samples [263]. With the parameter settings $mx = 50$, $my = 4$, $mz = 5$, and $\epsilon = 0.993$, the algorithm reported five tri-clusters. The biological meaning of these clusters was verified using the Gene Ontology [30]. The results showed that TRICLUSTER can find clusters with potential biological significance in genes, samples, time points, or any combination of these three dimensions.

7.5   Summary

As an emerging concept within the category of clustering techniques, pattern-based analysis has been widely applied to microarray data. Viewed through this paradigm, a gene-expression cluster is a "block" formed by a subset of genes and a subset of experimental conditions, where the genes in the block exemplify coherent expression patterns shaped by the context of the block. This approach is particularly apropos to applications in molecular biology, where only a small subset of genes typically participate in any cellular process of interest and, similarly, a cellular process typically takes place only in a subset of samples.

In general, pattern-based clustering approaches are conceptually either heuristic or deterministic. Approaches of both kinds employ some type of pattern-based cluster model to provide a quantitative description of the coherence within a cluster. These methods each offer a model-specific clustering algorithm to search the data set for pattern-based clusters.

The mining of coherent gene clusters from three-dimensional gene-sample-time microarray data sets has been discussed in detail in this chapter. Each cluster identified within these data sets contains a subset of genes and a subset of samples which are mutually coherent along the time series. These coherent gene clusters may indicate samples that correspond to particular phenotypes (e.g., diseases) and suggest candidate genes correlated to those phenotypes. The second half of this chapter was devoted to detailed presentations of two efficient algorithms, which have shown promise in mining complete sets of coherent gene clusters. Experimental results have shown that these approaches are both efficient and effective in identifying meaningful coherent gene clusters.

Chapter 8

Visualization of Microarray Data

Contributors: Daxin Jiang and Li Zhang

8.1 Introduction

As discussed in previous chapters, the primary goal of subjecting microarray data to pattern analysis is the potential of achieving a greater understanding of the data. However, this aim is often confounded by the inherent characteristics of microarray data, including structural complexities arising from the large number of variables (genes and experimental conditions), the relatively high noise ratio, and complex biological correlations among variables. Previous chapters have explored a series of numerical methods which can be employed for pattern analysis.

A radically different approach to gleaning insights from microarray data is based on the concept of information visualization. This term is loosely used to refer to a wide range of ideas, as indicated by the nearly fourteen million hits generated by a June 2005 Google search on the query "what is information visualization." The top hit provided the following definition: "A method of presenting data or information in non-traditional, interactive graphical forms. By using 2-D or 3-D color graphics and animation, these visualizations can show the structure of information, allow one to navigate through it, and modify it with graphical interactions" (dli.grainger.uiuc.edu/glossary.htm).

Here, we will use the term "visualization" to refer to an approach that enables exploration and detection of patterns and relationships in a complex data set by presenting the data in a graphical format in which the key characteristics become more apparent. Unlike conventional numerical methods such as those described in previous chapters, visualization is able to provide a direct examination of the data set being analyzed. This is especially important in the early stages of data analysis when qualitative analysis takes precedence over quantitative. In addition, an initial visual inspection of the complex data structure may greatly enhance the performance of numerical methods in the subsequent stages of analysis [97].

According to D. Keim [162], there are three aspects of visualization: explorative, confirmative, and presentative. *Explorative visualization* begins with data alone, without any preliminary hypotheses, and ends with a visualization which suggests hypotheses. *Confirmative visualization* takes as input both data and hypotheses and moves toward a visualization which confirms or rejects the hypotheses. *Presentative visualization* results in a visualization which uses an appropriate technique to present facts about the data with no hypotheses involved. Explorative and confirmative visualizations can usually be distinguished by the presence or absence of prior hypotheses regarding the data. The distinction between explorative and presentative visualization is sometimes more subtle, as it is not always readily apparent whether a particular visualization has generated hypotheses. In this chapter, we will assume that all visualizations will suggest some hypotheses and thus will be either explorative or confirmative.

An explorative visualization is closely related to the statistical concept of *exploratory data analysis* (EDA). EDA [287, 286, 288] is a strategy for data analysis that employs a variety of techniques, most of which are graphical, to achieve the following goals:

- Gain insights into a data set;
- Uncover underlying structures;
- Extract important variables;
- Detect outliers and anomalies; and
- Assist in formulating underlying assumptions [10, 208].

EDA is particularly well-suited for use with microarray data sets, which are characterized by complex data structures and poorly-understood biological mechanisms. EDA can be understood as an organizing philosophy or approach, rather than an actual technique of data analysis [208], while explorative visualization is a collection of techniques which embody the philosophy of EDA.

The key challenge of visualization is that human spatial imagination is restricted to three dimensions. Therefore, directly viewing relationships in higher dimensional spaces is beyond our capability. The task of visualization is to present high-dimensional data in a two- or three-dimensional space, using a particular arrangement to reflect important relationships among the original data points. There are many well-known techniques for visualizing multidimensional multivariate data sets[1]. Keim and Kriegel [163] divided these techniques into four general categories on the basis of the layout of the visualization:

- Icon-based techniques;
- Pixel-oriented techniques;
- Hierarchical and graph-based techniques; and
- Geometric projection techniques.

In this chapter, we will discuss several selected techniques for explorative and confirmative visualization of microarray data. These techniques will encompass examples of the four general categories of visualization layout listed above. Section 8.2 will present several simple methods, including *box plots*, *histograms*, *scatter plots*, and *gene pies*, for visualizing the measurements of a single array. Such methods are termed *single-array visualizations*. In Section 8.3, a series of more complicated approaches to visualizing an entire multiple-array gene expression data set will be discussed. These multi-array visualization approaches include *global visualization*, *optimal visualization*, and *projection visualization*. The chapter will end with the presentation in Section 8.4 of a theoretically-grounded visualization framework called VizStruct.

8.2 Single-Array Visualization

In this section, we will describe four methods (*box plots*, *histograms*, *scatter plots*, and *gene pies*) for visualizing the gene expression measurements from a single array. The first three methods are classical EDA graphics that summarize the statistical characteristics of the data set, while the gene pie is a new method developed specifically for microarray data. Viewed from another perspective, gene pies and scatter plots are representative of icon-based techniques and pixel-based techniques, respectively, while box plots and histograms do not easily fit into any of Keim and Kriegel's layout categories [163]. Nonetheless, all four methods are explorative techniques and are useful for comparing the two channels in cDNA arrays (see Section 3.3.1) or a control/sample pair of Affymetrix arrays.

---

[1]In traditional statistical society, *multidimensional* indicates independent dimensions and *multivariate* suggests correlated dimensions. Therefore, if it is not known in advance whether a particular attribute in the data is independent with others or not, the term *multidimensional multivariate information visualization* is more appropriate.

8.2.1   *Box Plot*

Box plots are simple graphical representations of several descriptive statistics, such as mean (or median) and variance, for given data sets. In general, the vertical axis of a box plot is formed by a response variable, while the horizontal axis corresponds to the factor of interest [208]. In the case of microarray data, the vertical axis represents expression measurement, while the horizontal axis may contain Cy3 and Cy5 channels (for cDNA arrays) or control and experimental samples (for Affymetrix arrays).

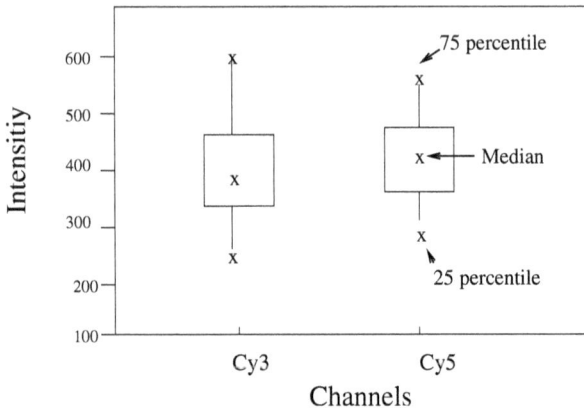

Fig. 8.1   The box plot for a representative data set. The vertical axis represents the expression levels, while the horizontal axis correspond to the channels of cDNA arrays or the Affymetrix arrays.

The box plot is characterized by a central box and two tails (Figure 8.1). To be specific, the layout is as follows [83]:

- The mark in the box indicates the position of the median (the value located halfway between the largest and smallest data value).
- The upper and lower boundaries of the box represent the locations of the *upper quartile* (UQ) and *lower quartile* (LQ), which are the 75th and 25th percentiles, respectively. Thus, the box will represent the interval that contains the central 50 percent of the data. The interval between the upper and lower quartiles is called the *interquartile distance* (IQD).
- The length of the tails is usually 1.5·IQD. Data points that fall beyond UQ+1.5·IQD or LQ−1.5·IQD are considered outliers.

The box plot is an important EDA tool for determining whether a factor has a significant effect on the response with respect to either location or variation [208]. For example, juxtaposing the distributions of measurements from Cy3 and Cy5 channels in the box plot may reveal whether the measurement of Cy3 is systematically higher or lower than that of Cy5. If this is the case, a normalization method such as those discussed in Section 3.6 may be needed to correct the data. The box plot can also assist in the identification of outliers.

8.2.2 *Histogram*

A *histogram* is a graphical summarization of the distribution of a given *univariate* data set. In essence, a histogram subdivides the range of the data into equal-sized intervals, called bins. The *frequency* of each bin B_i is then calculated by counting the number of points in the given data set that fall into B_i. The histogram plot represents the bins on the horizontal axis and the frequency of each bin on the vertical axis.

In comparison with box plots, histograms provide more detailed information about the data; the shape of a histogram (i.e., the height of the bins) is often used as an empirical estimation of the probability density function of the variable of interest. Like a box plot, a histogram reveals the mean (or median) and variance of the data, as well as outliers, but it additionally reflects the skewness of the data. For example, Figure 3.8 in Section 3.5 depicts (a) the histogram of the ratios of spot intensities in an microarray data set and (b) the histogram of the log transformed ratios. We can see that the data in Figure 3.8(a) are highly clumped, while, after log transformation, they are well dispersed. This figure guides the researcher to transform the data as a pre-processing step (see Section 3.5).

Histograms can be used as a "quality indicator" for microarray data. The Y-axis of such histograms represents the log ratios of the spot intensities measured in the two channels of cDNA arrays or the control/sample pair of Affymetrix arrays. The histogram of a quality expression data set is expected to exhibit the following two features [83]:

- The peak of the histogram is expected to be in the vicinity of value 0 if the experiment has been performed on an entire genome or on a large number of randomly-selected genes (see Figure 3.8). This reflects the expectation that most genes in an organism will remain unchanged, resulting in a preponderance of ratios around 1, or of log ratios around 0. A shift of the histogram away from these expected values might indicate the need for data preprocessing and normalization.

- The histogram is expected to be symmetrical, with a shape generally similar to that of a normal distribution (see Figure 3.8). Although the distribution can often depart from normality through unusually fat or thin tails, most such differences will not be detectable via a visual inspection. In most cases, large departures from a normal shape will indicate flaws in the data, strong artifacts, or the need for some normalization procedures.

Histograms can also guide users in selecting differentially-expressed genes in microarray data with a certain fold change (see Section 4.3). Again, the Y-axis in these histograms represents the log ratios between two channels (in a cDNA microarray) or the control/sample pair (in an Affymetrix array). In such histograms, selecting differentially-regulated genes can be simply accomplished by placing vertical bars (thresholds) on this axis and selecting the genes outside the bars. For instance, if we wish to select genes that have a fold change of two, we need only establish the bars at $+/-1$ ($log_2 2 = 1$) and select the genes outside the vertical bars (see Figure 4.1).

8.2.3 Scatter Plot

Scatter plots differ from box plots and histograms in two respects. First, unlike box plots and histograms, which provide a graphical presentation of the *statistical features* of the given data, scatter plots present the data points themselves. In fact, the scatter plot can be categorized as a "pixel-oriented technique" in that each data point corresponds to one pixel in the plot. Second, instead of characterizing univariate data as do box plots and histograms, scatter plots visualize data with two variables. In particular, a scatter plot reveals the relationships or associations between two variables in the data set.

Scatter plots have been widely used for the representation of microarray data. In these scatter plots, the two axes correspond to the two channels in cDNA arrays or the sample/control pairs in Affymetrix arrays. To facilitate the discussion of this technique, the following section will focus on only cDNA arrays.

Suppose gene g_i has an expression level x_i in the Cy3 channel and y_i in the Cy5 channel. Using the X-axis to represent the Cy3 channel and the Y-axis to represent the Cy5 channel, the point representing g_i will be plotted at coordinates $(log x_i, log y_i)$ in the scatter plot. In this way, all the genes in the data set can be plotted on the graph (see Figure 4.2(a) for an example of a scatter plot).

In a scatter plot, genes with similar expression levels in the two channels will lie on the diagonal line $y = x$; genes with different expression levels in the two channels will be located away from the diagonal. For example, using the X-axis to represent the Cy3 channel and the Y-axis to represent the Cy5 channel, genes having higher Cy3 than Cy5 values will appear below the diagonal line, while genes with the reversed situation will appear above the diagonal. Points located at greater distances from the diagonal have more significant discrepancies between the values of the two channels.

Although scatter plots using Cy3 and Cy5 intensities as coordinates can be effective in presenting gene values, it may be that the human eye and brain are better at perceiving deviations from horizontal and vertical lines rather than from diagonal lines [266]. An alternative scatter plot, the *R-I plot* (for *ratio-intensity plot*), uses the Y-axis to represent the difference between the log intensities of the Cy3 and Cy5 channels (i.e., $log\frac{Cy3}{Cy5}$) and the X-axis to represent the sum of the log intensities (i.e., $log(Cy3 * Cy5)$) [225]. In a R-I plot, a gene will lie on the horizontal line $y = 0$ if it has equal expression values in both channels(see Figure 4.2(b) for an example of a R-I plot).

In fact, the R-I plot is related to the Cy3/Cy5 plot; it can be obtained by rotating a Cy3/Cy5 plot by 45° and then scaling the two axes appropriately (see Figure 4.2(a) and (b)). Despite this simple difference, it is thought that the R-I plot is generally a more powerful tool for visualizing and quantifying the Cy3 and Cy5 channels [266].

Like histograms, scatter plots (including R-I plots) can be used as a "quality indicator" and "gene selector" [83] by observing the characteristics of the plot:

- The majority of genes are expected to fall in the vicinity of the diagonal $y = x$ (for scatter plots) or the horizontal line $y = 0$ (for R-I plots). If the plot does not exhibit this characteristic, then the data from one channel are consistently lower or higher than the data from the other channel, suggesting the need for some data pre-processing and normalization.
- Unusual plot shapes should be noted. For example, the specific shape of the data in Figure 3.9(a) in Section 3.6.2 is not an accident. Here, this shape indicates that the dyes have introduced a non-linear effect distorting the data which should be corrected through appropriate normalization procedures.

- In a scatter plot, the genes with a two-fold change will be at a distance of at least 1 from the diagonal $y = x$ ($log_2 2 = 1$). In general, given a threshold τ , the fold-change method reduces to drawing lines parallel to the diagonal at a distance $\pm log_2 \tau$ and selects the genes outside the lines (Figure 4.2(a)). Similarly, in a R-I plot, the genes with a particular fold change can be selected using horizontal lines (Figure 4.2(b)).

8.2.4   Gene Pies

The gene pie (http://www.biodiscovery.com/genesight.asp) is a visualization tool specifically designed for microarray data. It is often used to compare the gene expression data obtained from the two channels in cDNA arrays. The gene pie is an icon-based technique in the sense that each data point (gene) in the data is represented by an icon which takes the form of a pie.

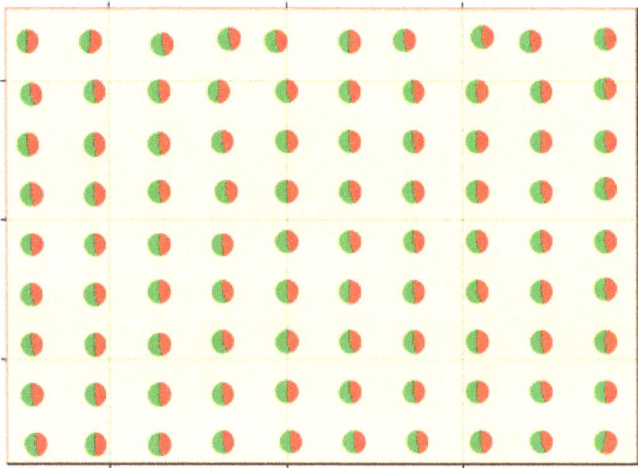

Fig. 8.2   An example of gene pies (Figure is adapted from http://www.biodiscovery. com/productcenters/images/genepie.gif).

In general, gene pies encode two types of information simultaneously: the size of a pie represents the absolute intensity, and the proportional color divisions correspond to the ratio between the Cy3 and Cy5 intensities (Figure 8.2). Although the ratio of Cy3 and Cy5 intensities is of greatest interest, since a particularly high or low ratio suggests that the gene may be up- or down-regulated, the absolute intensity values also may yield interesting information, such as [83]:

- If the intensities of the two channels are close to the background intensity, the ratio may be meaningless, as the ratio of two numbers close to zero is indeterminate.
- A gene with low intensity values may be of little biological interest, since it appears to be inactive in both mRNA samples.

For such reasons, it may be most fruitful to identify those "pies" which are of large size and contain a clearly-differentiated color division.

8.3   Multi-Array Visualization

In this section, we will discuss methods for the two-dimensional visualization of an entire microarray data set, which typically consists of a series of experiments. In particular, we will explore three methodologies which take different approaches to the treatment of dimension information: *global visualization*, *optimal visualization*, and *projection visualization*.

The visualization technologies will be illustrated in application to an iris data set which has been widely employed to assess discriminant and cluster-analysis algorithms. The iris data set contains 150 observations of four attributes (sepal length and width, petal length and width) from three iris species (50 observations each of iris Setosa, iris Versicolor, and iris Virginica). This data set was the result of the foundational work of Fisher [99].

8.3.1   *Global Visualizations*

Global visualization technologies simply encode each dimension uniformly according to a consistent visual clue. These technologies employ a system of *parallel coordinates* [103, 145, 146, 298], where the axes of a multidimensional space are defined with parallel vertical lines. A poly-line represents a point in Cartesian coordinates (Figure 8.3A). In a variation on this system, *circular parallel coordinates*, the axis radiates from the center of a circle and extends to the perimeter (Figure 8.3B). Unfortunately, the depiction of high-dimensional data sets via parallel coordinates results in a confusing display with many overlapping lines. Similarly, the radiating axes used in circular parallel coordinate depictions do not depict these data sets well.

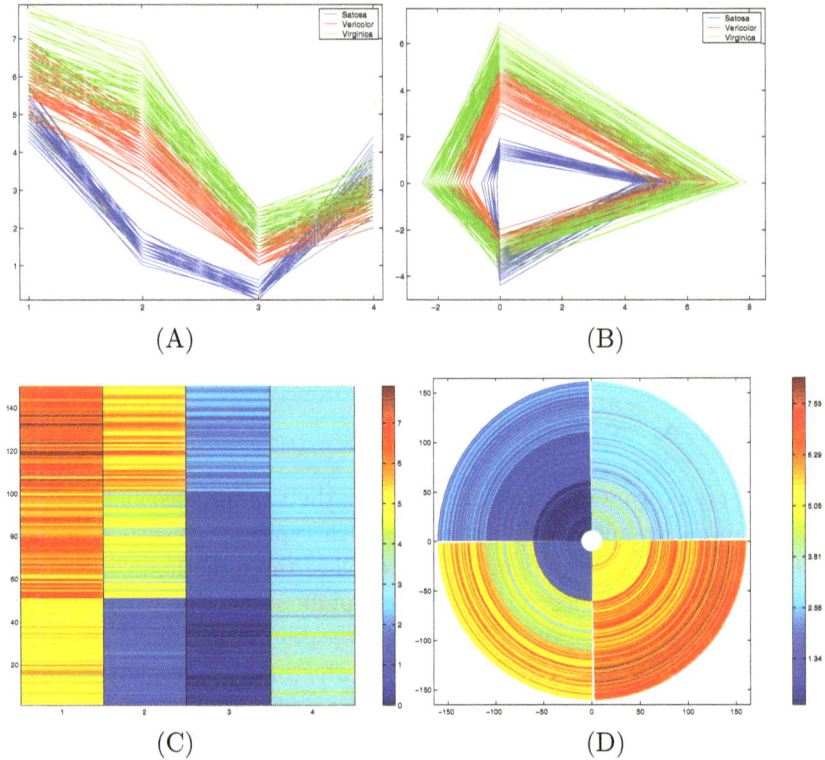

Fig. 8.3 Global visualization techniques on 3-class 150 × 4 iris data by Fisher. (A) Parallel coordinates. (B) Circular parallel coordinates. (C) Heat maps. (D) Circle segments.

Heat maps, in contrast, encode the dimensional values by a color map and arrange them as a flat graph. This precludes the overlapping lines which clutter parallel-coordinate depictions (Figure 8.3C). The heat map corollary to circular parallel coordinates is a *circle segment* [162]. Circle segments represent an entire data set as circles divided into segments, with each dimension corresponding to one segment (Figure 8.3D).

Eisen et al. [92] combine an agglomerative clustering algorithm with the heat map presentation (see Section 5.4.1). In this method (called *Tree-View*), each cell of the gene expression matrix is colored according to the measured intensity, and the rows and columns of the matrix are re-ordered based on the clustering result. Large contiguous patches of color represent groups of genes that share similar expression patterns over multiple conditions.

Figure 8.3 illustrates all four global visualization techniques. Global visualizations directly display all dimensional information and involve little or no computation. However, all these methods perform poorly with large data sets; parallel coordinates suffer from overlapping lines, and heat maps or circle segments require a large amount screen space to display. Also, the latter two methods require data to be sorted and similar data aggregated to create the desired visual effect.

8.3.2 *Optimal Visualizations*

Unlike global visualizations, other visualization techniques seek to use a point in two or three dimensions to represent an original high-dimensional point. Optimal visualizations consist of methods which estimate the parameters and assess the fit of various spatial-distance models to proximity data. Multidimensional scaling (MDS) [66] is a typical example of such a method. Of particular interest is Sammon's mapping technique [237], a classical form of MDS which maps high-dimensional data onto a two-dimensional space in attempt to replicate the relative Euclidean distance between the original points. For example, following stress function $\mathcal{E}$:

$$\mathcal{E} = \frac{1}{\sum_i \sum_{j<i} d_{ij}^*} \sum_i \sum_{j<i} \frac{(d_{ij}^* - d_{ij})^2}{d_{ij}^*}, \tag{8.1}$$

where d_{ij}^* is the distance between points i and j in the N-dimensional space and d_{ij} is the distance between i and j in the visualization. An illustration of Sammon's mapping for the iris data set is provided in Figure 8.4(A) and (B).

   Optimization of certain functions is a common feature of these methods. One advantage of optimal visualization is that concise layout and labeling of an individual data point is possible (Figure 8.4(B)). The visualization layout can optimize results while preserving the relative relationships in the original high-dimensional space. However, these methods do have several disadvantages. A heavy computational effort is involved, and the addition of new data points may require a rerun of the entire process. Each run is likely to generate a different display layout. Finally, the exact location of a data point cannot be predicted prior to visualization.

(A)

(B)

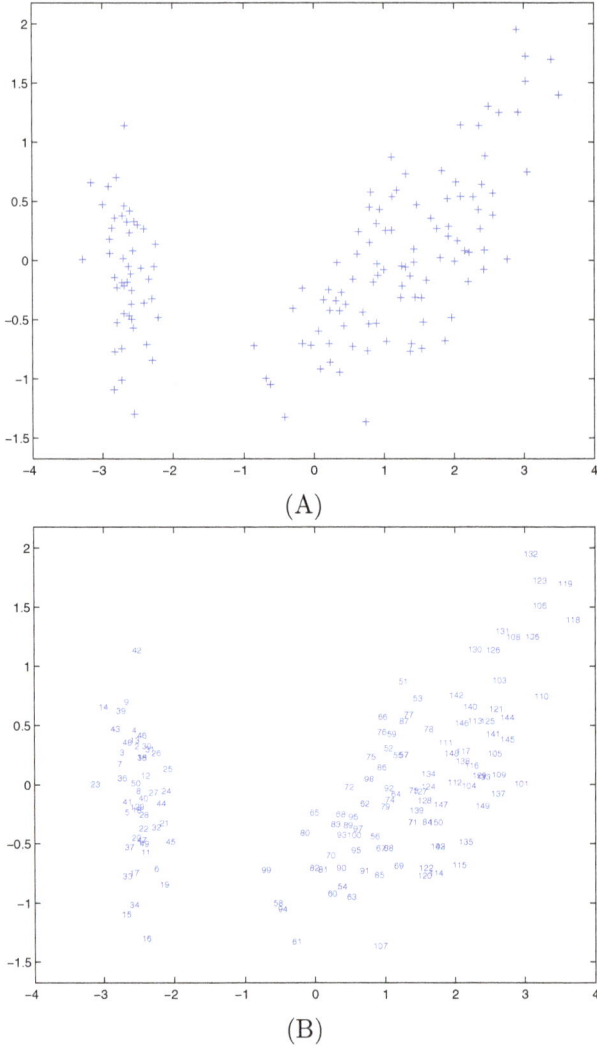

Fig. 8.4   Sammon's mapping of three-class 150 × 4 iris data from Fisher. (A) Each plus symbol represents a data point. (B) Each number represents a data point's number in the data set.

8.3.3   *Projection Visualization*

An alternative to optimal visualization is to use simple projection functions to achieve a low-dimensional (two- or three-dimensional) display. Unlike optimal visualizations, these projections produce a low-dimensional image

Table 8.1   Brief summary of advantages and disadvantages of three classes of visualization methods: global visualizations, optimal visualizations, and projection visualizations.

	Advantages	Disadvantages
Global visualizations	display all dimensional information, no computation	severe overlapping, large space to display
Optimal visualizations	achieve optimal result, sound theoretical basis	lack user interaction, heavy computation
Projection visualizations	concise display, little computation	lack rigorous proof, may not be optimal

with far less computation at the expense of possible non-optimal results. Recently, a group of methods called *radial coordinate visualization* has been developed; these include RadViz [137, 138], star coordinates [158], and VizCluster [26]. These methods share a common underlying concept, referred to as the "spring paradigm."

The spring paradigm is similar in spirit to circular parallel coordinates, in which N lines emanate radially from the center of the circle and terminate at the perimeter, where there are special endpoints. The equally-spaced points are called dimensional anchors. One end of a "spring" is attached to each dimensional anchor, while the other end is attached to a data point. The spring constant K_i has the value of the i-th coordinate of the data point. The data point values are typically normalized to have values between 0 and 1. Each data point is then displayed at the position that produces a spring force sum of 0. These methods achieve clear and useful results, but solid proof of their validity needs to be established. Figures 8.5(A) and (B) illustrate the radial coordinate visualization approach.

In summary, global visualizations, optimal visualizations, and projective visualizations are possible candidates for the visualization of gene expression data. Table 8.1 summarizes some of the advantages and disadvantages of these approaches.

8.4   VizStruct

DNA microarrays provide a broad snapshot of the state of the cell by measuring the expression levels of thousands of genes simultaneously. The visualization of this data faces a challenge in that gene expression matrices can be studied from either a sample-space or a gene-space orientation.

(A)

(B)

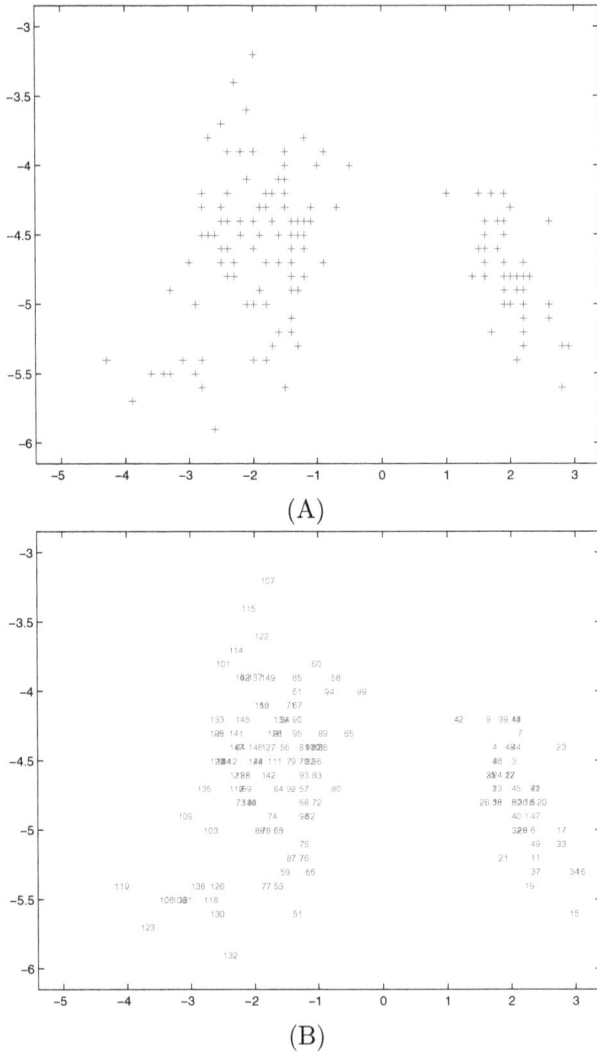

Fig. 8.5   RadViz projection of three-class 150×4 iris data from Fisher. (A) Each plus symbol represents a data point. (B) Each number represents a data point's number in the data set.

Both approaches to the data are characterized by very high and asymmetric dimensionality.

To address this challenge, the *VizStruct*[2] visualization framework pro-

---
[2]With permission from Oxford University Press.

posed in [327] re-envisions a two-dimensional visualization as a series of two-point characterizations of discrete-time signals. VizStruct also approaches information visualization through the frequency-domain representation of signals. *Fourier harmonic projections* (FHPs) are used to map the high-dimensional gene expression data onto a two-dimensional space, resulting in an intuitive scatter plot.

The VizStruct framework can generate either explorative or confirmative visualizations, as appropriate to the data under examination. In the explorative visualization process, the focus is on the sample space. The subsequent sections will present an interactive explorative visualization approach which classifies samples on the basis of gene expression variations. In contrast, in the confirmative visualization process, the gene space is the area of concentration. In this context, Fourier harmonic projections are applied in conjunction with global visualization techniques to investigate the structure of the data set, especially time-series data sets, to find patterns, outliers, and potential clusters.

The remainder of this chapter will be dedicated to an in-depth discussion of the VizStruct method. The application of Fourier harmonic projections will be explored in Section 8.4.1, and the properties of those projections will be explicated in Section 8.4.2. The application of FHPs to the sample and gene dimensions will be presented in Sections 8.4.4 and 8.4.5, respectively.

8.4.1 *Fourier Harmonic Projections*

8.4.1.1 *Discrete-time signal paradigm*

Given a data object in a N-dimensional data set $\mathcal{D}$, let's take a closer look when this data object is treated as a N point discrete time signal.

Definition 8.1 Discrete-time signal. A discrete-time signal is an indexed sequence of numbers denoted by $\mathbf{x}[n]$, where integer n is called a discrete time index ranging from $-\infty$ to ∞. Each term (or point) in $\mathbf{x}[n]$ is denoted by x_n. An N-point discrete-time signal uses the canonical index $n = 0, \ldots, N - 1$ which can be obtained by integer modulo by N [46, 210].

This paradigm reformulates the concept of two-dimensional projection visualization as the identification of two-dimensional point characterizations or estimations for signals (data points). The estimator should capture as fully as possible the characteristics of the original data, observing the following two rules:

Table 8.2 Mathematical notations and conventions used throughout this section.

Symbols	Definitions
N	number of dimensions
M	number of data points
i	imaginary unit, $i^2 = -1$
e	the base of natural logarithms
$\mathbf{x}[n], \mathbf{y}[n]$	discrete-time signals
$\mathbf{x}[n-d]$	signal $\mathbf{x}[n]$ with time point right shift by d
$\tilde{x}$	mean of signal $\mathbf{x}[n]$
x_n	n-th term of the signal $\mathbf{x}[n]$
$\mathbf{N}, \mathbf{R}, \mathbf{C}$	positive integer, real, and complex number
$\mathbf{0}$	the origin of complex plane, i.e., $0 + 0i$
$\| \; \|$	norm in complex space
$\bar{z}$	conjugate of complex number z
$[\;]$	sequence of numbers
$\mathbf{W}_N$	twiddle factor
$\mathcal{F}(\;)$	discrete-time Fourier transform function
$\mathcal{F}_k(\;)$	k-th harmonic of discrete Fourier transform

(1) Each estimator must maximize the approximation for each data point; and

(2) The visualization layout as a whole should reflect the original structure of the data.

The discrete-time paradigm for multidimensional multivariate information visualization can be summarized thus: For an N-dimensional data set, each data point is treated as an N-point discrete-time signal. Creating a two-dimensional projection visualization is considered equivalent to seeking a two-valued characterization for the N-point signal. The intent of this method is to identify appropriate candidates for this process.

Table 8.2 summarizes the descriptive notations and conventions used throughout this section. Unless otherwise indicated, all signals referred in the section are real-valued discrete-time signals[3].

8.4.1.2 *The Fourier harmonic projection algorithm*

The frequency-domain representation of signals (both continuous and discrete) provides an useful analysis and design tool. The most prominent frequency-domain representation is the Fourier transformation proposed

---

[3]Signals are allowed to take complex values. In this section, we only consider real-valued signals.

by Jean-Baptiste Joseph Fourier (1768-1830). The Fourier transformation, in essence, decomposes or separates a waveform or function into sinusoids of different frequency which sum to the original waveform.

Traditional, non-digital signals are continuous, existing in real time and with real amplitude. Due to constraints associated with the computer-encoding of information, a digital signal must be represented by a finite number of bytes. As a result, both the time and amplitude axes must be quantized, producing discrete-time signals.

The discrete-time Fourier transformation (DFT) [201] is a Fourier transformation tailored for use with discrete-time signals.

Definition 8.2 Discrete-time Fourier Transform. The discrete-time Fourier transform of an N-point signal $\mathbf{x}[n]$ is a frequency sequence with N complex values: $\mathcal{F}(\mathbf{x}[n]) = [\mathcal{F}_k(\mathbf{x}[n])]$, where each

$$\mathcal{F}_k(\mathbf{x}[n]) = \sum_{n=0}^{N-1} x_n \mathbf{W}_N^{nk}, \qquad k = 0, 1, \dots, N-1. \tag{8.2}$$

$\mathbf{W}_N = e^{-i2\pi/N}$ is termed a *twiddle factor*.

The discrete-time Fourier transformation is a transformation from an N-point of a discrete-time signal to the N-point of frequency-domain sample. The computational complexity of DFT is $O(N \log N)$ using the fast Fourier transform (FFT) or the Cooley-Tukey algorithm [63]. Each term in the discrete-time Fourier transform is defined as a *harmonic*.

Definition 8.3 Harmonics of DFT. Each $\mathcal{F}_k(\mathbf{x}[n])$ in Eq. (8.2) is referred as k-th harmonic of the discrete-time Fourier transform of the signal $\mathbf{x}[n]$, $k = 0, 1, \dots, N-1$. In particular, $\mathcal{F}_0(\mathbf{x}[n])$ is referred to as the zero harmonic or average and it is always a real number. $\mathcal{F}_1(\mathbf{x}[n])$ is referred to as the first harmonic. It is a complex number. Similarly, $\mathcal{F}_2(\mathbf{x}[n])$ is referred to as the second harmonic, and so forth.

Example 8.1 Let a signal $\mathbf{x}[n] = [0, 1, 2, 3]$, then $\mathcal{F}(\mathbf{x}[n]) = [6, -2 + 2i, -2, -2 - 2i]$. The zero harmonic is 6, and the first harmonic is $-2 + 2i$. By Eq. (8.2), $\mathcal{F}_1(\mathbf{x}[n]) = x_0 W_4^0 + x_1 W_4^1 + x_2 W_4^2 + x_3 W_4^3 = x_0(0) + x_1(-i) + x_2(0) + x_3(i) = 0 - i - 2 + 3i = -2 + 2i$. The remaining harmonics can be obtained similarly.

The discrete-time Fourier transformation (DFT) can be understood by considering the signal to be the sum of superposed sinusoid signals whose frequencies are the integer multiplication of a specific base frequency. Each harmonic $\mathcal{F}_k$ in the DFT is a measure of the kth sinusoidal frequency component in the signal. The first harmonic measures the base frequency component. The second harmonic measures a frequency double that of the base frequency, with subsequent harmonics measuring the pertinent frequencies. Fourier harmonics are typically complex numbers and thus can serve to provide the two-dimensional point estimate for mapping a multi-dimensional signal. For this reason, this mapping process can be referred to as the generation of Fourier harmonic projections (FHPs). In particular, we are interested in the projection derived by the first Fourier harmonic.

Definition 8.4   The First Harmonic of DFT. $\mathcal{F}_1(\mathbf{x}[n])$ is referred to as the first harmonic of the discrete-time Fourier transform of the signal $\mathbf{x}[n]$ or simply the first harmonic. $\mathcal{F}_1(\mathbf{x}[n])$ is a complex number

$$\mathcal{F}_1(\mathbf{x}[n]) = \sum_{n=0}^{N-1} x_n \mathbf{W}_N^n = \sum_{n=0}^{N-1} x_n e^{-i2\pi n/N}. \tag{8.3}$$

Definition 8.5   FHP, FFHP. The projection functions using the k-th harmonic of DFT in Eq. 8.2 are termed the k-th Fourier harmonic projections, or FHPs. A projection using the first harmonic of the DFT, as illustrated in Eq. 8.3, is termed the first Fourier harmonic projection, or FFHP.

The complex numbers representing the individual terms of the first Fourier harmonic projection in Equation 8.3 can be expressed in terms of magnitude r and phase θ to provide a useful geometric interpretation of the mapped projection. For a data point with N dimensions, the complex exponential divides a unit circle centered at the origin of the complex plane into N equally-spaced angles. The value of the first dimension is projected onto the radial line corresponding to $\theta = 0$. Similarly, the value of the kth dimension is projected onto the radial line corresponding to the $\theta = 2\pi(1 - k)/N$ radians. The overall two-dimensional FFHP mapping is the complex sum of all N projections from a data point. Figure 8.6 illustrates the geometric interpretation for a point containing six dimensions.

In this section, all figures representing FFHP visualizations are two-dimensional scatter plots. The x-axis is labeled $Re[F1]$ and represents the real component of the first Fourier harmonic, while the y-axis is labeled $Im[F1]$ and represents the imaginary component of the first Fourier harmonic.

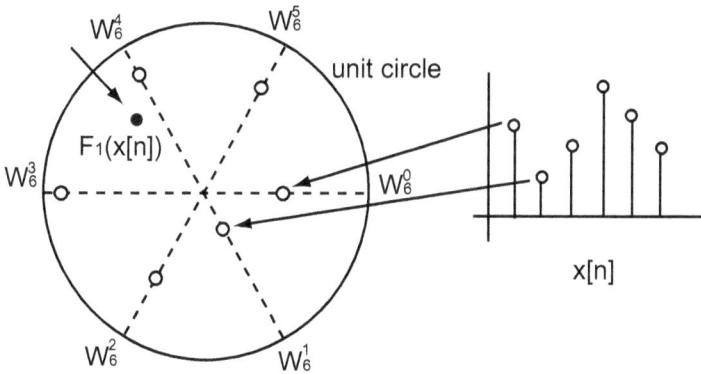

Fig. 8.6 Geometric interpretation of the first Fourier harmonic projection. A normalized six-dimensional data point is shown on the right by the stem plot. The "twiddle factor" divides the unit circle centered at the origin into six equal angles, and each dimension of the data point is projected onto a different radial angle (open circle). The projections are summed to give a two-dimensional image (solid dot).

8.4.2 Properties of FHPs

The ensuing subsections will provide a discussion of the properties of FHPs. First, operations upon signals and related properties will be introduced. We will then demonstrate that any higher k-th harmonic ($k > 1$) projection is equivalent to the first harmonic of the original signal being reordered.

8.4.2.1 Basic properties

Definition 8.6 Signal Operations. Let two signals be $\mathbf{x}[n] = [x_0, \ldots, x_{N-1}]$, $\mathbf{y}[n] = [y_0, \ldots, y_{N-1}]$, and $a \in \mathbf{R}$. (1) Amplitude shifting, $\mathbf{x}[n] + a = [x_0 + a, \ldots, x_{N-1} + a]$. (2) Amplitude scaling, $a\,\mathbf{x}[n] = [ax_0, \ldots, ax_{N-1}]$. (3) Sum/difference, $\mathbf{x}[n] \pm \mathbf{y}[n] = [x_0 \pm y_0, \ldots, x_{N-1} \pm y_{N-1}]$. (4) Conjugate, $\overline{\mathbf{x}}[n] = [\overline{x}_0, \ldots, \overline{x}_{N-1}]$. (5) Right shifting by d-unit incremental is by replacing time index n with $n - d$, denoted as $\mathbf{x}[n - d]$. Similarly, left shifting is denoted by $\mathbf{x}[n+d]$. (6) Transposing a given signal is by replacing its time index n with $-n$, denoted by $\mathbf{x}[-n]$.

A discrete-time signal system is by definition "cyclic by modulo N." A canonical index can be obtained after shifting or transposing by performing modulo N on time indices.

Example 8.2   Let a signal $\mathbf{x}[n] = [0, 1, 2, 3]$, then $\mathbf{x}[n-1] = [3, 0, 1, 2]$, $\mathbf{x}[n-2] = [2, 3, 0, 1]$, and $\mathbf{x}[-n] = [0, 3, 2, 1]$. To see why, the original sequence of $\mathbf{x}[n]$ is $[x_0, x_1, x_2, x_3]$ and the right shifting sequence of $\mathbf{x}[n-1]$ becomes $[x_{-1}, x_0, x_1, x_2]$. Performing time index modulo by 4 results in $[x_3, x_0, x_1, x_2]$.

Proposition 8.1   Constant Cancellation. *An N-dimensional point with equal dimension values will be mapped onto the origin. If* $\mathbf{x}[n] = [a, \ldots, a]$, *then* $\mathcal{F}_1(\mathbf{x}[n]) = \mathbf{0}$.

Proposition 8.2   Amplitude Shifting. *Two N-dimensional points with constant amplitude shifting will be mapped onto the same point. If* $\mathbf{y}[n] = \mathbf{x}[n] + a$, *then* $\mathcal{F}_1(\mathbf{y}[n]) = \mathcal{F}_1(\mathbf{x}[n])$.

Proposition 8.3   Amplitude Scaling. *Two N-dimensional points with constant amplitude scaling will be mapped onto two points on a line through the origin. If* $\mathbf{y}[n] = a\,\mathbf{x}[n]$, *then* $\mathcal{F}_1(\mathbf{y}[n]) = a\,\mathcal{F}_1(\mathbf{x}[n])$.

Figure 8.7(A) below illustrates the amplitude-shifting and scaling properties. Figure 8.7(B) illustrates the time-shifting property. Propositions 8.2 and 8.3 are special cases of of a more general proposition.

Proposition 8.4   Homomorphism. *The first Fourier harmonic projection is homomorphic, i.e.,* $\mathcal{F}_1(a\,\mathbf{x}[n] + b\,\mathbf{y}[n]) = a\,\mathcal{F}_1(\mathbf{x}[n]) + b\,\mathcal{F}_1(\mathbf{y}[n])$.

One consequence of Proposition 8.4 is that the first Fourier harmonic projection preserves linear property of the original space; two data points with linear correlation will be mapped onto a line.

Theorem 8.1   Line. *Under the first Fourier harmonic projection, an N-dimensional line will be mapped onto a two dimensional line.*

8.4.2.2   Advanced properties

Proposition 8.5   Time Shifting. *Two N-dimensional points with constant time shifting will be mapped onto a circle concentric to the unit circle. The angle between two images is* $d \cdot 2\pi/N$. *If* $\mathbf{y}[n] = \mathbf{x}[n-d]$, *then* $\mathcal{F}_1(\mathbf{y}[n]) = \mathcal{F}_1(\mathbf{x}[n])W_N^d$.

Proposition 8.6   Transposing. *Two N-dimensional points transposing each other will be mapped onto two points conjugate each other, i.e., symmetric to the real axis. If* $\mathbf{y}[n] = \mathbf{x}[-n]$, *then* $\mathcal{F}_1(\mathbf{x}[-n]) = \overline{\mathcal{F}_1(\mathbf{x}[n])}$.

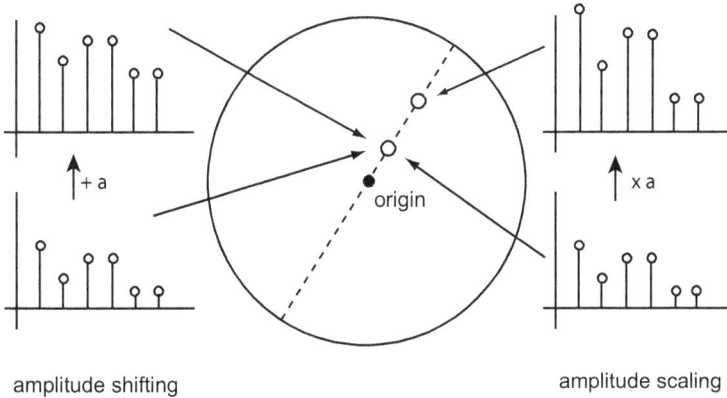

amplitude shifting   amplitude scaling

(A)

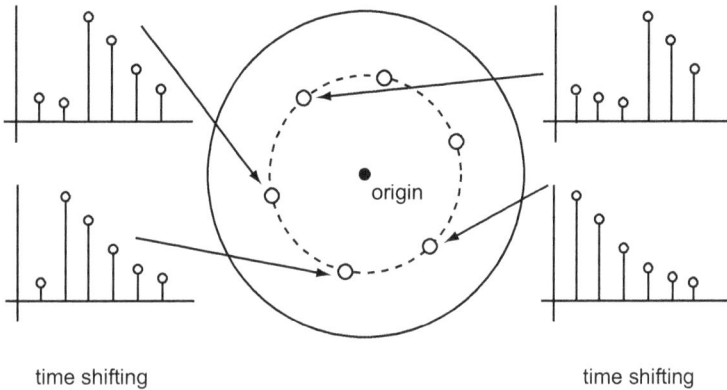

time shifting   time shifting

(B)

Fig. 8.7   Illustration of the effect of adding a constant, scaling each value by a constant factor, and time-shifting. (A) illustrates the effects of adding a constant to each dimension value (left) and of scaling each value by a constant factor (right). The two data points differ by the constant mapping to the same point in the visualization. In the visualization, the scaled points lie on a radial line. (B) illustrates the effect of time-shifting, a circular shift of each dimension value to the right. The four data points are each shifted by one dimension and are mapped to a circle concentric to the unit circle.

The next property, termed *periodic cancellation*, is not intuitively obvious. A data point is said to have a complete periodic pattern if its dimensional values are the concatenation of one or more identical subsequences. Data points such as $(1, 2, 1, 2)$, $(1, 2, 1, 2, 1, 2, 1, 2)$, or $(1, 3, 2, 1, 3, 2, 1, 3, 2)$ will be mapped onto the origin. Proposition 8.1 is a special case with only one subsequence.

Proposition 8.7   Periodic Cancellation. *An N-dimensional point with dimension values having a complete periodic pattern will be mapped onto the origin.*

8.4.2.3 Harmonic equivalency

The preceding discussion has focused exclusively on the first Fourier harmonic projection. The first projection is of particular significance, due to the property of *harmonic equivalency*. It can be shown that for any harmonic (> 1), there exists an equivalent first harmonic resulting from a proper reordering of the original discrete signal. The proof of this property is based on the concept of the *harmonic twiddle power index*.

Definition 8.7   Twiddle Power Index. For an N-point signal, the k-th harmonic twiddle power index (HTPI in short) is a permutation of the N time indices from $0, \ldots, N-1$. It corresponds to the sequence that a particular time index mapped on the ascending sorted powers of twiddle factor $\mathbf{W}_N^0, \ldots, \mathbf{W}_N^{N-1}$, by the k-th harmonic.

Example 8.3   Given an arbitrary 5-point signal, the first harmonic twiddle power index (HTPI) is $[0, 1, 2, 3, 4]$, the second HTPI is $[0, 3, 1, 4, 2]$, and the third HTPI is $[0, 2, 4, 1, 3]$. Take a closer look at the second harmonic twiddle power index. Since $\mathcal{F}_2(\mathbf{x}[n]) = \sum_{n=0}^{N-1} x_n \mathbf{W}_N^{2n}$, we have $\mathcal{F}_2(\mathbf{x}[n]) = x[0]\mathbf{W}_5^0 + x[1]\mathbf{W}_5^2 + x[2]\mathbf{W}_5^4 + x[3]\mathbf{W}_5^6 + x[4]\mathbf{W}_5^8 = x[0]\mathbf{W}_5^0 + x[1]\mathbf{W}_5^2 + x[2]\mathbf{W}_5^4 + x[3]\mathbf{W}_5^1 + x[4]\mathbf{W}_5^3$. Sort by the power of $\mathbf{W}_5$, we have $\mathcal{F}_2(\mathbf{x}[n]) = x[0]\mathbf{W}_5^0 + x[3]\mathbf{W}_5^1 + x[1]\mathbf{W}_5^2 + x[4]\mathbf{W}_5^3 + x[2]\mathbf{W}_5^4$. The sequence of the time index $[0, 3, 1, 4, 2]$ is the second HTPI.

The harmonic twiddle power index (HTPI) is always a permutation of $0, \ldots, N-1$ even if certain harmonics do not use all powers of $\mathbf{W}_N$ for the calculation.

Example 8.4   Let $N = 4$, $\mathcal{F}_2(\mathbf{x}[n]) = x[0]\mathbf{W}_4^0 + x[1]\mathbf{W}_4^2 + x[2]\mathbf{W}_4^0 + x[3]\mathbf{W}_4^2 = x[0]\mathbf{W}_4^0 + x[2]\mathbf{W}_4^0 + x[1]\mathbf{W}_4^2 + x[3]\mathbf{W}_4^2$. The second harmonic twiddle power index is $[0, 2, 1, 3]$.

The relationship between the k-th harmonic of the original signal and the first harmonic of the reordered signal can be readily determined.

Theorem 8.2   Harmonic Equivalence. *Any k-th harmonic of a signal ($1 < k < N$) is equivalent to the first harmonic of the original signal whose time index be rearranged by the k-th harmonic twiddle power index.*

Owing to the properties of symmetry, over half of all harmonic projections do not have a distinct layout.

Proposition 8.8 Symmetry of Harmonics. *$\mathcal{F}_k(\mathbf{x}[n])$ and $\mathcal{F}_{N-k}(\mathbf{x}[n])$ are conjugate each other. In other words, they are symmetric to the real axis, $\mathcal{F}_k(\mathbf{x}[n]) = \overline{\mathcal{F}_{N-k}(\mathbf{x}[n])}$.*

8.4.2.4 *Effects of harmonic twiddle power index*

Each harmonic is actually a measurement of the corresponding frequency component in the signal. The "shape" of a signal determines the contribution of its harmonics. It is intuitively obvious that a very flat signal will be represented by a zero harmonic, while a signal with two cycles (two peaks) is evidence of the dominance of the second harmonic. The reordering of time indices effectively rearranges the order of the dimension of data objects. This reordering will reshape the signal and thus redistribute the contribution of its harmonics.

A theorem advanced by Parseval states that the total energy of a signal will be preserved through a discrete-time Fourier transformation [199]:

$$\sum_{n=0}^{N-1} |x[n]|^2 = \frac{1}{N} \sum_{k=0}^{N-1} \|\mathcal{F}_k(\mathbf{x}[n])\|^2. \tag{8.4}$$

Since $\sum_{n=0}^{N-1} |x[n]|^2$ is invariant to the order of n, Parseval's theorem suggests that signal's total sum of harmonic norms is fixed regardless of the arrangement of its time indices. The arrangement of time indices only affects the proportion of each harmonic.

Harmonic equivalency suggests that higher harmonics can be considered as a systematical "reshuffling" of the dimensions rather than a physical rearrangement. Figure 8.8 offers an insightful view of the effects of the second harmonic twiddle power index. In Figure 8.8A, a 25-point signal simulating $\cos(t)$ for $t = [0, 4\pi]$ is depicted by the blue curve with circular points. This signal has a dominant second harmonic component, which is depicted in Figure 8.8B. A reordering of this signal according to the second harmonic twiddle index results in a signal resembling $\cos(t/2)$; this is depicted as a red curve with star-shaped points. This newly-reordered signal is dominated by the first harmonic, as indicated in Figure 8.8C. Since $\cos(t) = \sin(t + \pi/2)$, similar relationships hold true for $\sin(t)$.

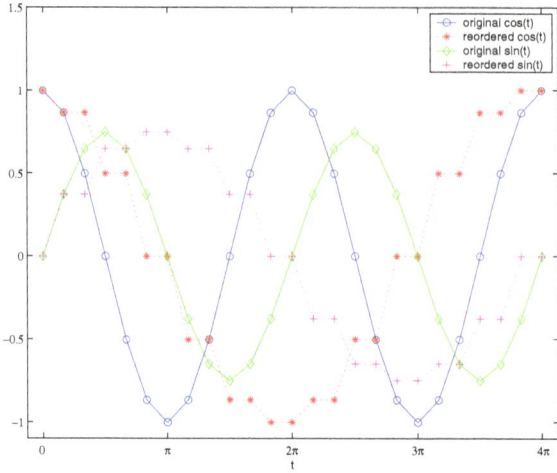

(A)

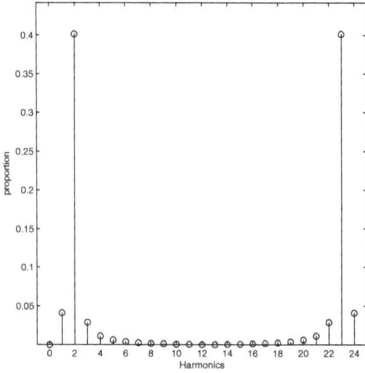

(B)

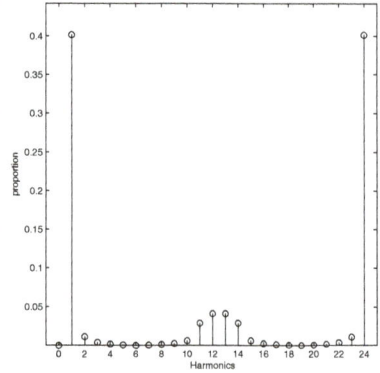

(C)

Fig. 8.8   The effect of the second harmonic twiddle power index. (A) Two signals $\cos(t)$ and $\sin(t)$ for $t = [0, 4\pi]$ before and after reordering by the second harmonic twiddle power index. (B) The proportion of each of the harmonic for the original $\cos(t)$ signal; the second harmonic is the dominant component. (C) The proportion of each of the harmonics after the $\cos(t)$ signal has been reordered by the second harmonic twiddle power index; the first harmonic is dominant.

Figures 8.8B and C provide a concise view of the contribution made by each harmonic in a single signal. The harmonic distribution for a set of signals, termed the *harmonic spectrum*, can be arrived at via a histogram of these distributions.

8.4.3 *Enhancements of Fourier Harmonic Projections*

By the harmonic equivalency demonstrated in the previous section, we may focus on the first Fourier harmonic Projection and create enhancements to give users more options for visualization presentations and also help users to gain better understanding of the data set.

Like all projection functions, possible information loss of the first Fourier harmonic projections is a major concern. Despite preserving some crucial relationships among the original data set, little individual dimension information is left by the projection. It is unlikely that a single two-dimensional projection can reveal all of the interesting relationships in the original high dimensional space. One solution is to create distortions in the visualization, i.e., to create different layouts by transforming the data. Since users are often involved in this process, it is part of the user interactions. The transformation of data usually includes scaling, rotation, and reordering. By distortions, static visualization can become dynamic.

The dimension tour is a feature that allows the user to interact with the data via dynamic animations. It is analogous to the grand tour[4] and the user interacts with the visualization by changing the dimension parameter w_n, for each dimension in the following variation of the first Fourier harmonic projection:

$$\mathcal{F}_1(\mathbf{x}[n]) = \sum_{n=0}^{N-1} w_n\, x[n] \mathbf{W}_N^n = \sum_{n=0}^{N-1} w_n\, x[n] e^{-i2\pi n/N}. \tag{8.5}$$

The default value for each w_n is 0.5 and the individual parameters can be changed over a given range either manually or systematically using the program[5]. Because no two-dimensional mapping can capture all the interesting properties of the original multi-dimensional space, two points that are close in the visualization can theoretically be far apart in the multi-dimensional space. The dimension tour, which creates animations that explore dimension parameter space, can reveal structures in the multi-dimensional input that were hidden due to overlap with other points in the visualization.

Figure 8.9 illustrates the capabilities of the dimension tour using a syn-

---

[4]The grand tour is a method for viewing multivariate statistical data via orthogonal projections onto a sequence of two dimensional subspace [12, 45, 62].

[5]One systematic approach is to choose the eigen values of corresponding dimensions calculated using PCA as the parameters to give different weights on dimensions.

(A)

(B)

(C)

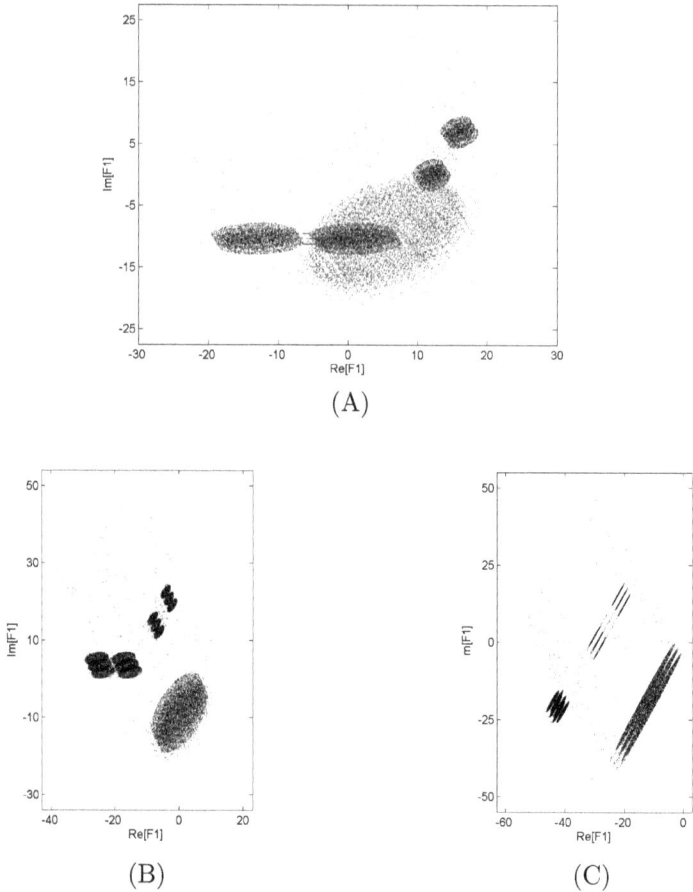

Fig. 8.9   Three snapshots from a dimension tour through a synthetic three-dimensional data set containing 25,000 points. The parameter settings for A-C were $\langle 0.5, 0.5, 0.5 \rangle$, $\langle 0.3, 0.5, 1 \rangle$, and $\langle 0.05, 1, 1 \rangle$, respectively. The x-axis is labeled $Re[F1]$ and represents the real part of the first Fourier harmonic and the y-axis is labeled $Im[F1]$ because it represents the imaginary part of the first Fourier harmonic. The units for both axes are those of the input data values.

thetic data set containing 25,000 points in three dimensions. At the default settings of dimensional parameters (Figure 8.9A), 5 clusters are apparent. However, during the course of the animations (Figures 8.9B and 8.9C), the multi-layered structures of the original 5 clusters become increasingly apparent.

8.4.4 *Exploratory Visualization of Gene Profiling*

The preceding subsections provided a thorough discussion of the technical aspects of the VizStruct (FFHP) approach to information visualization. Results from the application of VizStruct to several real data sets will be presented in this section. The iris data set [99] mentioned in Section 8.3 was deployed as an initial benchmark to obtain a preliminary assessment of the usefulness of the FFHP approach. The system was then applied to several more challenging high-dimensional microarray data sets. The visualizations discussed in this subsection are exploratory in character, as they are intended to classify samples by phenotype structure on the basis of the their genetic profiles.

8.4.4.1 *Microarray data sets for visualization*

Visualizations were generated using a multiple sclerosis (MS) data set [206] which contained two pair-wise group comparisons of interest. The first data subset, "MS vs. controls," contained array data from fifteen MS patients and fifteen age- and sex-matched controls. The second subset, "MS-IFN," contained array data obtained from fourteen MS patients prior to and 24 hours after interferon-β (IFN) treatment (see Section 6.4.3.3). The remaining gene-array data sets were obtained from the publicly-available sources listed in Table 8.3.

8.4.4.2 *Identification of informative genes*

Informative genes are genes whose expression levels vary significantly across different sample classes (see Section 6.1). Neighborhood analysis [114] (see Section 4.4.3) was used to create subsets of informative genes for the MS vs. controls, MS-IFN, and leukemia-A data sets. The significance analysis of microarrays (SAM) algorithm [289] (see Section 4.6.4) was used to identify subsets of informative genes in the remaining data sets.

8.4.4.3 *Classifier construction and evaluation*

Classifiers for these visualizations were constructed using the oblique decision tree approach, which constructs straight lines of arbitrary slope to separate classes of data [204]. The OC1 (Oblique Classifier 1) algorithm [203], which combines hill-climbing with two forms of randomization, was used to identify valid oblique splits at each node.

Table 8.3   Sources of the published data sets used.

Data Set	Size	Reference	URL from which data set was obtained
Leukemia-A	7129×72	[114]	http://www-genome.wi.mit.edu/cgi-bin/cancer/datasets.cgi
Colon	2000×62	[7]	http://microarray.princeton.edu/oncology/affydata/index.html
Colorectal	7464×36	[209]	http://microarray.princeton.edu/oncology
Liver	3964×156	[50]	http://genome-www.stanford.edu/hcc/supplement.shtm
Ovarian	3363×125	[239]	http://genome-www.stanford.edu/ovarian cancer/data.shtml
Lymphoma	2984×42	[6]	http://llmpp.nih.gov/lymphoma
Leukemia-B	7129×27	[293]	http://thinker.med.ohio-state.edu/projects/aml8/index.html
BRCA	3226×22	[129]	http://www.nhgri.nih.gov/DIR/Microarray/NEJM_Supplement/
Malenomas	8067×38	[29]	http://www.nature.com/nature/journal/v406/n6795/suppinfo/40653 6a0.html
Pancreatic	1492×36	[144]	http://genome-www.stanford.edu/pancreatic1/data.shtml
SRBCT	2308×82	[167]	http://www.nhgri.nih.gov/DIR/Microarray/Supplement/index.html
Soft–tissue	5878×46	[207]	http://genome-www.stanford.edu/sarcoma/download.shtml

The accuracy of the classifier was evaluated using the holdout and leave-one-out cross-validation techniques. The holdout technique, used for the larger data sets, involves constructing the classifiers with a subset of training samples and assessing accuracy using a mutually-exclusive subset of the test samples. The leave-one-out method was employed with the smaller data sets. Here, the classifiers were constructed using all the data points except one, which was withheld and used for testing. The process of classifier construction and testing was repeated for all the data points in the data set.

The accuracy of the classifier predictions was assessed using the overall error rate (the ratio of the number of errors to the total number of test samples) and the κ-coefficient. The κ-coefficient [59] is defined as:

$$\kappa = \frac{p_0 - p_c}{1 - p_c}. \tag{8.6}$$

Here, p_0 is the proportion of cases in which the raters agree. The term p_c represents the proportion of cases for which agreement may occur by chance, assuming that the marginal distributions of ratings are independent.

8.4.4.4 Dimension arrangement

Parseval's theorem [199] states that, while the total energy of a signal will be preserved through a discrete Fourier transformation, the contribution of each harmonic may change. As the result, the first Fourier harmonic projection is sensitive to the order of dimensions in the data. This is both a weakness and a strength; it can hinder class and cluster detection but provides advantages for data sets in which the dimensions are naturally ordered, such as time-series, dose, and concentration-effect curves.

Testing of the VizStruct method using the data sets described above was designed to capitalize upon this property. Prior to application of the FFHP, the dimensions were reordered to enhance the separation of classes upon visualization. The separation is accomplished by reordering the genes according to their similarity to the predefined class pattern in a manner that makes the first harmonic more dominant. The canonical dimension-reordering algorithm employed is provided in Algorithm 8.1.

Algorithm 8.1   : Canonical Dimension-Reordering

Let data be x_{ij}, genes x_i, $i = 1, 2, \ldots, p$ and samples y_j, $j = 1, 2, \ldots, n$. The total number of sample classes is K. $C_k = \{j : y_j = k\}$ for $k = 1, 2, \ldots, K$. Let $|C_k|$ be the size of C_k, $\bar{x}_{ik} = \sum_{j \in C_k} x_{ij}/|C_k|$. Let B be the set of $n!$ sequences of all permutations of $1, \ldots, K$.

(1) For each $b \in B$, find set of genes $x^b = \{x_i | \bar{x}_{ik(1)} \leq \bar{x}_{ik(2)}, \ldots, \leq \bar{x}_{ik(K)}$ and $b = k(1), k(2), \ldots, k(K)\}$.

(2) Let $b^* = \mathrm{argmax}_{b \in B} |x^b|$. It is some permutation of $1, 2, \ldots, K$, denoted as $b_{k(1)}, \ldots, b_{k(K)}$.

(3) Create a sample class pattern q based on b^*
$$q = \{\underbrace{b_{k(1)}, \ldots, b_{k(1)}}_{|C_1|}, \underbrace{b_{k(2)}, \ldots, b_{k(2)}}_{|C_2|}, \ldots, \underbrace{b_{k(K)}, \ldots, b_{k(K)}}_{|C_K|}\}.$$

(4) For each gene x_i, compute class coefficient $r_i = \sigma_{x_i q}/\sqrt{\sigma_{x_i}\sigma_q}$, i.e., Pearson's correlation coefficient with class pattern q. Then sort r_i's.

(5) The canonical order is defined as: $i_{(1)}, i_{(2)}, \ldots, i_{(p)}$ where $r_{i(1)} \leq r_{i(2)}, \ldots, \leq r_{i(p)}$.

The canonical dimension-reordering algorithm is based upon the observation that the ideal informative gene for samples grouped by class will be visualized as a rectilinear wave with consistent gene expressions for each sample class. This rectilinear wave signal causes the first harmonic to be the dominant component among all harmonics, a situation illustrated in Figure 8.10.

This can be understood by assuming a case with two sample classes, one comprising nine samples and the other seven. The corresponding signal of an ideal informative gene is depicted in Figure 8.10A, and Figure 8.10B illustrates the harmonic distribution of this signal. The x-axis displays the harmonic from 0 to 15, while the complex norm of each harmonic is shown on the y-axis. The first harmonic is dominant. Figure 8.10C illustrates a visualization of the same informative gene with the sample randomly rearranged. In this instance, the first harmonic is not the dominant component, as shown in Figure 8.10D. A similar visualization of a case with three sample classes is provided in Figures 8.10 E through H.

The canonical dimension-ordering algorithm is used to reorder genes according to their similarity to a predefined sample-class pattern. The intent of this reordering is to form a more rectilinear wave and thus cause the first harmonic to be dominant.

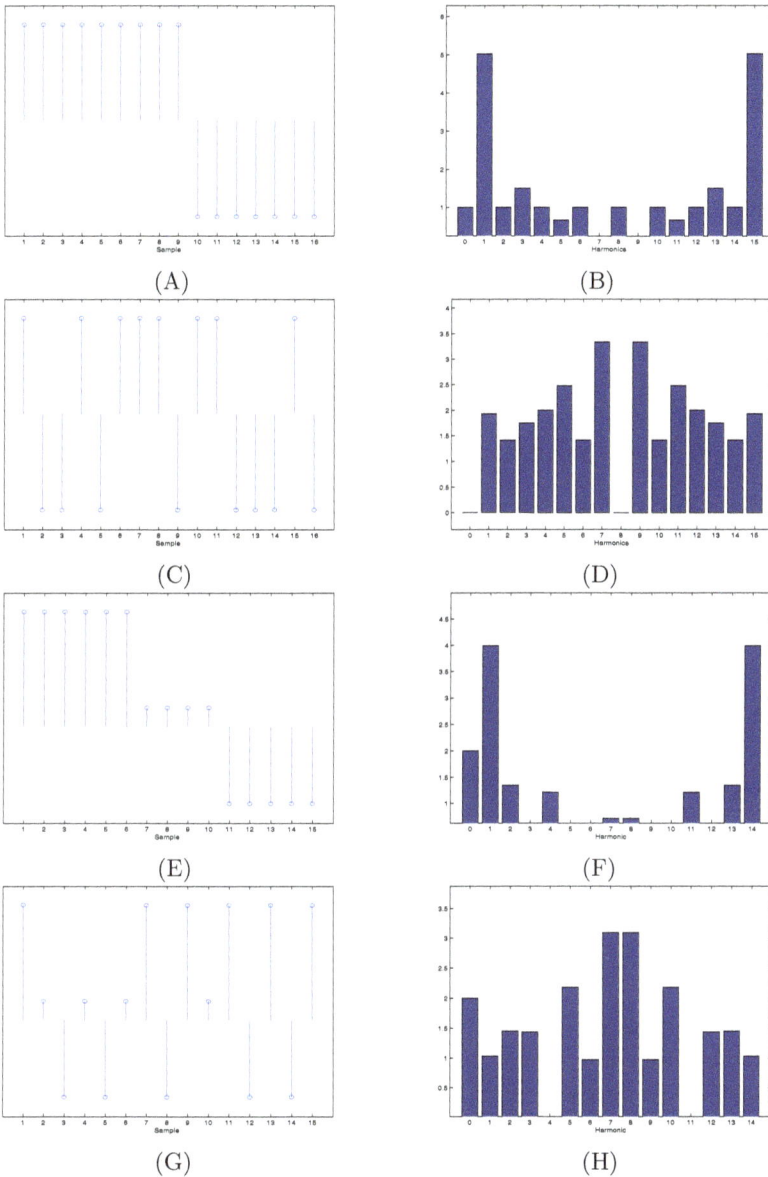

(A)

(B)

(C)

(D)

(E)

(F)

(G)

(H)

Fig. 8.10 Harmonic norm $\|\mathcal{F}_k(\mathbf{x}[n])\|$ distributions affected by the ordering of dimensions. (A) Stem plot of datum with a rectangular like wave form, (B) Harmonic norm distribution of (A), showing the first harmonic as the dominant one. (C) Same datum in (A) with random dimension ordering, (D) Harmonic norm distribution of (C), in which the first harmonic is not dominant. (E) through (H) is a three-class situation.

8.4.4.5 *Visualization of various data sets*

The iris data set contains 150 observations of four attributes (sepal length and width, petal length and width) from three iris species (50 observations each of iris setosa, iris versicolor, and iris virginica). Figures 8.11A-B show the data set with two-dimensional parameter settings. The three clusters are clearly apparent from the visualization and segregate by species. The iris setosa cluster (indicated by plus symbols) is clearly separated from the cluster formed by iris virginica (circles) and iris versicolor (stars). The separation between iris virginica and iris versicolor is more evident in Figure 8.11B than in Figure 8.11A.

These findings demonstrated the feasibility of applying FFHP mapping to array-derived gene expression profiles. We will below systematically show the visualization of two-class and multi-class data sets.

Visualization of Two-class Data Sets
Figure 8.12(A) presents the visualization results from the MS-IFN data set, which contains gene expression profiles from fourteen multiple sclerosis (MS) patients before and 24 hours after IFN treatment. Figure 8.12(B) presents the results from the MS vs. controls data set, which compares fifteen MS patients to age- and sex-matched controls. Subsets of 88 informative genes were identified from the 4,132 genes in each data set. The informative genes were mapped using VizStruct; oblique classifiers then were constructed and evaluated using the leave-one-out method (see Section 6.5.1). The pre-treatment group (indicated by plus symbols) in Figure 8.12(A) segregated to upper half of the visualization field, while the post-treatment group (circles) occupied the lower half. The results for MS patients (plus symbols) and controls (circles) in Figure 8.12(B) were generally similar. The pre-treatment and post-treatment groups in Figure 8.12(A) were clearly separated, while the separation between MS patients and controls in Figure 8.12(B) was less distinct. Table 8.4 indicates that the accuracy of the prediction was 100% (the κ-coefficient was 1.00) for the MS-IFN data set and 90% (the κ-coefficient was 0.80) for the MS vs. controls data set.

The leukemia-A data set [114] consists of expression values of 7,129 genes from 72 patients. Fifty informative genes were identified using the neighborhood-analysis approach of the original report. The training set consists of 27 acute lymphoblastic leukemia (ALL) and eleven acute myeloid leukemia (AML) patient samples, indicated by plus symbols and circles, respectively, in Figure 8.13(A). An oblique classifier was constructed based

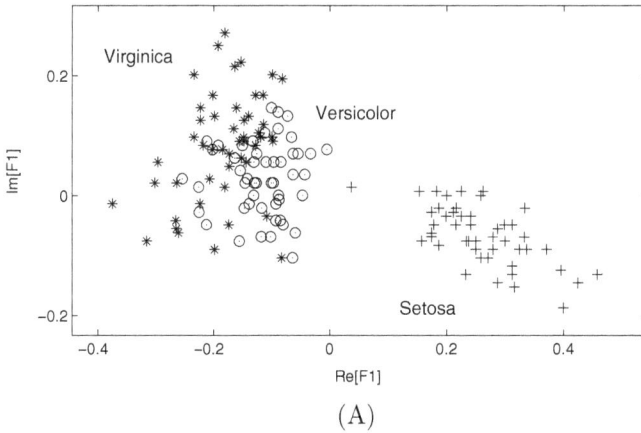

(A)

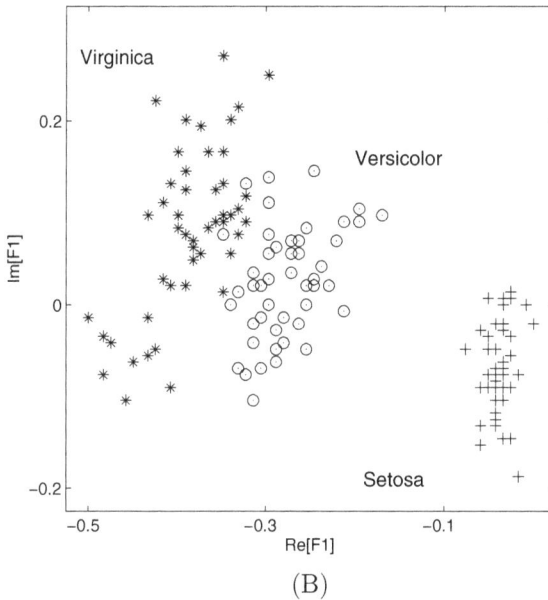

(B)

Fig. 8.11 Visualization of the iris data set generated by VizStruct. The iris setosa cluster (indicated by plus symbols) was clearly separated from the clusters formed by iris virginica (circles) and iris versicolor (stars) in both panels. The dimension parameters for (A) and (B) were $\langle 0.5, 0.5, 0.5, 0.5 \rangle$ and $\langle 0, 0.5, 0.5, 0.5 \rangle$, respectively.

on the visualization of the training set, and the holdout method was used for classifier evaluation. The testing set contained 20 ALL and fourteen AML samples (indicated by triangles and stars in Figure 8.13(A)). The

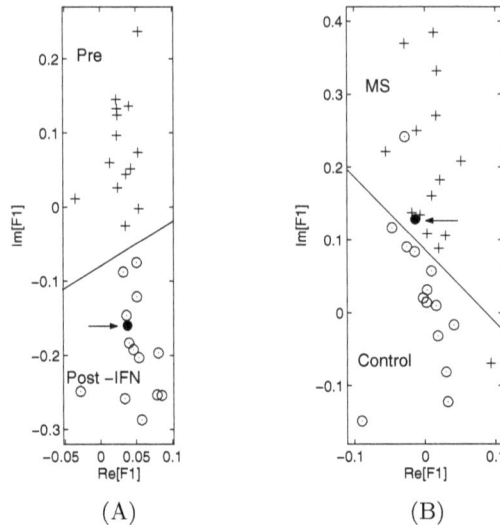

Fig. 8.12   Two-class classifications. (A) provides the results from mapping the MS-IFN data set. The pre-treatment samples are indicated by plus symbols, the post-treatment samples by circles, and a test sample is denoted by a dot highlighted with an arrow. (B) provides the results from mapping the MS vs. control data set. The MS patient samples are indicated by plus symbols, the post-treatment samples by circles, and a test sample is denoted by a dot highlighted with an arrow. On all panels, the solid dark line is the oblique classifier used. There were two misclassified samples in (B) which occurred during training and were not counted toward classifier evaluation.

overall accuracy of the classifier was 88%, and the κ-coefficient was 0.75.

The visualization of the colon cancer data set [7] consists of gene expression profiles for 2,000 genes from 22 normal samples and 39 cancerous colon tissues. The result from mapping 50 informative genes is summarized in Figure 8.13(B). The upper part of the visualization field was dominated by normal colon tissue, while cancerous colon tissues occupied the lower half. The overall accuracy was 89%, and the κ-coefficient was 0.75.

Figure 8.13(C) provides results from mapping the 100 informative-gene subset of the lymphoma data set [6]. The germinal-center lymphomas were relatively well separated from the activated B-cell-like lymphomas. The robustness of the separation contributed to 100% accuracy upon classification; the κ-coefficient was 1.00.

Visualization of Multi-class Data Sets

The performance of VizStruct in application to multi-class array-derived data sets was examined in a subsequent series of experiments. The

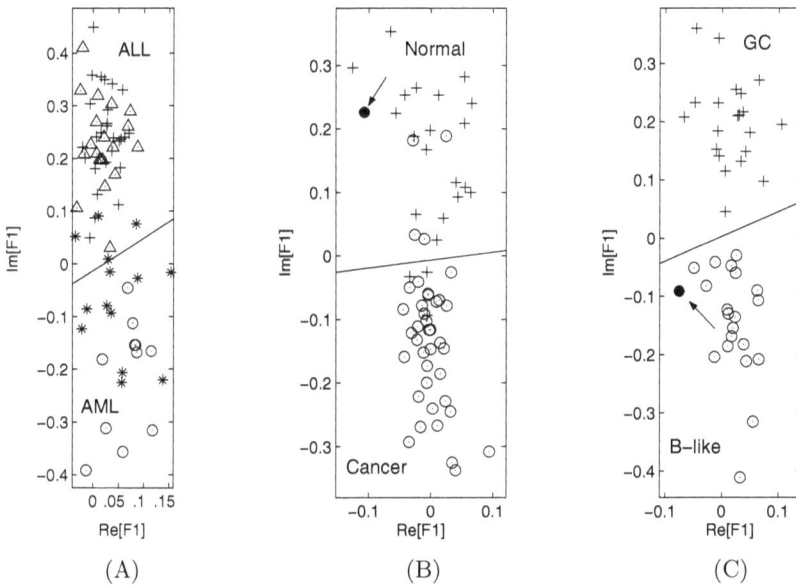

Fig. 8.13 Two-class classifications. (A) provides results from mapping the leukemia-A data set. Test samples are overlaid with training samples. The plus and circle symbols represent ALL and AML samples, respectively, during the training phase. The triangles and stars represent test ALL and test AML samples, respectively. (B) provides results from mapping the colon cancer data set. The plus and circle symbols indicate normal and cancerous colon tissue, respectively. The test cancerous sample is indicated by a dot highlighted by an arrow; it was misclassified. (C) provides results from mapping the lymphoma data set. The plus and circle symbols represent germinal center (GC) lymphoma samples and activated B cell-like lymphomas, respectively. The test-activated B cell-like lymphoma sample is indicated by a dot highlighted by an arrow; it was correctly classified.

leukemia-B data set [293] contains three classes, and a 50-gene informative subset was mapped using VizStruct (Figure 8.14A). The normal CD34+ samples segregated to the northwestern region of the visualization and were well separated from the AML samples. The samples from cytogenetically normal AML (AML-CN) patients and trisomy 8 AML (AML+8) patients also segregated, albeit less distinctly. This was reflected in the lower accuracy of 70%; the κ-coefficient was 0.55.

The FFHP mapping of a 50-gene informative subset of the BRCA data set [129] resulted in a clear separation of the BRCA1, BRCA2, and sporadic classes (Figure 8.14B).The accuracy was 100%, and the κ-coefficient was 1.00.

For the four-class SRBCT data set [167], a 100-gene informative subset

(A)

(B)   (C)

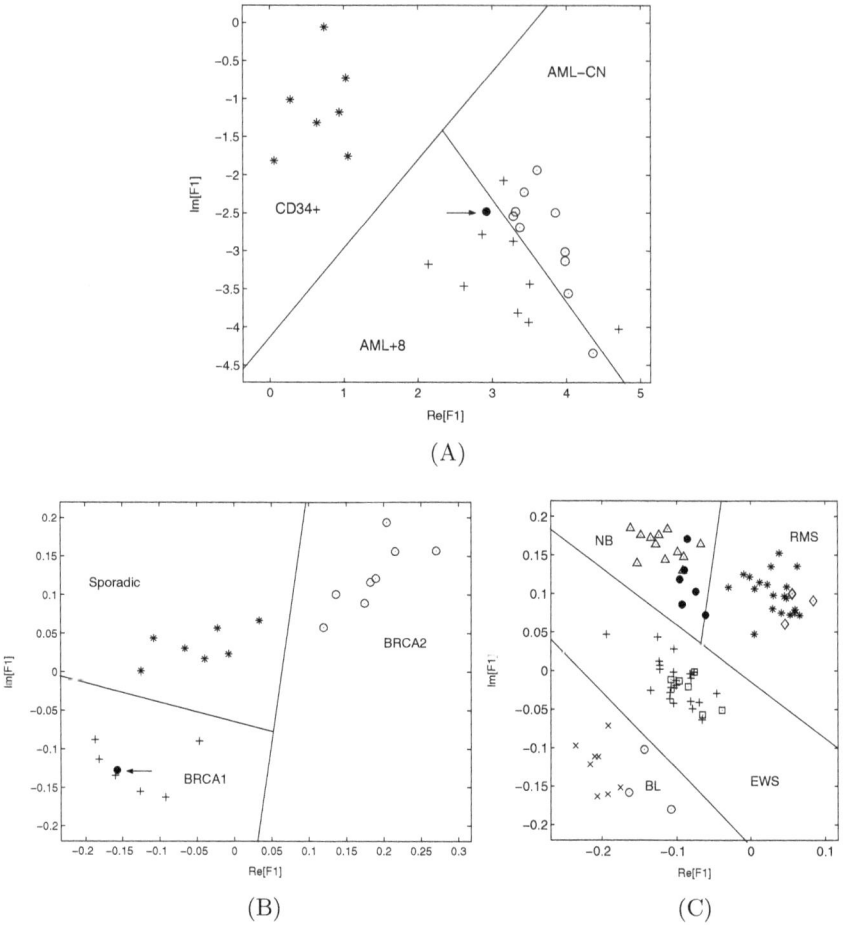

Fig. 8.14   Multi-class classifications.   (A) provides results from mapping the
leukemia-B data set.   The plus symbols are samples from AML patients with
trisomy 8 (AML+8), the circles are samples from AML patients with normal cy-
togenetics (AML-CN), and the stars are CD34+ cells from controls. There were
three training errors (1 AML-CN and 2 AML+8) that were not counted toward
classifier evaluation. (B) illustrates the mapping of the BRCA data set (plus symbols: patients with BRCA1 mutations; circles: patients with BRCA2 mutations;
stars: sporadic cases). The dots highlighted with arrows (in (A) and (B)) are a
representative test sample. (C) provides the results of mapping the informative
genes from the SRBCT data set. The triangles represent neuroblastomas (NB),
the stars are rhabdomyosarcomas (RMS), the plus symbols are Ewing's family
of tumors (EWS), and the crosses are Burkitt's lymphomas (BL). These symbols
represent the training set; for the corresponding test samples, four sets of symbols
were used: dot, diamond, square, and open circle. In all panels, the solid dark
lines represent the classifiers.

was mapped (Figure 8.14C). The four classes of small round blue cell tumors segregated to different regions of the visualization field. The accuracy of the oblique classifiers was of 95%, and the κ-coefficient was 0.93.

Table 8.4 shows the summary of results from sample classification (see [325] for all figures of experiments). As a general rule, κ values between 0.41 and 0.60 indicate moderate agreement, values in the range of 0.61-0.80 indicate substantial agreement, and values in the range of 0.81-1.0 indicate almost perfect agreement[174]. The summary in Table 8.4 indicates that, for the MS-IFN, MS-control, lymphoma, BRCA, and SRBCT data sets, almost perfect agreement was obtained using VizStruct. For the leukemia-A and colon cancer data sets, substantial agreement was obtained. However, only moderate agreement was achieved for the leukemia-B data set. Taken together, these experimental results highlight the effectiveness of the FFHP mapping technique for visualization-driven sample space classification of both binary and multi-class gene array data sets.

8.4.4.6   *Comparison of FFHP to Sammon's mapping*

As mentioned in Section 8.3.2, multi-dimensional scaling (MDS) offers an alternative approach to the visualization of multidimensional data. Tests have been performed to compare the accuracy of FFHP to Sammon's mapping method [237], a variant of MDS.

Figure 8.15A illustrates the results obtained from applying Sammon's mapping method to the iris data set. A comparison of Figure 8.15A to Figure 8.11A reveals the extensive similarities between FFHP and Sammon's mapping method. For example, the iris setosa cluster is clearly separated from the iris virginica and iris versicolor clusters, and the latter cluster is closer to the first than the second. The relative locations of individual samples are also remarkably similar (data not shown).

Figure 8.15B illustrates the results obtained from applying Sammon's mapping method to the SRBCT data set. A comparison of Figure 8.15B to Figure 8.14C reveals extensive similarities. For example, the relative locations of various tumor clusters are quite similar, and the cluster of Burkitt's lymphoma samples is distant from the rhabdomyosarcomas.

These comparisons indicate that, despite their substantial conceptual differences, the FFHP method and Sammon's mapping technique yield consistently similar mapping results.

Table 8.4 Summary of results from sample classifications.

Data Set	Classes	Genes in Classifier	Evaluation	Training Set Size	Testing Set Size	Errors Count	Accuracy	κ
MS-IFN	2	88	leave-one-out	28	28	0	100%	1.00
MS vs. Control	2	88	leave-one-out	30	30	3	90%	0.80
Leukemia-A	2	50	holdout	38	34	4	88%	0.75
Colon	2	50	leave-one-out	62	62	7	89%	0.75
Colorectal	2	240	leave-one-out	36	36	0	100%	1.00
Liver	2	106	holdout	120	36	2	94%	0.89
Ovarian	2	101	holdout	90	35	2	94%	0.89
Lymphoma	2	100	leave-one-out	42	42	0	100%	1.00
Leukemia-B	3	50	leave-one-out	27	27	8	70%	0.55
BRCA	3	50	leave-one-out	22	22	0	100%	1.00
Melanomas	3	103	leave-one-out	38	38	4	89%	0.82
Pancreatic	3	101	leave-one-out	36	36	2	94%	0.89
SRBCT	4	100	holdout	63	19	1	95%	0.93
Soft-tissue	5	102	leave-one-out	46	46	4	91%	0.88

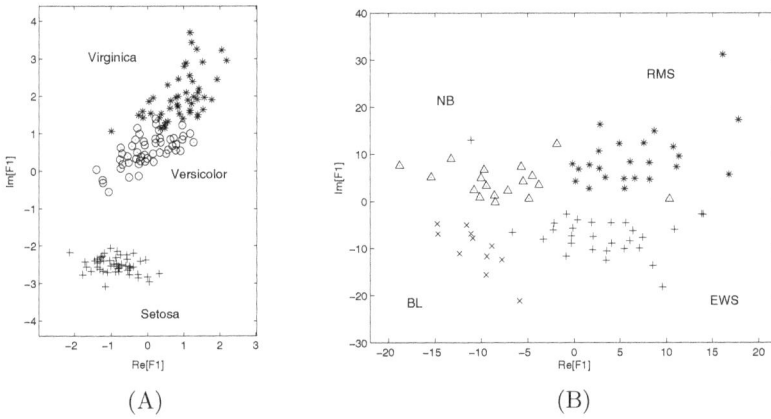

(A)   (B)

Fig. 8.15   Sammon's mapping method. (A) provides the results from Sammon's mapping of the iris data set (iris setosa is indicated by plus symbols, iris versicolor by circles, and iris virginica by stars). (B) provides results from Sammon's mapping of the informative genes from the SRBCT data set. The triangles represent neuroblastomas (NB), the stars are rhabdomyosarcomas (RMS), the plus symbols are Ewing's family of tumors (EWS), and the crosses are Burkitt's lymphomas (BL). The symbols represent both the training and test data sets.

8.4.5   *Confirmative Visualization of Gene Time-series*

The VizStruct technique was also tested in application to several time-series gene expression data sets. These visualizations were intended to graphically confirm the clustering of genes. In theory, genes belonging to a given cluster would be visualized in the same area, while genes from different clusters would be relatively more separated. Such visualizations can provide a useful tool for validation of the results of clustering algorithms.

8.4.5.1   *Data sets for visualization*

VizStruct was tested using three published array-derived data sets. The rat kidney array data set generated by Stuart et al. [269] contains measurements of gene expressions during rat kidney organogenesis. The data were downloaded from http://organogenesis.ucsd.edu/data.html. This data set contains 873 genes which vary significantly during kidney development at seven different time points.

The yeast-A data set provided by Alter et al. [8] is the result of a study of the yeast *S. cerevisiae* over two cell-cycle periods taken at seven-minute intervals over a period of 119 minutes, for a total of 18 time points. VizStruct testing used a subset of 77 genes classified by traditional methods into five

cell-cycle stages: MyG1, G1, S, SyG2, and G2yM. The data set was downloaded from http://genomewww.stanford.edu/GSVD/htmls/pnas.html.

The yeast-B data set taken from the report of Cho et al. [54] provides
a genome-wide characterization of mRNA transcript levels during the cell
cycle of the budding yeast S. cerevisiae. It consists of 416 genes at 17 time
points. Five groups of genes are reported by the author. The data were
downloaded from http://171.65.26.52/yeast_cell_cycle/cellcycle.html.

8.4.5.2   *The harmonic projection approach*

As mentioned in Section 8.4.4.4, it is useful to apply the first Fourier harmonic projection if the proportion of the first harmonic is significant. In
such cases, a canonical dimension ordering should be adopted to ensure
the dominance of the first harmonic. With time-series data, the order of
dimensions is fixed. These tests used the following three-step approach to
selecting the harmonic:

(1) Calculate the harmonic spectrum of the data set;
(2) Locate the dominant harmonic k; and
(3) Apply the kth Fourier harmonic project for the visualization.

8.4.5.3   *Rat kidney data set*

A seven-time point rat kidney data set was used to illustrate the visualization results generated by FFHP and the roles played by each of the
propositions discussed above. This data set was characterized by five discrete patterns or substructures of gene groups. Figures 8.16A-B illustrate
the idealized and actual gene expression profiles. Figure 8.16C provides
the FFHP-generated visualization in the form of a colorized scatter plot
reflecting the structure of the data. A colored symbol is used to indicate
each of the five gene groups, with each symbol representing one gene across
seven time points. As suggested by Proposition 8.6, genes from each group
appear as aggregated.

Two large clusters symmetric to the real axis are clearly apparent from
the visualization. The propositions underlying FFHP suggest the reason for
the formation of these clusters. Groups 1 and 2 contain genes which have
very high relative levels of expression in early development. In contrast,
groups 3, 4, and 5 contain genes that demonstrated a relatively steady
increase in expression throughout development. Temporal profiles of groups
1 and 4 suggest that they are somewhat symmetric to the middle time

(A)

(B)

(C)

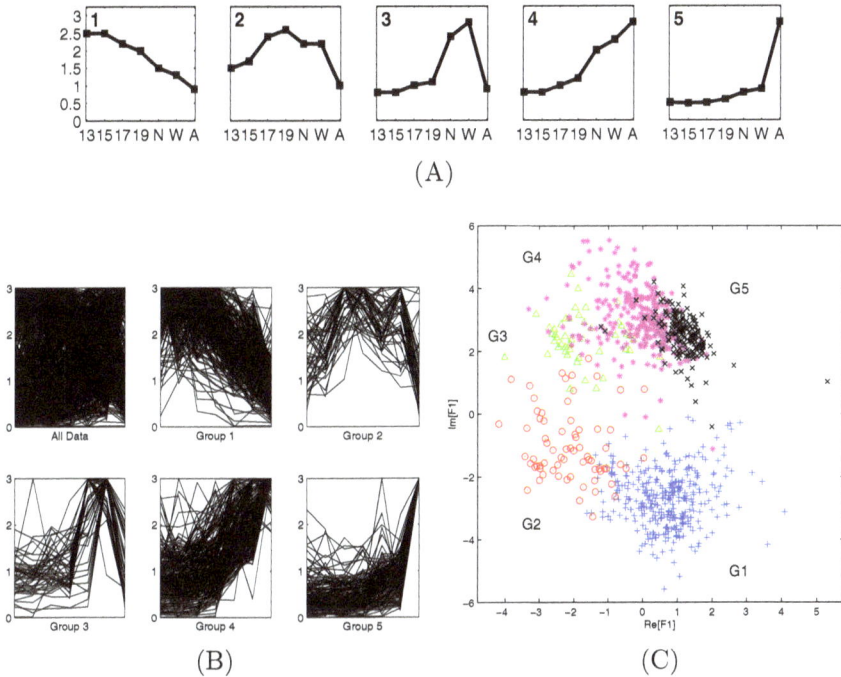

Fig. 8.16 The rat kidney data set. (A) Idealized temporal gene expression profiles. The groups were numbered 1 through 5 based on the timing of their peak expression during development. Time points were established at 13, 15, 17, and 19 embryonic days; N (newborn); W (one week after birth); and A (adult). (B) Visualization in parallel coordinates for the entire data set and for each of the gene groups. Patterns of genes in each group conform to the profiles depicted in (A). (C) Visualization generated by FFHP. The five gene groups are represented by blue plus symbols, red circles, green triangles, magenta stars, and black crosses.

point (gestational day 19). Following Proposition 4, these groups should be mapped to points symmetric to the real axis. On the other hand, groups 4 and 5 are mapped closely, reflecting the similarity of their profiles (with the exception of the significantly up-regulated gene at the last time point). Similar analyses explain the relative positions of other pairs of groups.

8.4.5.4 *Yeast-A data set*

When the first harmonic is not the dominant component, applying FFHP may yield undesirable results. This is illustrated in Figure 8.17A, where the visualization included overlapping groups, and the separation of different temporal patterns was very poor. The harmonic spectrum depicted in

(A)   (B)

(C)

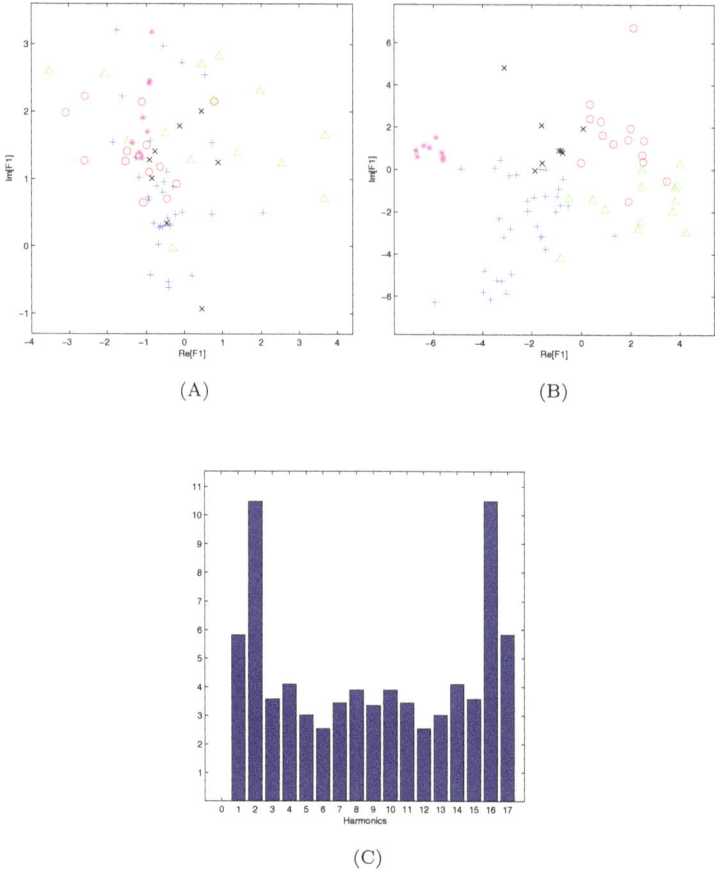

Fig. 8.17   The yeast-A data. (A) Visualization using the first Fourier harmonic projection. Five temporal patterns are represented by blue plus symbols, red circles, green triangles, magenta stars, and black crosses. (B) Visualization using the second Fourier harmonic projection. (C) The harmonic spectrum reveals the dominance of the second harmonic.

Figure 8.17C indicated that the dominant component was the second harmonic. A closer inspection of the temporal patterns in Figure 8.18A reveals evenly-spread two-peak shapes, particularly in the first and fourth panels. This is the signature of a signal dominated by the second harmonic. The observation was confirmed by inspection of the harmonic spectrum of those gene groups (Figure 8.18B).

Guided by this understanding of the harmonic spectrum, the second Fourier harmonic projection was applied to the yeast-A dataset, and these

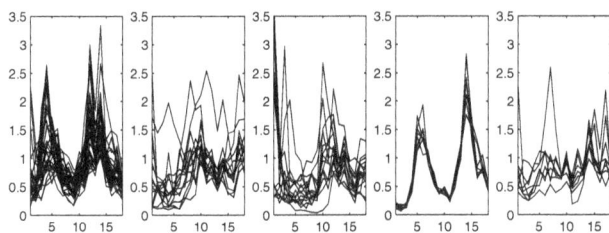

(A)

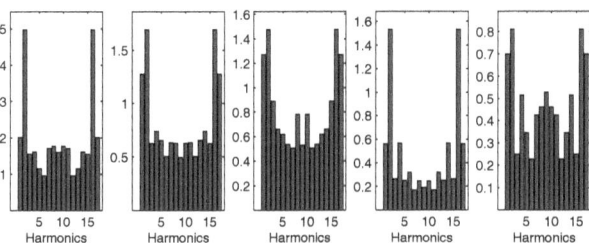

(B)

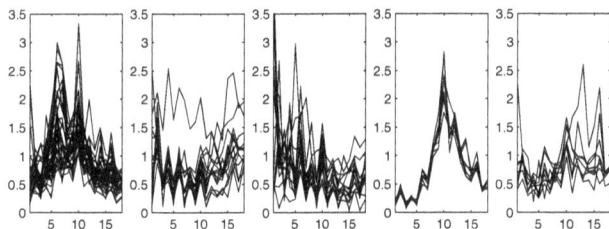

(C)

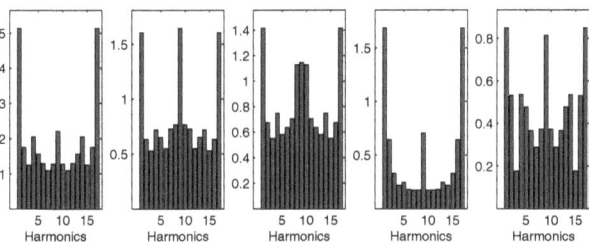

(D)

Fig. 8.18   Detail of the yeast-A data. (A) Parallel coordinates for the five gene groups of the original data set. (B) The harmonic spectrum of these gene groups in (A). (C) Parallel coordinates for the gene groups in (A) reordered by the second harmonic twiddle power index. (D) Harmonic spectrum of corresponding gene groups in (C).

results are shown in Figure 8.17B. The separation between groups improved significantly over the visualization shown in Figure 8.17A. Figures 8.18B and C provide an insightful view of the underlying structure. The discussion in Section 8.4.2 showed that applying the second harmonic projection is equivalent to applying the first harmonic projection to a dataset which has been dimensionally reordered by the second HTPI. Figure 8.18B illustrates the yeast-A data set with dimensions (time points) rearranged by the second HTPI. Shapes which were previously two-peak have been reconfigured by this process into single-peak shapes, a characteristic of dominance by the first harmonic. As confirmed in Figure 8.18D, the first harmonic indeed became much more dominant as a result of this reordering.

8.4.5.5 *Yeast-B data set*

Application of these techniques to the yeast-B data set generated similar results. This data set also exhibited second-harmonic dominant patterns, illustrated in Figure 8.19A and confirmed by the harmonic-spectrum graph provided in Figure 8.19(B). As expected, application of the second harmonic projection greatly improved the gene-group separation, as illustrated in Figures 8.19(C) and (D).

8.5 Summary

Visualization can be an effective tool for summarizing and interpreting data sets, describing their contents, and revealing significant features. Although many techniques have been developed to visualize multivariate data, their effectiveness in application to gene expression microarray data is limited by the high dimensionality of these data sets. Little research addressing the visualization of gene expression profiles has been published to date, and this field is still in its incipient stages. At present, visualization serves principally as an adjunct tool or graphical presentation mode for major clustering methods. The most prominent visualization-enhanced analysis tool for gene expression data is TreeView [92], which provides a user-friendly computational and graphical environment for assessing the results obtained from hierarchical clustering.

A recent innovation in this field had been the development of the VizStruct, a dynamic, interactive visualization environment for analyzing gene expression data. This environment employs innovative visualization methods to reveal the underlying data patterns. VizStruct uses Fourier

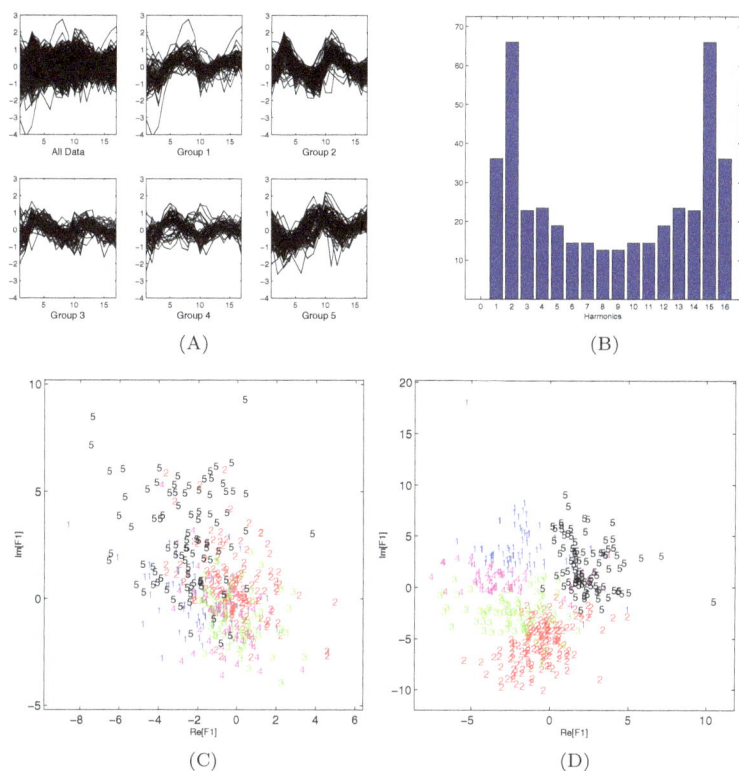

Fig. 8.19 The yeast-B data. (A) Parallel coordinates for the entire data set and each of the groups. (B) The harmonic spectrum reveals the dominance of the second harmonic. (C) Visualization using the first Fourier harmonic projection. Five colored letters represent genes in each of the five groups. (D) Visualization using the second Fourier harmonic projection. Five colored letters represent genes in each of the five groups.

harmonic projections (FHPs) to map multi-dimensional gene expression data onto two dimensions to create an intuitive scatter plot. Fourier harmonic projections preserve the key characteristics of multidimensional data through the reduction to two dimensions in a manner that facilitates visualization. They are computationally efficient and can be used for cluster detection and class prediction within gene expression data sets.

Visualization should not be considered a substitute for quantitative analysis. Rather, it is a qualitative means of focusing analytic approaches and guiding users toward the most appropriate parameters for quantitative techniques. The performance of VizStruct suggests that visualization may be capable of a larger and richer role than is currently appreciated. It is appropriate, however, to be mindful of the "curse of dimensionality" [20] that attends multi-dimensional gene expression data. Any experimental findings revealed by two-dimensional mapping should be rigorously confirmed through appropriate quantitative techniques.

Chapter 9

New Trends in Mining Gene Expression Microarray Data

9.1 Introduction

The existing mining techniques may not be sufficient for mining complex genomic data sets, as these methods do not adequately incorporate domain knowledge into the mining process. Biomedical experimentation is typically well-controlled and hypothesis-driven and therefore represents a particularly promising context for the integration of domain knowledge. In addition, the mining process would also be enhanced by the integration of data from various sources available in this rapidly-evolving field. In this chapter, we will briefly discuss several new approaches which pursue these directions.

9.2 Meta-Analysis of Microarray Data

In the past few years, a myriad of microarray experiments have been conducted, and large quantities of microarray data have been made available in public data repositories such as the Stanford Microarray Database (http://genome-ww5.stanford.edu/MicroArray/SMD) [113], the ArrayExpress repository (http://www.ebi.ac.uk/arrayexpress) [40], and the Gene Expression Ominibus (GEO, http://www.ncbi.hlm.nih.gov/geo/) [108]. A meta-analysis which combines multiple microarray data sets may provide more reliable insights into the functional relationships between genes than can be gained by analysis of individual data sets (e.g., [177, 200, 268]). However, integration of microarray data also poses significant technical and computational challenges. Data generated by individual investigators may be governed by distinct protocols and utilize different microarray platforms (such as cDNA or short and long oligonucleotides). Normalization and comparison of such variable data poses technical difficulties. These technical issues will not be discussed here; more information on these matters can be found in [200]. On the computation side, novel approaches are needed to combine and facilitate meta-analysis of differentially expressed genes (see Chapter 4) and co-expressed genes (see Chapter 5) across multiple microarray data sets. This issue will be addressed briefly in the following section.

9.2.1   *Meta-Analysis of Differential Genes*

Most analyses of multiple microarray data sets have focused on differential expression, comparing two or more related data sets to identify genes that distinguish different groups of samples (see Chapter 4) [43, 55, 75, 228, 231, 261, 307, 322]. For example, Rhodes et al. [231] combined four prostate cancer data sets and proposed a robust statistical method to determine those genes that are differentially expressed between benign prostate tissue and clinically-localized prostate cancer. The meta-analysis included the following major steps:

- *Statistical analysis of each individual data set.* The t-statistic (see Section 4.4) for each gene was calculated within each individual data set. To determine the statistical significance of the t-statistic, the value was compared with 10,000 t-statistics generated by randomly assigning the sample labels to the expression values of the gene. The p-value was then evaluated as the fraction of random t-statistics that were greater than or equal to the actual t-statistic.
- *Combination of the p-values from individual data sets.* For each gene that was present in all of the data sets involved in the meta-analysis, a *summary statistic* S was computed using the p-values from individual data sets:

$$S = -2logp_1 - \cdots - 2logp_k,$$

where k is the number of data sets and p_i ($1 \le i \le k$) is the p-value from the ith data set. The p-value of the summary statistic S was then obtained by a comparison to 100,000 summary statistics generated by randomly selecting a p-value from each individual data set contributing to the meta-analysis. The p-value of the summary statistic was the fraction of random summary statistics that were greater than or equal to the actual p-value.

- *Estimation of the gene-specific false discovery rate (FDR).* Since the microarray data sets involved in the meta-analysis typically contain thousands of genes, the p-values of the S statistic require correction to address the errors arising from multiple testing (see Section 4.6). The false discovery rate (FDR) (see Section 4.6.2) for gene i was calculated as follows:

$$FDR_i = \frac{p_i \times n}{m_i},$$

where p_i is the p-value of the summary statistic for the ith gene, n is the total number of genes in the meta-analysis, and m_i is the number of genes with a p-value equal to or smaller than p_i.

- *Selection of differentially expressed genes.* A threshold for the FDR was established, and genes below this threshold were reported as differentially expressed.

Rhodes et al. [231] reported an interesting discrepancy between results obtained from the analysis of individual data sets and from meta-analysis. The former identified 758, 665, 0, and 1,194 genes as over-expressed in each of the four data sets, while the meta-analysis recognized only 50 genes as consistently over-expressed; the FDR-adjusted value was held constant at 0.1. Moreau et al. [200] noted that the method used by Rhodes et al. is highly conservative because of the choice of null hypothesis. In their re-analysis of the data, 233 genes were determined to be reliably over-expressed at the same FDR-adjusted value [200].

9.2.2 Meta-Analysis of Co-Expressed Genes

Another approach to meta-analysis of multiple microarray data is to foreground gene co-expression rather than differential expression. For example, in [177], Lee et al. collected 60 large human data sets containing a total of 3,924 microarrays. They sought pairs of genes that were reliably co-expressed in multiple data sets. Every pair of gene expression profiles within each data set was compared with the value of Pearson's correlation coefficient (see Section 5.2.2). The significance (p-value) of each correlation was assessed by assuming that the distribution of correlations under the null hypothesis of no correlation follows a t-distribution with $n-2$ degrees of freedom (see Section 4.4), where n is the number of measurements in the expression profile (the number of samples). P-values were then corrected using the Bonferroni correction (see Section 4.6.1.1) for the number of genes tested, and the family-wise error rate (see Section 4.6.1) was set at $\alpha = 0.01$ per data set. In addition, pairs were only considered for further study if they were among the upper or lower 0.5% of correlations in the data set. Such pairs of genes are termed "co-expression links."

Lee et al. [177] established a high-confidence network of 8,805 genes connected by 220,649 "co-expression links" that were observed in at least three of the 60 data sets. Links demonstrating a higher level of confirmation through their appearance in multiple data sets were more likely to represent gene pairs already known to have a functional relationship. An analysis of clusters within the high-confidence network revealed functionally coherent groups of genes. These findings demonstrated that the large body of accumulated microarray data can be synthesized to increase the reliability of inferences about gene function.

In another study by Stuart et al. [268], 3,182 DNA microarrays taken from four different species (humans, flies, worms, and yeasts) were collected to identify co-expression relationships which have persisted throughout evolution. This persistence implies that the co-expression of these gene pairs confers a selective advantage and therefore that these genes are functionally related. Stuart et al. constructed a multiple-species co-expression network by identifying pairs of genes (or their orthologs) that were co-expressed not only in a single experiment and organism but also in diverse experiments in multiple organisms. They showed that the multiple-species co-expression network provides better functional predictions than single-species networks because it uses evolution to filter out gene interactions that are not functionally relevant.

9.3   Semi-Supervised Clustering

Clustering algorithms are generally recognized as "unsupervised" learning approaches, in that no prior knowledge is assumed at the start of the mining process. However, in the case of genomic data sets, some prior domain knowledge is often available. For example, some genes are known to be function-related, while others may be irrelevant to any biological process. Integrating such domain knowledge into the mining process may effect substantial improvement in the mining results. The following section will introduce several general semi-supervised clustering approaches which have been proposed in the data-mining literature. We will then present a framework for "seed generation" which was specifically designed for pattern analysis in microarray data.

9.3.1 General Semi-Supervised Clustering Algorithms

Recently, a series of semi-supervised clustering algorithms based on various applications have been proposed. In general, these approaches can be divided into two categories [18, 116]:

- *Constraint-based approaches.* With these methods, the clustering algorithm is modified so that user-provided constraints or labels can be deployed to bias the search for an appropriate partition. Approaches include performing a transitive closure of the constraints and using them to initialize clusters [17], including in the cost function a penalty for lack of compliance with the specified constraints [68], or requiring constraints to be satisfied during the cluster-assignment step of the clustering process [294].
- *Distance-based approaches.* These approaches take as a basis an existing clustering algorithm which uses some similarity measure; typical algorithms include the hierarchical *single-link* [28], *complete-link* [170], or *k-means algorithm* [28, 309]). The similarity measure is adapted so that specified constraints can be more readily satisfied. Similarity measures which have been employed in this way include the *Jensen-Shannon divergence* trained with a gradient descent [60], the Euclidean distance modified by a shortest-path algorithm [170], and *Mahalanobis distances* adjusted by a convex optimization [28, 309].

Basu et al. [18] combined the constraint-based and distance-based approaches in a probabilistic framework based on *hidden Markov random fields (HMRFs)*. Given a data set $\mathcal{X} = \{x_i\}_{i=1}^{N}$ where N is the number of data objects, an HMRF model has the following two components [18]:

- A *hidden field* $\mathcal{L} = \{l_i\}_{i=1}^{N}$ of random variables with unobservable values. In the clustering framework, the hidden variables are the unobserved labels which indicate object cluster assignments. Every hidden variable l_i takes its values from the set $\{1, \ldots, K\}$ which contains the indices of the clusters (K is the number of clusters in the data set).
- An *observable* set $\mathcal{X} = \{x_i\}_{i=1}^{N}$ of random variables, where every random variable x_i is generated from a conditional probability distribution $Pr(x_i|l_i)$ determined by the corresponding hidden variable l_i. The random variables $\mathcal{X}$ are conditionally independent from the hidden vari-

ables $\mathcal{L}$, i.e., $Pr(\mathcal{X}|\mathcal{L}) = \prod_{x_i \in \mathcal{X}} Pr(x_i|l_i)$. In their framework, the set of observable variables for the HMRF corresponds to the given data points.

To incorporate domain knowledge into this framework, supervision is modeled as a set of *must-link* constraints $\mathcal{M}$ with a set of associated violation costs $\mathcal{W}$ and a set of *cannot-link* constraints $\mathcal{C}$ associated with a set of violation costs $\overline{\mathcal{W}}$. The clustering process is then redefined as the identification of the *maximum a-posteriori* (MAP) configuration of the HMRF which maximizes $Pr(\mathcal{X}|\mathcal{L})$. This can also be defined as minimization of the following objective function:

$$\mathcal{J}_{obj} = \sum_{x_i \in \mathcal{X}} D(x_i, \mu_{l_i}) + \sum_{(x_i, x_j) \in \mathcal{M}} f_{\mathcal{M}}(x_i, x_j) + \sum_{(x_i, x_j) \in \mathcal{C}} f_{\mathcal{C}}(x_i, x_j) + logZ,$$

where $D(x_i, \mu_{l_i})$ is the distance between data object x_i and the centroid μ_{l_i} of the cluster to which x_i belongs, $f_{\mathcal{M}}(x_i, x_j)$ is the penalty function for violating a must-link constraint, $f_{\mathcal{C}}(x_i, x_j)$ is the penalty function for violating a cannot-link constraint, and Z is a constant.

The EM algorithm [69] (see Section 5.3.4) can be applied to minimize the objective function. In the E-step, assignments of data points to clusters are updated using the current estimates of the cluster representatives $\{\mu_k\}_{k=1}^{K}$. In the M-step, the objective function is decreased by re-estimation of the cluster representatives from their currently-assigned objects. If a parameterized distance measure is used, the measure parameters are also updated to decrease the objective function in this M-step. The E-step and M-step iterate until the objective function converges.

Basu et al. applied their semi-supervised approach to three subsets of the *20-Newsgroups collection* (http://www.ai.mit.edu/people/jrennie/20Newsgroups). These trials indicated that the application of the algorithm derived from the HMRF framework to these complex textual data sets led to improved cluster quality when compared with unsupervised clustering approaches. Microarray data share a number of characteristics with textual data, including sparsity, high dimensionality, and a disproportionate relationship between the number of objects and the dimensionality of the shape. It may therefore be inferred that the semi-supervised approach would achieve a better performance in sample-based analysis (see Chapter 6) than other existing methods. Further empirical studies should be conducted to verify this conjecture.

9.3.2 A Seed-Generation Approach

In semi-supervised approaches, the greatest challenge lies in appropriately selecting the domain knowledge of greatest pertinence to the mining process. The following subsections will explore the concepts of *seed generation* and *pattern selection rules* as a means for formulating domain knowledge.

9.3.2.1 Seed-generation methods

Input from domain experts identifying significant or irrelevant genes can be used as initial "seeds" for the generation of potential candidate clusters. Three conceptual approaches to this task are described below.

- Method 1: *Domain expert selected seeds.* Here, the mean expression pattern of selected genes identified by the domain expert is used as the seed. These genes have typically been previously identified as having a role in the cellular or physiological processes of interest. The seed is then deployed as a template for identification of other genes with similar patterns; these may be novel or poorly-characterized genes that may merit additional experimental investigation. For example, Figure 9.1 shows the temporal profiles of a set of four genes, $H38522$, $AA704613$, $N92443$ and $N75595$, that are known to be induced by interferon-β, the drug used to treat multiple sclerosis patients in clinical studies.
- Method 2: *User defined patterns.* This method is similar to the previous but uses a seed as an original pattern rather than the temporal profile of a specific gene. These patterns may not have been observed in the data set, but the user may be interested in identifying unknown genes that share similarities with the pattern of interest. Figure 9.2 provides two examples of such patterns; the first increases over the first several time points and then declines, while the second reverses this profile.
- Method 3: *Seed searching with constraints.* Unlike the first two methods of seed generation, which are supervised and require substantial domain knowledge, Method 3 can be employed when the user has only partial or minimal knowledge about the pattern of interest. With this method, the seeds are identified randomly from the data in a process directed by certain pattern-selection rules. Such rules can be provided by users and are independent from other seed-generation processes. These rules and their implementation will be discussed in detail in the next subsection.

A. H38522

B. AA704613

C. N92443

D. N75595

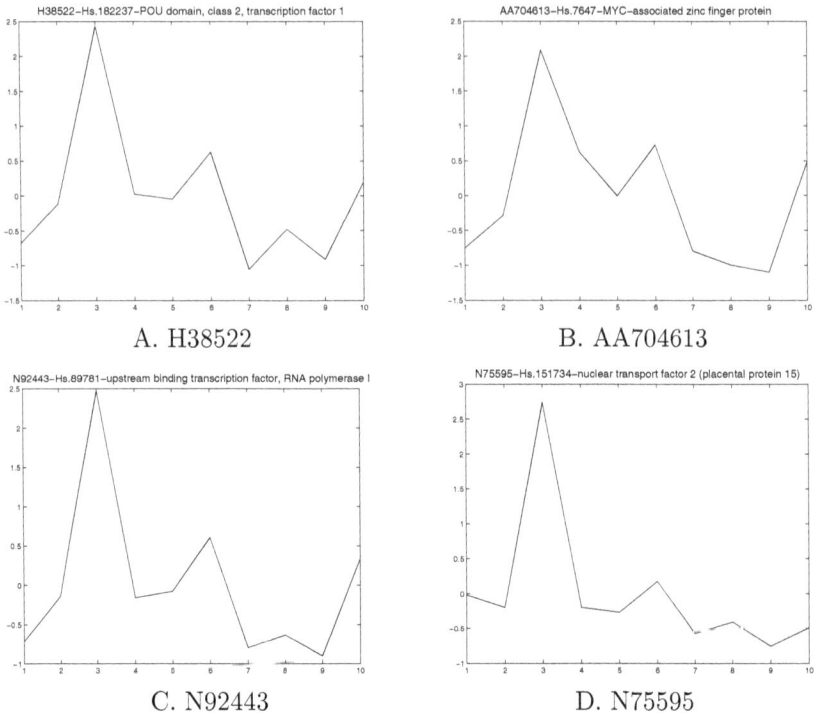

Fig. 9.1   Four seeds identified with domain knowledge.

9.3.2.2   *Pattern-selection rules*

Pattern-selection rules are intended to allow users to pre-define the features of the patterns to be selected as candidate seeds. The subsequent scan of the data matrix will eliminate from consideration of all patterns which fail to meet the specified seed-exclusion rules. Seed-exclusion rules consist of a set of parameters, each of which is defined by the user. Some examples of these parameters are:

- *Trend.* The trend defines a threshold number of consecutive time points in a pattern over which values monotonically increase or decrease. This eliminates patterns with frequent fluctuations, which are likely to represent noise. Figure 9.3 displays two patterns that differ in trend length. The pattern in Figure 9.3(A) has a trend of length 1, while that of the pattern in Figure 9.3(B) is 3. The subsequent search of the microarray data matrix will seek additional patterns having comparable trend

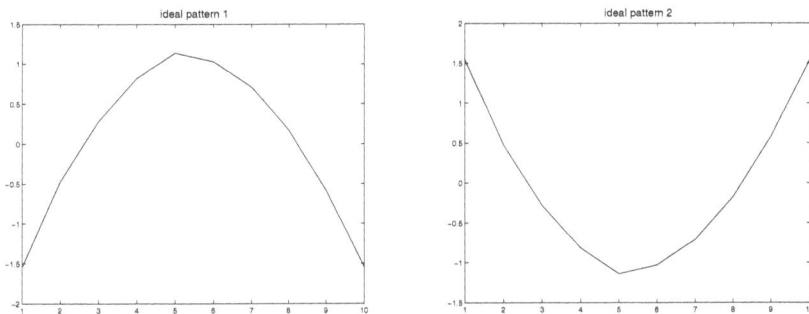

Fig. 9.2   Two examples of the synthetic patterns defined by user.

values. Patterns with trends below the specified threshold will not be selected as seeds.

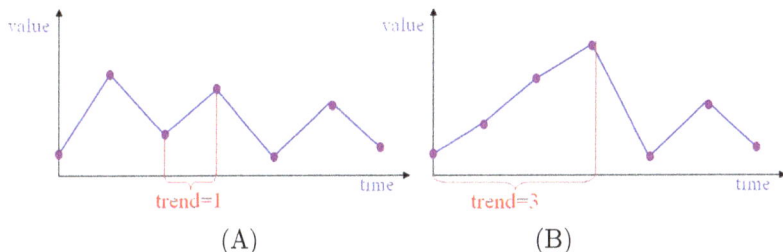

(A)   (B)

Fig. 9.3   Two patterns differ in the length of the trends.

- *Pick point.* The pick point specifies the relative time point at which the pattern reaches its highest value or lowest value. This is a relative value only, since the data-normalization and standardization process has stripped the patterns of absolute values. Figure 9.4 provides an example of a pattern which reaches its highest value at time 1 and its lowest value at time 4. It can therefore be described as *"highpickpoint"* = 1 and *"lowpickpoint"* = 4. If the pick points of a pattern differ from the user-set values, the pattern will not be selected as a seed.

- *Pick interval.* The pick interval defines the number of consecutive time points between the first two local high pick points. This feature reveals the duration of expression of a gene after a drug treatment. A carefully-selected combination of pick point and pick interval can provide insight into the mechanisms of diseases and therapies. Figure 9.5 displays a

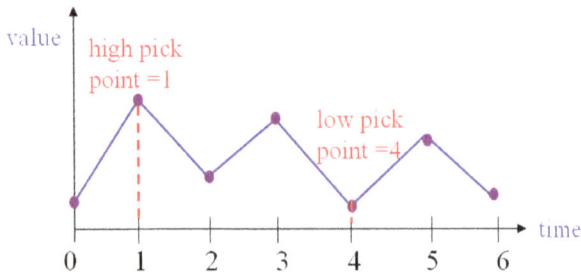

Fig. 9.4   The high pick point and low pick point of a pattern.

pattern which reaches its first local high pick value at time 1 and its second local high value at time 6. It can therefore be described as having *"pickinterval"* = 5. The low pick interval can be similarly defined. If the pick interval of a pattern is below the user-specified threshold, it will not be selected as a seed.

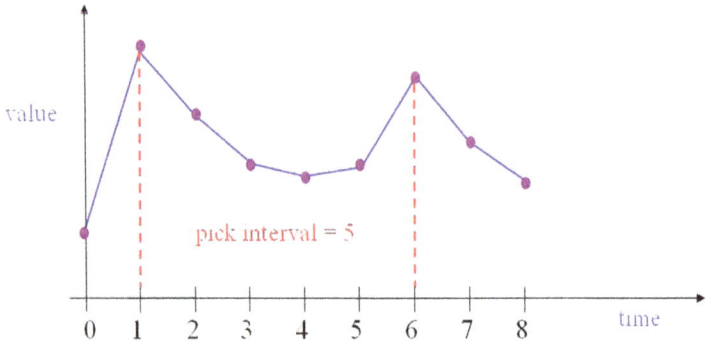

Fig. 9.5   The pick interval between the first two local high pick points.

- *Period.* The period value indicates the duration of repetition of a pattern. Figure 9.6 illustrates a pattern which repeats after every 4 time slots. This pattern can be described by $w(i + p) = w(i)$, and p is the period value. This feature is pertinent to long-term time-series data. Only patterns which conform to the user-defined period value will be selected as potential seeds.

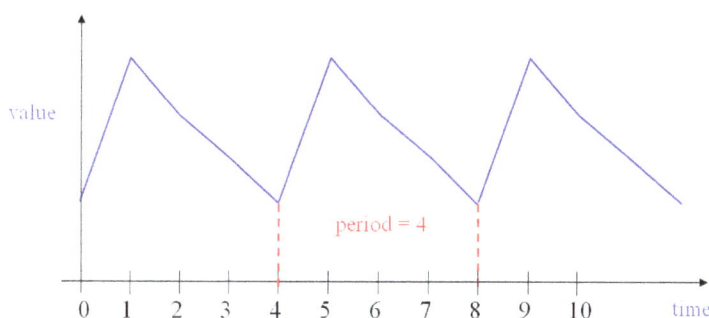

Fig. 9.6 The "period" of a pattern.

- *Sparsity.* It is usually desirable to avoid seeds with closely-related patterns. The user can define a threshold of similarity *sparsity threshold* which will exclude patterns which resemble or are identical to a previously-selected seed.

9.3.2.3 *The framework for the seed-generation approach*

Pattern-selection rules allow sophisticated users to define thresholds for some or all parameters and to add features or re-define the existing rules for specific applications. These features also can be incorporated into other pattern-related retrieval or mining frameworks.

Users can select different sets of rules to generate a range of candidate clusters. For example, variant seeds can be selected with thresholds set first at *"pickinterval"* = 5 and *"trend"* = 4 and then at *"highpickpoint"* = 1 and *"lowpickpoint"* = 5. These seeds can then serve as input to the cluster-detection phase.

Figure 9.7 illustrates the detailed framework of the seed-generation phase corresponding to the GST data set discussed in Section 7.4.

In Figure 9.7, the seeds generated using Method 1 (domain expert selected seeds) are gene patterns from the original gene-sample-time series data set. The seeds generated using Method 2 (user defined patterns) are original patterns which may not exist in the original data set. With Method 3 (seed searching with constraints), different seeds may have varying constraints. For example, $seed_{31}$ is constrained by two parameters (trend and pick interval), $seed_{34}$ is defined by all parameters, while $seed_{35}$ is governed only by the sparsity rule.

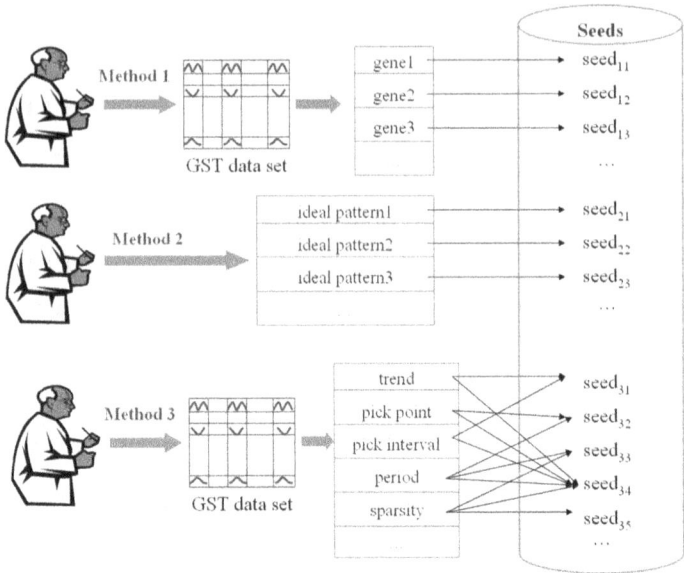

Fig. 9.7   The detailed framework of the seed generation.

9.4   Integration of Gene Expression Data with Other Data

The increasing quantities of high-throughput biological data have made biology an information-rich science. Various types of functional genomic data sets are now available. Many of these, such as DNA/protein expression profiles, protein-protein interactions, protein structures, and DNA binding data, are key to predicting protein functions and recovering functional modules. Due to the high noise ratio of high-throughput data sets, an integrated analysis of multiple data sources often generates more reliable results.

In Section 9.2, we discussed some approaches to the combination of multiple microarray data sets. In this section, we will focus on the combination of microarray data with other genomic or proteomic data. Such integration is often able to provide a more comprehensive view of the biological system. The value of combining groupings of genes obtained through different methods has been illustrated by several studies in which functional predictions were made based on several types of data [13, 71, 164, 190, 215, 246, 247, 272, 285].

An examination of gene expression data can reveal co-expressed genes, groups of genes that demonstrate coherent patterns within samples or in reaction to stimuli. From the inspection of protein-protein interaction data, groups of proteins which frequently interact can be identified. Since proteins are the products of genes, it would be fruitful to integrate both data types to locate those gene clusters that are both co-expressed and demonstrate protein interactions.

Such cross-data-set clusters revealed by data integration are often interesting and meaningful. This approach assists in cross-confirming the validity of identified clusters, since both gene expression and protein data are frequently noisy. Identification of a cluster in both data sets strongly substantiates the correlation and connection among their constituent genes. Furthermore, although they are closely related, gene expression and protein interaction data convey different biological meanings, and a coincidence of co-expressed genes and interacting proteins is therefore biologically significant. As indicated in [246], many pathways exhibit two properties: their genes exhibit a similar gene expression profile, and the protein products of the genes often interact.

In general, approaches to data integration fall within the following categories, which are differentiated by the performance of the integration before, during, or after the mining process:

- *Pre-integration.* Pre-integration approaches typically combine the "correlation" or "distance" between data objects in different data sets into a combined distance function. This function can then be incorporated into conventional mining algorithms which target a single data source.

  For example, Hanisch et al. [122] constructed a distance function which combines information from expression data and biological networks. The combined distance function is intended to: (1) assign a small distance value to pairs of objects which are proximate in the biological network and show similar expression patterns; (2) assign a larger distance value to objects which are far apart in the network or are not co-expressed in the gene expression data; and (3) assign the highest distance value to objects which are far apart as defined by both individual distance measures δ_{exp} and δ_{net}. Given two data objects o_i, o_j which correspond to genes g_i, g_j and network nodes v_i, v_j, the combined function Δ is defined as

  $$\Delta(o_i, o_j) = 1 - 0.5 \times (\lambda_{exp}(g_i, g_j) + \lambda_{net}(v_i, v_j)),$$

  where $\lambda_\Psi(x_i, x_j) = \frac{1}{1 + e^{-s_\Psi(\delta_\Psi(x_i, x_j) - \nu_\Psi)}}$ for $\Psi \in \{exp, net\}$, and

ν_Ψ, s_Ψ are parameters normalizing the individual distance measures. In [122], an *average-link* agglomerative hierarchical clustering algorithm is adopted to derive clusters on the basis of the combined function.

The incorporation of such a function allows any existing mining algorithm designed for single data sets to be applied directly to multiple data sources. In pre-integration approaches, it is often necessary to assign weights to each data source to avoid excessive influence on the combined distance function by a small number of data sets. These weights are usually difficult to determine, and their assignment is often somewhat arbitrary.

• *Post-integration.* Post-integration approaches analyze several data sets separately and combine the results obtained.

For example, in [283], Tornow et al. applied the super-paramagenetic [31] algorithm (see Section 5.4.2.2) to protein-protein interaction networks. Subsets of proteins which interact closely in the network are recognized as clusters by the algorithm. The *correlation strength* of these clusters is then calculated in relationship to the expression profiles of their corresponding genes. Higher correlation-strength values indicate greater coherence on the part of the expression profiles of the genes within the cluster. This approach tests the relationship between two types of data, since correlation strength is calculated in the context of gene expression data, while the clusters are distilled from the protein-protein interaction data set. Therefore, if the correlation strength of a cluster is significantly greater than that which might have occurred by chance, the proteins in that cluster are highly likely to perform the same function.

Post-integration approaches consider the data sets separately during the clustering stage and then combine the results from individual data sets in the integration stage. The major drawback of this strategy is the loss of information regarding correlations among different data sets. Therefore, each data set will be mined independently, uninformed by the characteristics of the data distribution in other data sets.

• *Joint mining.* Joint mining approaches integrate disparate genomic data sets during the mining process and exploit the inherent correlations among the sets to give unity and coherence to the process. For this reason, they may produce more meaningful results than either pre- and post-integration approaches. However, most existing algorithms designed for use with single data sets are not extensible for joint-mining applications. More sophisticated models and mining algorithms are needed to maximize the benefits of integrating heterogeneous data sets. In the following subsections, we will present two joint mining approaches which represent advances in that direction.

9.4.1 A Probabilistic Model for Joint Mining

Recently, the concept of "joint mining" has been proposed to suggest a mining process that explores various data sets in parallel and is guided by the inherent correlations across different data spaces [246, 247]. In [246], Segal et al. proposed a unified probabilistic model to combine gene expression profiles E with protein-protein interactions Υ. The combined model defines a joint distribution over the hidden variables $C = \{c_1, \ldots, c_n\}$, as follows:

$$P(C, E|\Upsilon) = \frac{1}{Z} (\prod_{i=1}^{n} P(c_i) \prod_{j=1}^{m} P(e_{i,j}|c_i)) \cdot (\prod_{[g_i - g_j] \in \Upsilon} \phi_2(c_i, c_j)),$$

where Z is a normalizing constant that ensures that P sums to 1. The hidden variable $c_i \in C$ represents the cluster to which gene g_i belongs. Given the number of clusters k, the variable c_i is associated with a multinomial distribution with parameters $\Theta = \{\theta_1, \ldots, \theta_k\}$. $P(e_{i,j}|c_i)$ is the *conditional probability distribution* (CPD) which represents the probability of observing expression value $e_{i,j}$ at the j-th condition of the expression profile of gene g_i, assuming g_i belongs to cluster c_i. The model assumes that the CPD $P(e_{i,j}|c_i = p)$ follows a Gaussian distribution $\mathcal{N}(\mu_{p,j}, \sigma_{p,j}^2)$. Finally, each pair of genes g_i, g_j that interact in Υ are associated with a compatibility potential $\phi_2(c_i, c_j)$, defined as follows:

$$\phi_2(c_i = p, c_j = q) = \begin{cases} \alpha & p = q \\ 1 & \text{otherwise} \end{cases},$$

where α is a parameter and required to be greater than or equal to 1. This model is based on the observation that, if g_i and g_j interact, they are more likely to belong to the same cluster.

The framework described in [246] uses the *EM* algorithm [69] (see Section 5.3.4) to learn the probabilistic model. The EM algorithm contains two major steps. In the *Expectation step*, the current estimate of the parameters Θ, $\{\mu_{p,j}\}$, and $\{\sigma_{p,j}\}$ are used to compute the distribution of the hidden variables C. In the *Maximization step*, the algorithm estimates the parameters using the posterior distribution on the basis of current C. This iterative process usually converges after several steps, and the clusters can be derived from C.

In [247], Segal et al. applied a similar probabilistic model to find clusters of genes which share common motif profiles (from DNA sequence data) and similar expression patterns (from gene expression data). Although this approach has generated promising experimental results, it does have some limitations. First, the user is required to specify k, the number of pathways in the data set. However, the value of k is usually unknown in advance. Second, each gene is permitted to belong to one and only one pathway. However, it is frequently the case that a gene may participate in several cellular processes under different conditions, and some genes in the data set may not participate in any process. Third, the approach assumes that the expression levels of the genes in a given pathway will follow a Gaussian distribution, while the real data may poorly fit the Gaussian model [320].

9.4.2   *A Graph-Based Model for Joint Mining*

Gene expression and protein interaction data can each be modeled using a gene coherence graph G_g and a protein interaction graph G_p. In the gene coherence graph G_g, vertices are genes, and two genes are connected if their expression patterns conform to a user-specified similarity measure such as Euclidean distance, Pearson's correlation coefficient, KL-distance [160], or pattern-based similarity measures [52, 296]. In the protein interaction graph G_p, the vertices are proteins, and two proteins are connected if they interact with each other. A subjective (i.e., onto) mapping $m(p) = g$ from proteins to genes is used to model the relationship between genes and proteins.

Given a gene expression data set G_g and a protein-protein interaction data set G_p, assume that G is a group of genes in G_g and P is the group of protein products of G in G_p. A group of genes G is relevant if each gene $g_i \in G$ is connected with most other genes $g_j \in G$ in the gene coherence graph G_g, and, at the same time, each protein $p_i \in P$ is connected with most other proteins $p_j \in P$ in the protein interaction graph G_p.

The core issue of mining relevant patterns from multiple data graphs can be reduced to the identification of "cross-graph quasi-cliques" [217]. Let us assume that there is a set of objects which is modeled in multiple graphs, and these objects form the vertices of each graph. The goal of the mining process is to identify groups of objects within each graph which satisfy a certain level of inter-group connectivity. Specifically, in every graph, each object in a group should be connected to at least a portion γ ($0 < \gamma \leq 1$) of the other objects in the same group, where γ is a user-specified parameter. In this instance, the induced subgraph of each group of objects in each graph approximates a clique and is thus termed a quasi-clique.

This understanding permits the formulation of a generic graph-based model for cross-graph quasi-clique mining. A cross-graph quasi-clique is a maximal set of vertices that are quasi-complete subgraphs in multiple graphs. In [217], only simple graphs which do not contain internal loops or multi-edges were considered.

Within this general graph-based model, we can formalize the definition of a joint cluster which incorporates both gene expression and protein-protein interaction data. Each gene in a gene expression data set corresponds to a vertex in the gene expression graph, and there exists an edge between two genes g_i and g_j if those genes have similar expression profiles. Similarly, each protein in a protein-protein interaction data set corresponds to a vertex in the protein-protein interaction graph, and there is an edge connecting two proteins p_i and p_j if and only if p_i and p_j interact. A mapping function f is used to map the genes to the proteins. Given a gene expression graph $G_E(V_E, E_E)$, a protein-protein interaction graph $G_P(V_P, E_P)$, and a mapping function $f : V_E \to V_P$, a joint cluster can be defined as a cross-graph quasi-clique Q such that (1) the induced graph of Q in G_E is a clique; (2) the induced graph of Q in G_P is a γ-quasi-clique; and (3) $|Q| = min_s$, where γ and min_s are two thresholds which control the connectivity and size of cross-graph quasi-cliques. Clearly, a cross-graph quasi-clique Q has two salient properties: the genes in Q are co-expressed genes, and the proteins in Q frequently interact. The task of mining cross-graph quasi-cliques is therefore reducible to the computation of the complete set of cliques in a graph, a problem which is NP-hard. The efficiency of identification of joint clusters can be improved by exploiting the characteristics of the data sets and formulating appropriate pruning rules.

One proposed approach [217] to the computation of a complete set of cross-graph quasi-cliques proceeds via the following two steps:

- *Graph reduction step.* The edges in the protein-protein interaction graph G_P are scanned, and the edge (u, v) is removed if there is no edge connecting the mapped vertices u' and v' in the gene expression graph G_E. This operation is termed *edge reduction*. In a similar *vertex reduction* process, the vertices in G_P and G_E are scanned, and the vertex u is removed if the degree of u falls below the connectivity threshold.

This operation is iterated until no additional vertices can be pruned. Since the protein-protein interaction graph is typically sparse, the sizes of the graphs can be reduced substantially. In a preliminary study of this method, when $\gamma = 0.5$, these graph reduction steps (edge and vertex reductions) effected at least a 30% improvement in efficiency.

- *Graph mining step.* In this step, two algorithms are applied to the reduced graphs. The first algorithm, a rudimentary algorithm, conducts a recursive, depth-first search of the vertex set. It enumerates the subsets of vertices and applies several pruning rules to compute the complete set of γ-quasi-cliques in one graph. It then determines whether a γ-quasi-clique is a γ-quasi-clique in the other graphs. The second algorithm, *Crochet*, also applies a depth-first search to enumerate the vertex set. It further improves the efficiency by pruning the subsets of vertices which are unlikely to become maximal cross-graph quasi-cliques.

This approach permits the integration of several genomic/proteomic data sets and facilitates a nuanced exploration of the data. Experimental results have indicated that this method is able to identify gene groups that correspond well to functional groups and which contain entire protein complexes. It can be readily applied to the joint mining of multiple genomic data sets.

An example of cross-graph quasi-clique. In [217], cross-graph quasi-cliques were mined from gene expression data and protein-protein interaction data. Figure 9.8 illustrates a cross-graph quasi-clique Q ($\gamma_E = 1$ and $\gamma_P = 0.4$) with a diameter of 3. The induced graph of G_E (the gene expression graph) for Q is a perfect clique, so only the induced graph of G_P (the protein interaction graph) for Q is provided here. The cross-graph quasi-clique contains eleven vertices. ORF (Open Reading Frame, see Section 2.5.2) names are used to represent the corresponding genes and proteins.

This cross-graph quasi-clique is of particular biological significance, since these eleven genes are highly coherent, and the corresponding eleven proteins demonstrate intensive interactions.

Additional examples of joint mining using multiple sources have been provided by Page and Craven [211] in their survey of the biological applications of mining multiple tables. Some applications include pharmacophore discovery, gene regulation, and information extraction from text and sequence analysis.

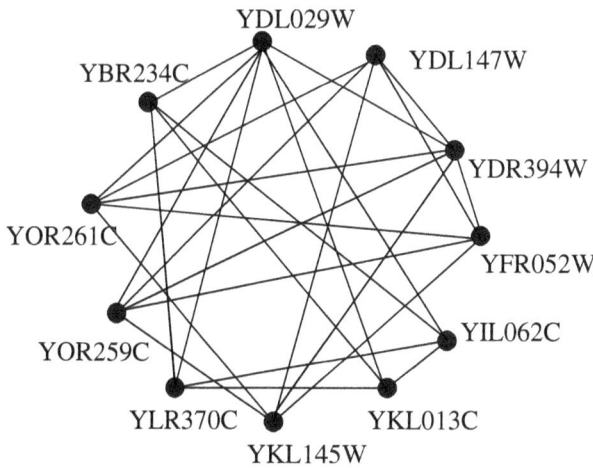

Fig. 9.8   A cross-graph quasi-clique of 11 proteins.

More advanced mining techniques based on multi-relational biological data should also be pursued. For example, a database containing the operational details of a single-cell organism would need to encode at least the following information:

- Genome: DNA sequence and gene locations.
- Proteome: the organism's full complement of proteins, not necessarily a direct mapping from its genes.
- Metabolic pathways: linked biochemical reactions involving multiple proteins, small molecules and protein-protein interactions.
- Regulatory pathways: the mechanism by which the expression of some genes into proteins, such as transcription factors, influences of the expression of other genes, including protein-DNA interactions.

This collection of data may draw from a variety of high-throughput techniques, including gene expression microarrays, proteomics (primarily mass spectrometry), metabolomics, and single-nucleotide polymorphism (SNP) measurements. Although it is possible to gain insights from any one of these data sources, it is now understood that more revealing results can be obtained by synthesizing several types of data. It will therefore behoove the field of biomedical data-mining to effectively incorporate the full spectrum of data sources.

9.5   Summary

The community of researchers involved in mining gene expression data sets has come to understand the value of incorporating detailed biological expertise to inform the mining process. This chapter has discussed some of the latest approaches being taken to that integration, including semi-supervised analysis and the joint mining of multiple data sources. Incorporation of gene ontology into the mining process has also recently been considered [168]. These new approaches show much promise in improving the effectiveness of gene expression microarray data analysis.

Chapter 10

Conclusion

The generation of gene expression microarray data is proceeding at a rapid and accelerating pace, heightening the demand for advances in the analytical methods used to seek and detect patterns and relationships in these complex data sets. This book has offered a systematic presentation of a variety of advanced data-mining approaches which are currently available or in development. These data-mining techniques have been presented as belonging to four basic conceptual categories:

- *Gene-based analysis.* These analytical methods deploy various clustering algorithms to identify co-expressed genes and coherent patterns. The approach taken by these clustering algorithms can be partition-based, hierarchical, or density-based. In particular, promising results have been achieved with an interactive clustering approach, GPX, which integrates concepts taken from both the hierarchical and density-based approaches. The arbitrary setting of cluster borders which compromises the results of other methods is avoided, and enhanced performance is obtained even with data sets containing many intermediate genes or significant noise. Experimental studies have indicated that GPX can identify most of the coherent patterns in a data set with higher accuracy than comparable clustering methods.
- *Sample-based analysis.* Sample-based methods use a variety of clustering and classification techniques to identify phenotype structures. Since a microarray experiment typically involves many more genes than samples, meaningful sample-based analysis must effectively reduce the extremely high dimensionality of the genes in a data set. Both supervised and unsupervised methods for the reduction of dimensionality have been discussed, and a series of approaches to disease classification and discovery has been surveyed. Detailed presentations were provided

of two novel approaches, CLIFF and ESPD, which utilize the relationship between the genes and samples to perform gene selection and sample clustering simultaneously in an iterative paradigm.

- *Pattern-based analysis.* These methods seek to ascertain the relationship between genes and samples through examining subsets of each. Association rules characterizing this relationship are at the heart of these methods. Our discussion reviewed several classical approaches to mining frequent itemsets, as well as a post-mining method, to identify association rules which may be applicable to gene expression microarray data. We then explored various heuristic and deterministic pattern-based clustering algorithms specifically developed for the detection of coherent patterns embedded in gene and sample subsets. Several novel approaches were presented which offer a sophisticated and efficient means for detecting the inherent correlations among genes, samples, and time-series.

- *Analysis by visualization.* Chapter 8 explored the use of visualization tools to detect patterns and relationships in gene expression microarray data sets. Classical visualization approaches vary in their ability to transform a gene expression data set from high-dimensional space into a more easily understood two- or three-dimensional space. A recently-developed approach, VizStruct, offers a dynamic, interactive visualization environment for analyzing gene expression data. The VizStruct method uses Fourier harmonic projections (FHPs) to map multi-dimensional gene expression data onto two dimensions to create an intuitive scatter plot. Fourier harmonic projections preserve the key characteristics of multidimensional data through the reduction to two dimensions in a manner that facilitates visualization. They are computationally efficient and can be used for cluster detection and class prediction within gene expression data sets.

The book concludes with a discussion of the latest trends in data-mining for the analysis of gene expression data. It has become clear that incorporation of the knowledge and expertise of biologists into the mining process can be of significant benefit. The integration of data from other sources also offers promising results, as a combination of heterogeneous data is often able to provide a more comprehensive view of the biological system. It is hoped that further exploration into these novel conceptual approaches will bring us to a fuller understanding of our genetic constitution and thus to a more sustainable and healthier future.

Bibliography

[1] Adams, M.D., Celniker, S.E., Holt, R.A., Evans, C.A., Gocayne, J.D., Amanatides, P.G., Scherer, S.E., Li, P.W., Hoskins, R.A., Galle, R.F., et al. The genome sequence of Drosophila melanogaster. *Science*, 287(5461):2185–2195, Mar 2000.

[2] Aggarwal, C.C. and Yu, P.S. Finding generalized projected clusters in high dimensional spaces. In *SIGMOD 2000, Proceedings ACM SIGMOD International Conference on Management of Data*, pages 70–81, 2000.

[3] Aggarwal, C.C., Wolf, J.L., Yu, P.S., Procopiuc, C. and Park, J.S. Fast algorithms for projected clustering. In *SIGMOD 1999, Proceedings ACM SIGMOD International Conference on Management of Data*, pages 61–72, Philadephia, Pennsylvania, USA, 1999.

[4] Agrawal, R. and Srikant, R. Fast algorithms for mining association rules. *In proc. 1994 Int. Conf. Very Large DataBases (VLDB'94)*, pages 487–499, Sept. 1994.

[5] Agrawal, R., Gehrke, J., Gunopulos, D. and Raghavan, P. Automatic subspace clustering of high dimensional data for data mining applications. In *SIGMOD 1998, Proceedings ACM SIGMOD International Conference on Management of Data*, pages 94–105, 1998.

[6] Alizadeh, A.A., Eisen, M.B., Davis, R.E., Ma, C., Lossos, I.S., Rosenwald, A., Boldrick, J.C., Sabet, H., Tran, T., Yu, X., Powell, J.I., Yang, L., Marti, G.E., Moore, T., Hudson, J.Jr., Lu, L., Lewis, D.B., Tibshirani, R., Sherlock, G., Chan, W.C., Greiner, T.C., Weisenburger, D.D., Armitage, J.O., Warnke, R., Levy, R., Wilson, W., Grever, M.R., Byrd, J.C., Botstein, D., Brown, P.O., and Staudt, L.M. Distinct types of diffuse large b-cell lymphoma identified by gene expression profiling. *Nature*, Vol.403:503–511, February 2000.

[7] Alon, U., Barkai, N., Notterman, D.A., Gish, K., Ybarra, S., Mack, D. and Levine, A.J. Broad patterns of gene expression revealed by clustering analysis of tumor and normal colon tissues probed by oligonucleotide array. *Proc. Natl. Acad. Sci. USA*, Vol. 96(12):6745–6750, June 1999.

[8] Alter, O., Brown, P. O., and Botstein, D. Generalized Singular Value Decomposition for Comparative Analysis of Genome-Scale Expression

Data Sets of Two Different Organisms. *Proc. Natl. Acad. Sci. USA*, Vol. 100(6):3351–3356, March 2003.

[9] Alter, O., Brown, P.O. and Bostein, D. Singular value decomposition for genome-wide expression data processing and modeling. *Proc. Natl. Acad. Sci. USA*, Vol. 97(18):10101–10106, Auguest 2000.

[10] Amaratunga, D. and Cabrera, J. *Exploration and analysis of DNA microarray and protein array data.* Wiley Series in Probability and Statistics, 2003.

[11] Ankerst, M., Breunig, M.M., Kriegel, H., Sander, J. OPTICS: Ordering Points To Identify the Clustering Structure. *Sigmod*, pages 49–60, 1999.

[12] Asimov, D. The Grand Tour: A Tool for Viewing Multidimensional Data. *SIAM Journal of Scientific and Statistical Computing*, 6(2):128–143, 1985.

[13] Bader, G.D. and Hogue, C.W. Analyzing yeast protein-protein interaction data obtained from different sources. *Nat. Biotechnol.*, 20(10):991–997, Oct. 2002.

[14] Balakrishnama, S., Ganapathiraju, A. Linear Discriminant Analysis - A Brief Tutorial. Technical report, Institute for Signal and Information Processing, Department of Electrical and Computer Engineering, Mississippi State University.

[15] Bar-Joseph, Z., Demaine, E.D., Gifford, D.K., Srebro, N., Hamel, A.M. and Jaakkola, T.S. K-ary clustering with optimal leaf ordering for gene expression data. *Bioinformatics*, 19(9):1070–1078, 2003.

[16] Bar-Joseph, Z., Gerber, G., Gifford, D.K., Jaakkola, T.S. and Simon, I. A new apporach to analyzing gene expression time series data. In *Proc. 6th Annual International Conference on Computational Molecular Biology*, pages 39–48, 2002.

[17] Basu, S., Banerjee, A. and Mooney, R.J. Semi-supervised clustering by seeding. In *Proceedings of 19th International Conference on Machine Learning (ICML'02)*, pages 19–26, 2002.

[18] Basu, S., Bilenko, M. and Mooney, R.J. A Probabilistic Framework for Semi-Supervised Clustering. In *Proceedings of the 10th ACM SIGKDD International Conference on Knowledge Discovery and Data Mining (KDD'04)*, pages 59–68, Seattle, WA, August 2004.

[19] Beer, D.G., Kardia, S.L., Huang, C.C., Giordano, T.J., Levin, A.M., Misek, D.E., Lin, L., Chen, G., Gharib, T.G., Thomas, D.G., Lizyness, M.L., Kuick, R., Hayasaka, S., Taylor, J.M., Iannettoni, M.D., Orringer, M.B. and Hanash, S. Gene-expression profiles predict survival of patients with lung adenocarcinoma. *Nat Med.*, 8(8):816–824, Aug 2002.

[20] Bellman, R. *Adaptive Control Processes: A Guided Tour.* Princeton University Press, 1961.

[21] Ben-Dor, A., Bruhn, L., Friedman, N., Nachmann, I., Schummer, M. and Yakhini, Z. Tissue Classification with Gene Expression Profiles. *J. Comput. Biol.*, 7:99–111, 2000.

[22] Ben-Dor, A., Shamir, R. and Yakhini, Z. Clustering gene expression patterns. *Journal of Computational Biology*, 6(3/4):281–297, 1999.

[23] Benjamini, Y. and Hochberg, Y. Controlling the false discovery rate: a

practical and powerful approach to multiple testing. *J. Roy Stat Soc.*, Ser B. 57:289–300, 1995.

[24] Benson, D.A., Karsch-Mizrachi, I., Lipman, D.J., Ostell, J., Rapp, B.A., and Wheeler, D.L. Genbank. *Nucl. Acids Res.*, 30:17–20, 2002.

[25] Beyer, K., Goldstein, J., Ramakrishnan, R. and Shaft, U. When is nearest neighbors meaningful. In *Proceedings of the International Conference of Database Theories*, pages 217–235, 1999.

[26] Bhadra, D. and Garg, A. An interactive visual framework for detecting clusters of a multidimensional dataset. Technical Report 2001-03, Dept. of Computer Science and Engineering, University at Buffalo, NY., 2001.

[27] Bhattacharjee, A., Richards, W.G., Staunton, J., Li, C., Monti, S., Vasa, P., Ladd, C., Beheshti, J., Bueno, R., Gillette, M., Loda, M., Weber, G., Mark, E.J., Lander, E.S., Wong, W., Johnson, B.E., Golub, T.R., Sugarbaker, D.J. and Meyerson, M. Classification of human lung carcinomas by mRNA expression profiling reveals distinct adenocarcinoma subclasses. *Proc Natl Acad Sci U S A*, 98(24):13790–13795, Nov 2001.

[28] Bilenko, M. and Mooney R.J. Adaptive duplicate detection using learnable string similarity measures. In *Proceedings of 9th International Conference on Knowl-edge Discovery and Data Mining (KDD'03)*, pages 39–48, Washington, DC, 2003.

[29] Bittner, M., Meltzer, P., Chen, Y., Jiang, Y., Seftor, E., Hendrix, M., Radmacher, M., Simon, R., Yakhini, Z., Ben-Dor, A., Sampas, N., Dougherty, E., Wang, E., Marincola, F., Gooden, C., Lueders, J., Glatfelter, A., Pollock, P., Carpten, J., Gillanders, E., Leja, D., Dietrich, K., Beaudry, C., Berens, M., Alberts, D., and Sondak, V. Molecular Classification of Cutaneous Malignant Melanoma by Gene Expression Profiling. *Nature*, 406(6795):536–540, August 2000.

[30] Blake, J.A. and Harris, M. *Current Protocols in Bioinformatics*, chapter The Gene Ontology Project: Structured vocabularies for molecular biology and their application to genome and expression analysis. Wiley and Sons, Inc., 2003.

[31] Blatt, M., Wiseman, S. and Domany, E. Superparamagnetic Clustering of Data. *Physical Review Letters*, 76(18):3251–3254, Apr. 1996.

[32] Bo, T. and Jonassen, I. New feature subset selection procedures for classification of expression profiles. *Genome Biol.*, 3(4):research 0017.1–0017.11, Mar 2002.

[33] Boguski, M.S., Lowe, T.M., Tolstoshev, C.M. dbEST—database for "expressed sequence tags". *Nat. Genet.*, 4(4):332–333, Aug 1993.

[34] Bolstad, B.M., Irizarry, R.A., Astrand, M. and Speed, T.P. A comparison of normalization methods for high density oligonucleotide array data based on variance and bias. *Bioinformatics*, 19(2):185–193, 2003.

[35] Bonferroni, C.E. *Il calcolo delle assicurazioni su gruppi di teste*, chapter Studi in Onore del Professore Salvatore Ortu Carboni, pages 13–60. Rome, 1935.

[36] Bonferroni, C.E. Teoria statistica delle classi e calcolo delle probabilità. *Pubblicazioni del Istituto Superiore di Scienze Economiche e Commerciali*

di Firenze, 8:3–62, 1936.

[37] Boser, B.E., Guyon, I.M. and Vapnik, V.N. A training algorithm for optimal margin classifiers. In *Proceedings of the fifth annual workshop on Computational learning theory table of contents*, pages 144–152, Pittsburgh, PA, 1992.

[38] Bowtell, D.D. Options available from start to finish for obtaining expression data by microarray. *Nature Genetics*, 21(Suppl.):25–32, 1999.

[39] Brazma, A. and Vilo, J. Minireview: Gene expression data analysis. *Federation of European Biochemical societies*, 480:17–24, June 2000.

[40] Brazma, A., Parkinson, H., Sarkans, U., Shojatalab, M., Vilo, J., Abeygunawardena, N., Holloway, E., Kapushesky, M., Kemmeren, P., Lara, G.G., Oezcimen, A., Rocca-Serra, P. and Sansone, S.A. ArrayExpress–a public repository for microarray gene expression data at the EBI. *Nucleic Acids Res.*, 31(1):68–71, Jan 2003.

[41] Breiman, L. Baggin predictors. *Machine Learning*, 24:123–140, 1996.

[42] Breiman, L., Friedman, J.H., Olshen, R.A. and Stone C.J. *Classification and Regaression Trees*. Wadsworth, Monterey, CA, 1984.

[43] Breitling, R., Sharif, O., Hartman, M.L. and Krisans, S.K. Loss of compartmentalization causes misregulation of lysine biosynthesis in peroxisome-deficient yeast cells. *Eukaryot Cell*, 1(6):978–986, Dec. 2002.

[44] Brown, M.P.S., Grundy, W.N., Lin, D., Cristianini, N., Sugnet, C.W., Furey, T.S., Ares, M.Jr. and Haussler, D. Knowledge-based analysis of microarray gene expression data using support vector machines. *Proc. Natl. Acad. Sci.*, 97(1):262–267, January 2000.

[45] Buja, A., Cook, D. and Swayne, D. Interactive High-Dimensional Data Visualization. *Journal of Computational and Graphical Statistics*, 5(1):78–99, 1996.

[46] Cadzow, J. A., Landingham, H. F. *Signals, Systems, and Transforms*. Prentice-Hall, Inc., Englewood Cliffs, NJ, 1985.

[47] Chambers, J., Angulo, A., Amaratunga, D., Guo, H., Jiang, Y., Wan, J.S., Bittner, A., Frueh, K., Jackson, M.R., Peterson, P.A., Erlander, M.G. and Ghazal, P. DNA microarrays of the complex human cytomegalovirus genome: profiling kinetic class with drug sensitivity of viral gene expression. *J. Virol.*, 73(7):5757–5766, 1999.

[48] Šidák, Z. Rectangular confidence regions for the means of multivariate normal distributions. *Journal of the American Statistical Association*, 62:626–633, 1967.

[49] Chen, J.J., Wu, R., Yang, P.C., Huang, J.Y., Sher, Y.P., Han, M.H., Kao, W.C., Lee, P.J., Chiu, T.F., Chang, F., Chu, Y.W., Wu, C.W. and Peck, K. Profiling expression patterns and isolating differentially expressed genes by cDNA microarray system with colorimetry detection. *Genomics*, 51:313–324, 1998.

[50] Chen, X., Cheung, S. T., So, S., Fan, S. T., Barry, C., Higgins, J., Lai, K. M., Ji, J., Dudoit, S., Ng, I. O., Van De Rijn M., Botstein, D., and Brown, P. O. Gene Expression Patterns in Human Liver Cancers. *Mol Biol Cell.*, 13(6):1929–1939, June 2002.

[51] Cheng, C.H., Fu, A.W. and Zhang, Y. Entropy-based subspace clustering for mining numerical data. In *Proceedings of the fifth ACM SIGKDD international conference on Knowledge discovery and data mining*, pages 84–93, San Diego, CA, 1999.

[52] Cheng, Y. and Church, G.M. Biclustering of expression data. *Proceedings of ISMB'00*, 8:93–103, 2000.

[53] Cheung, V.G., Morley, M., Aguilar, F., Massimi, A., Kucherlapati, R. and Childs, G. Making and reading microarrays. *Nature Genetics*, 21(Suppl.):15–19, 1999.

[54] Cho, R.J., Campbell, M.J., Winzeler, E.A., Steinmetz, L., Conway, A., Wodicka, L., Wolfsberg, T.G., Gabrielian, A.E., Landsman, D., Lockhart, D.J., and Davis, R.W. A Genome-Wide Transcriptional Analysis of the Mitotic Cell Cycle. *Molecular Cell*, Vol. 2(1):65–73, July 1998.

[55] Choi, J.K., Yu, U., Kim, S. and Yoo, O.J. Combining multiple microarray studies and modeling interstudy variation. *Bioinformatics*, 19(Suppl 1):i84–90, 2003.

[56] Chu, S., DeRisi, J., Eisen, M., Mulholland, J., Botstein, D., Brown, P.O., et al. The transcriptional program of sporulation in budding yeast. *Science*, 282(5389):699–705, 1998.

[57] Chung, C.H., Bernard, P.S. and Perou, C.M. Molecular portraits and the family tree of cancer. *Nature Genetics*, 32(Suppl):533–540, 2002.

[58] Cleveland, W.S. Robust locally weighted regression and smoothing scatterplots. *J. Amer. Stat. Assoc.*, 74:829–836, 1979.

[59] Cohen, J. A Coefficient of Agreement for Nominal Scales. *Educational and Psychological Measurement*, 20(1):37–46, 1960.

[60] Cohn, D., Caruana, R. and McCallum, A. Semi-supervised clustering with user feedback. In *Technical Report TR2003-1892, Cornell University.*, 2003.

[61] Collins, F.S. Microarrays and macroconsequences. *Nature genetics*, 21(1):2, 1999.

[62] Cook, C., Buja, A., Cabrera, J. and Hurley, C. Grand Tour and Projection Pursuit. *Journal of Computational and Graphical Statistics*, 2(3):225–250, 1995.

[63] Cooley, J. W. and Tukey, J. W. An Algorithm for The Machine Calculation of Complex Fourier Series. *Mathematics of Computation*, 19(90):297–301, 1965.

[64] Cover, T.M. and Thomas J.A. *Elements of Information Theory.* John Wiley & Sons, Inc., New York, 1991.

[65] Crick, F. Central dogma of molecular biology. *Nature*, 227(5258):561–563, Aug. 1970.

[66] Davision, M. L. *Multidimensional Scaling.* Krieger Publishing, Inc., Malabar, FL, 1992.

[67] Debouck, C. and Goodfellow, P.N. DNA microarrays in drug discovery and development. *Nat. Genet.*, 21(1 Suppl):48–50, 1999.

[68] Demiriz, A., Bennett, K. and Embrechts, M. *Intelligent Engineering Systems Through Artificial Neural Networks 9*, chapter Semi-supervised clustering using genetic algorithms., pages 809–814. ASME Press, 1999.

[69] Dempster, A.P., Laird, N.M. and Rubin, D.B. Maximal Likelihood from Incomplete Data Via the EM Algorithm. *Journal of the Royal Statistical Society*, Ser B(39):1–38, 1977.

[70] Deng, L., Pei, J., Ma J. and Lee D.L. A Rank Sum Test Method for Informative Gene Discovery. In *Proceedings of the 10th ACM SIGKDD International Conference on Knowledge Discovery and Data Mining (KDD'04)*, pages 410–419, Seattle, WA, Aug. 2004.

[71] Deng, M., Chen, T. and Sun, F. An integrated probabilistic model for functional prediction of proteins. In *Recomb'03*, pages 95–103, 2003.

[72] Der, S.D., Zhou, A., Williams, B.R. and Silverman, R.H. Identification of genes differentially regulated by interferon alpha, beta, or gamma using oligonucleotide arrays. *Proc Natl Acad Sci*, 95(26):15623–15628, 1998.

[73] DeRisi, J., Penland, L., Brown, P.O., Bittner, M.L., Meltzer, P.S., Ray, M., Chen, Y., Su, Y.A. and Trent, J.M. Use of a cDNA microarray to analyse gene expression patterns in human cancer. *Nature Genetics*, 14:457–460, 1996.

[74] DeRisi, J.L., Iyer, V.R. and Brown, P.O. Exploring the metabolic and genetic control of gene expression on a genomic scale. *Science*, pages 680–686, 1997.

[75] Detours, V., Dumont, J.E., Bersini, H. and Maenhaut, C. Integration and cross-validation of high-throughput gene expression data: comparing heterogeneous data sets. *FEBS Lett.*, 546(1):98–102, Jul. 2003.

[76] Devore, J. and Peck, R. *Statistics: the Exploration and Analysis of Data*. Duxbury, Pacific Grove, California, third edition.

[77] D'haeseleer, P., Wen, X., Fuhrman, S. and Somogyi, R. Mining the Gene Expression Matrix: Inferring Gene Relationships From Large Scale Gene Expression Data. *Information Processing in Cells and Tissues*, pages 203–212, 1998.

[78] Dhanasekaran, S.M., Barrette, T.R., Ghosh, D., Shah, R., Varambally, S., Kurachi, K., Pienta, K.J., Rubin, M.A. and Chinnaiyan, A.M. Delineation of prognostic biomarkers in prostate cancer. *Nature*, 412(6849):822–826, 2001.

[79] Ding, C. Analysis of gene expression profiles: class discovery and leaf ordering. In *Proc. of International Conference on Computational Molecular Biology (RECOMB)*, pages 127–136, Washington, DC., April 2002.

[80] J. Dopazo and JM. Carazo. Phylogenetic reconstruction using an unsupervised growing neural network that adopts the topology of a phylogenetic tree. *Journal of Molecular Evolution*, 44:226–233, 1997.

[81] Dopazo, J., Zanders, E., Dragoni, I., Amphlett, G. and Falciani, F. Methods and approaches in the analysis of gene expression data. *Immunol. Meth.*, 250:93–112, 2001.

[82] Draghici, S. Statistical intelligence: effective analysis of high-density microarray data. *Drug Discov Today*, 7(11 Suppl):S55–63, 2002.

[83] Drăghici, S. *Data analysis tools for DNA microarrays*. Chapman & Hall/CRC, 2003.

[84] Dubes, R. and Jain, A. *Algorithms for Clustering Data*. Prentice Hall, 1988.

[85] Duda, R.O., Hart, P.E. and Stork, D.H. *Pattern Classification.* Wiley Interscience, 2000.

[86] Dudoit, S., Fridlyand J. and Speed, T.P. Comparison of Discrimination Methods for the Classification of Tumors Using Gene Expression Data. *Journal of the American Statistical Association,* pages 77–87, 2002.

[87] Dudoit, S., Yang, Y.H., Callow, M. and Speed, T. Statistical methods for identifying differentially expressed genes in replicated cDNA microarray experiments. Technical Report 578, University of California, Berkeley, 2000.

[88] Duggan, D.J., Bittner, M., Chen, Y., Meltzer, P. and Trent, J.M. Expression profiling using cDNA microarrays. *Nat. Genet.,* 21(1 Suppl):10–14, 1999.

[89] Dunham, I., Shimizu, N., Roe, B.A., Chissoe, S., Hunt, A.R., Collins, J.E., Bruskiewich, R., Beare, D.M., Clamp, M., Smink, L.J., Ainscough, R., Almeida, J.P., Babbage, A., Bagguley, C., Bailey, J., Barlow, K., Bates, K.N., Beasley, O., Bird, C.P., Blakey, S., Bridgeman, A.M., Buck, D., Burgess, J., Burrill, W.D., O'Brien, K.P. The DNA sequence of human chromosome 22. *Nature,* 402(6761):489–95, Dec 1999.

[90] Dysvik, B. and Jonassen, I. J-Express: exploring gene expression data using Java. *Bioinfomatics,* 17(4):369–370, 2001. Applications Note.

[91] Efron, B. The Jackknife, the Bootstrap, and Other Resampling Plans. *CBMS-NSF Regional Conference Series in Applied Mathematics,* 38, 1982.

[92] Eisen, M.B., Spellman, P.T., Brown, P.O. and Botstein, D. Cluster Analysis and Display of Genome-wide Expression Patterns. *Proc. Natl. Acad. Sci. USA,* 95(25):14863–14868, December 1998.

[93] Ekins, R.P. and Chu, R.W. Microarrays: their origins and applications. *Trnds in Biotechnology,* 17:217–218, 1999.

[94] Ermolaeva, O., Rastogi, M., Pruitt, K.D., Schuler, G.D., Bittner, M.L., Chen, Y., Simon, R., Meltzer, P., Trent, J.M. and Boguski, M.S. Data management and analysis for gene expression arrays. *Nature Genetics,* 20:19–23, 1998.

[95] Ester, M., Kriegel, H., Sander, J., Xu, X. A Density-Based Algorithm for Discovering Clusters in Large Spatial Databases with Noise. In *Proceedings of 2nd International Conference on KDD,* pages 226–231, Portland, OR, 1996.

[96] Farabee, M.J. *On-Line Biology Book. http://www.emc.maricopa.edu/ faculty/farabee/BIOBK/BioBookTOC.html.*

[97] Fayyad, U., Grinstein, G., Wierse, A., editor. *Information Visualization in Data Mining and Knowledge Discovery.* Morgan Kaufmann Publishers, San Diego, CA, 2002.

[98] Ferrando, A.A., Neuberg, D.S., Staunton, J., Loh, M.L., Huard, C., Raimondi, S.C., Behm, F.G., Pui, C.H., Downing, J.R., Gilliland, D.G., Lander, E.S., Golub, T.R. and Look A.T. Gene expression signatures define novel oncogenic pathways in T cell acute lymphoblastic leukemia. *Cancer Cell,* 1(1):75–87, Feb 2002.

[99] Fisher, R. A. The Use of Multiple Measurements on Taxonomic Problems. *Annals of Eugenics,* 7(2):179–188, 1936.

[100] Fleischmann, R.D., Adams, M.D., White, O., Clayton, R.A., Kirkness, E.F., Kerlavage, A.R., Bult, C.J., Tomb, J.F., Dougherty BA, Merrick JM, et al. Whole-genome random sequencing and assembly of Haemophilus influenzae Rd. *Science*, 269(5223):496–512, Jul 1995.

[101] Fodor, S.P., Read, J.L., Pirrung, M.C., Stryer, L., Lu, A.T. and Solas, D. Light-directed, spatially addressable parallel chemical synthesis. *Science*, 251(4995):767–773, 1991.

[102] Fraley, C. and Raftery, A.E. How Many Clusters? Which Clustering Method? Answers Via Model-Based Cluster Analysis. *The Computer Journal*, 41(8):578–588, 1998.

[103] Fua, Y. H., Ward, M. O. and Rundensteiner, E. A. Hierarchical Parallel Coordinates for Exploration of Large Datasets. In *IEEE Visualization '99*, pages 43–50, 1998.

[104] Furey, T.S., Cristianini, N., Duffy, N., Bednarski, D.W., Schummer, M., and Haussler, D. Support Vector Machine Classification and Validation of Cancer Tissue Samples Using Microarray Expression Data. *Bioinformatics*, Vol.16(10):909–914, 2000.

[105] Garber, M.E., Troyanskaya, O.G., Schluens, K., Petersen, S., Thaesler, Z., Pacyna-Gengelbach, M., van de Rijn, M., Rosen, G.D., Perou, C.M., Whyte, R.I., Altman, R.B., Brown, P.O., Botstein, D. and Petersen, I. Diversity of gene expression in adenocarcinoma of the lung. *Proc Natl Acad Sci U S A*, 98(24):13784–13789, Nov. 2001.

[106] Gasch, A.P., Huang, M., Metzner, Botstein, D., S., Elledge, S.J. and Brown, P.O. Genomic Expression Responses to DNA-damaging Agents and the Regulatory Role of the Yeast. *Molecular Biology of the Cell*, 12:2987–3003, Oct. 2001.

[107] Gasch, A.P., Spellman, P.T., Kao, C.M., Carmel-Harel, O., Eisen, M.B., Storz, G., Botstein, D., and Brown, P.O. Genomic Expression Programs in the Response of Yeast Cells to Environmental Changes. *Molecular Biology of the Cell*, 11:4241–4257, 2000.

[108] Genetics-Editorial. Coming to terms with microarrays. *Nature Genetics*, 32(Suppl.):333–334, 2002.

[109] Getz, G., Levine, E. and Domany, E. Coupled two-way clustering analysis of gene microarray data. *Proc. Natl. Acad. Sci. USA*, Vol. 97(22):12079–12084, October 2000.

[110] Getz, G., Levine, E., Domany, E. and Zhang, M.Q. Super-paramagnetic clustering of yeast gene expression profiles. *Phsica A*, 279:457–464, 2000.

[111] Ghosh, D. and Chinnaiyan, A.M. Mixture modelling of gene expression data from microarray experiments. *Bioinformatics*, 18:275–286, 2002.

[112] Goffeau, A., Barrell, B.G., Bussey, H., Davis, R.W., Dujon, B., Feldmann, H., Galibert, F., Hoheisel, J.D., Jacq, C., Johnston, M., Louis, E.J., Mewes, H.W., Murakami, Y., Philippsen, P., Tettelin, H., Oliver, S.G. Life with 6000 genes. *Science*, 274(5287):546,563–567, Oct 1996.

[113] Gollub, J., Ball, C.A., Binkley, G., Demeter, J., Finkelstein, D.B., Hebert, J.M., Hernandez-Boussard, T., Jin, H., Kaloper, M., Matese, J.C., Schroeder, M., Brown, P.O., Botstein, D. and Sherlock, G. The Stanford

Microarray Database: data access and quality assessment tools. *Nucleic Acids Res.*, 31(1):94–96, Jan 2003.

[114] Golub, T.R., Slonim, D.K., Tamayo, P., Huard, C., Gassenbeek, M., Mesirov, J.P., Coller, H., Loh, M.L., Downing, J.R., Caligiuri, M.A., Bloomfield, D.D. and Lander, E.S. Molecular classification of cancer: Class discovery and class prediction by gene expression monitoring. *Science*, Vol. 286(15):531–537, October 1999.

[115] Gray, N.S., Wodicka, L., Thunnissen, A.M., Norman, T.C., Kwon, S., Espinoza, F.H., Morgan, D.O., Barnes, G., LeClerc, S., Meijer, L., Kim, S.H., Lockhart, D.J. and Schultz, P.G. Exploiting chemical libraries, structure, and genomics in the search for kinase inhibitors. *Science*, 281:533–538, 1998.

[116] Grira, N., Crucianu, M. and Boujemaa, N. Unsupervised and Semi-supervised Clustering: A Brief Survey.
http://www.ercim.org/pub/bscw.cgi/d28935/Muscle-WP8-grira.pdf.

[117] Gruvberger, S., Ringner, M., Chen, Y., Panavally, S., Saal, L.H., Borg, A., Ferno, M., Peterson, C. and Meltzer, P.S. Estrogen receptor status in breast cancer is associated with remarkably distinct gene expression patterns. *Cancer Res.*, 61(16):5979–5984, Aug. 2001.

[118] Hakak, Y., Walker, J.R., Li, C., Wong, W.H., Davis, K.L., Buxbaum, J.D., Haroutunian, V. and Fienberg, A.A. Genome-Wide Expression Analysis Reveals Dysregulation of Myelination-Related Genes in Chronic Schizophrenia. *Proc. Natl. Acad. Sci. USA*, Vol. 98(8):4746–4751, April 2001.

[119] Halkidi, M., Batistakis, Y. and Vazirgiannis, M. On Clustering Validation Techniques. *Intelligent Information Systems Journal*, 17(2-3):107–145, 2001.

[120] Han, J. and Kamber, M. *Data Mining: Concept and Techniques*. The Morgan Kaufmann Series in Data Management Systems. Morgan Kaufmann Publishers, August 2000.

[121] Han, J., Pei, J. and Yin, Y. Mining frequent patterns without candidate generation. *In proc. 2000 ACM-SIGMOD Int. Conf. Management of Data (SIGMOD'00)*, pages 1–12, May 2000.

[122] Hanisch, D., Zien, A., Zimmer, R. and Lengauer, T. Co-clustering of biological networks and gene expression data. *Bioinformatices*, 18(Suppl 1): S145–54, 2002.

[123] Hartigan, J.A. *Clustering Algorithm*. John Wiley and Sons, New York., 1975.

[124] Hartuv, E. and Shamir, R. A clustering algorithm based on graph connectivity. *Information Processing Letters*, 76(4–6):175–181, 2000.

[125] Hastie, T., Tibshirani, R., Boststein, D. and Brown, P. Supervised harvesting of expression trees. *Genome Biology*, Vol. 2(1):0003.1–0003.12, January 2001.

[126] Hastie, T., Tibshirani, R., Eisen, M.B., Alizadeh, A., Levy, R., Staudt, L., Chan, W.C., Botstein, D. and Brown, P. 'Gene shaving' as a method for identifying distinct sets of genes with similar expression patterns. *Genome Biology*, 1:1–21, 2000.

[127] Hattori, M., Fujiyama, A., Taylor, T.D., Watanabe, H., Yada, T., Park, H.S., Toyoda, A., Ishii, K., Totoki, Y., Choi, D.K., Soeda, E., Ohki, M., Takagi, T., Sakaki, Y., Taudien, S., Blechschmidt, K., Polley, A., Menzel, U., Delabar, J., Kumpf, K., Lehmann, R., Patterson, D., Reichwald, K., Rump, A., Schillhabel, M., Schudy, A. The DNA sequence of human chromosome 21. The chromosome 21 mapping and sequencing consortium. *Nature*, 405(6784):311–319, May 2000.

[128] Hearst, M.A., Schlkopf, B., Dumais, S., Osuna, E. and Platt. J. Trends and controversies - support vector machines. *IEEE Intelligent Systems*, 13(4):18–28, 1998.

[129] Hedenfalk, I., Duggan, D., Chen, Y. D., Radmacher, M., Bittner, M., Simon, R., Meltzer, P., Gusterson, B., Esteller, M., Kallioniemi, O. P., Wilfond, B., Borg, A., and Trent, J. Gene-expression profiles in hereditary breast cancer. *The New England Journal of Medicine*, 344(8):539–548, February 2001.

[130] Heller, R.A., Schena, M., Chai, A., Shalon, D., Bedilion, T., Gilmore, J., Woolley, D.E. and Davis, R.W. Discovery and analysis of inflammatory disease-related genes using cDNA microarrays. *Proc. Natl. Acad. Sci. USA*, 94:2150–2155, 1997.

[131] Heller, R.A., Schena, M., Chai, A., Shalon, D., Bedilion, T., Gilmore, J., Woolley, D.E. and Davis, R.W. Discovery and analysis of inflammatory disease-related genes using cDNA microarrays. *Proc. Natl. Acad. Sci. USA*, 94:2150–2155, 1997.

[132] Herrero, J., Valencia, A. and Dopazo, J. A hierarchical unsupervised growing neural network for clustering gene expression patterns. *Bioinformatics*, 17:126–136, 2001.

[133] Heyer, L.J., Kruglyak, S. and Yooseph, S. Exploring expression data: identification and analysis of coexpressed genes. *Genome Res*, 9(11):1106–1115, 1999.

[134] Hinneburg, A. and Keim, D.A. An efficient approach to clustering in large multimedia database with noise. *Proc. 4th Int. Con. on Knowledge discovery and data mining*, pages 58–65, 1998.

[135] Hipp, J., Gntzer, U. and Nakhaeizadeh, G. Algorithms for association rule mining a general survey and comparison. *ACM SIGKDD Exploration*, 2(Issue 1):58–64, 2000.

[136] Hippo, Y., Taniguchi, H., Tsutsumi, S., Machida, N., Chong, J.M., Fukayama, M., Kodama, T. and Aburatani, H. Global gene expression analysis of gastric cancer by oligonucleotide microarrays. *Cancer Res.*, 62(1):233–240, Jan 2002.

[137] Hoffman, P., Grinstein, G. G., and Pinkney, D. Dimensional Anchors: A Graphic Primitive for Multidimensional Multivariate Information Visualizations. In *Workshop on New Paradigms in Information Visualization and Manipulation (NPIVM '99), in conjunction with CIKM '99*, pages 9–16, Kansas City, Missouri, November 1999. ACM.

[138] Hoffman, P.E., Grinstein, G.G., Marx, K., Grosse, I., and Stanley, E. Dna visual and analytic data mining. In *IEEE Visualization '97*, pages 437–441,

Phoenix, AZ, 1997.

[139] Hofmann, W.K., de Vos, S., Elashoff, D., Gschaidmeier, H., Hoelzer, D., Koeffler, H.P. and Ottmann, O.G. Relation between resistance of Philadelphia-chromosome-positive acute lymphoblastic leukaemia to the tyrosine kinase inhibitor STI571 and gene-expression profiles: a gene-expression study. *Lancet*, 359(9305):481–486, Feb 2002.

[140] Holland, J.H. *Adaptation in Natural and Artificial Systems*. The University of Michigan Press, Ann Arbor, IL, 1975.

[141] Hollander, M. and Wolfe D.A. *Nonparametric Statistical Method*. New York: Wiley, second edition, 1999.

[142] Holm, S. A simple sequentially rejective multiple test procedure. *Scandinavian Journal of Statistics*, 6:65–70, 1979.

[143] Holter, N.S., Mitra, M., Maritan, A., Cieplak, M., Banavar, J.R. and Fedoroff, N.V. Fundamental patterns underlying gene expression profiles: simplicity from complexity. *Proc Natl Acad Sci*, 97(15):8409–8414, 2000.

[144] Iacobuzio-Donahue, C. A., Maitra, A., Olsen, M., Lowe, A. W., van Heek, N. T., Rosty, C., Walter, K., Sato, N., Parker, A., Ashfaq, R., Jaffee, E., Ryu, B., Jones, J., Eshleman, J. R., Yeo, C. J., Cameron, J. L., Kern, S. E., Hruban, R. H., Brown, P. O., and Goggins, M. Exploration of Global Gene Expression Patterns in Pancreatic Adenocarcinoma Using cDNA Microarrays. *Am J Pathol*, 162(4):1151–1162, April 2003.

[145] Inselberg, A. N-dimensional graphics, part I – lines and hyperplanes. Technical Report G320-2711, IBM Los Angeles Scientific Center, IBM Scientific Center, Los Angeles, CA, 1981.

[146] Inselberg, A. The plane with parallel coordinates. *The Visual Computer*, 1:69–91, 1985.

[147] Iyer, V.R., Eisen, M.B., Ross, D.T., Schuler, G., Moore, T., Lee, J.C.F., Trent, J.M., Staudt, L.M., Hudson, Jr. J., Boguski, M.S., Lashkari, D., Shalon, D., Botstein, D. and Brown, P.O. The transcriptional program in the response of human fibroblasts to serum. *Science*, 283:83–87, 1999.

[148] Jagadish, H.V., Madar, J. and Ng, R. Semantic compression and pattern extraction with fasicicles. In *VLDB*, pages 186–196, 1999.

[149] Jain, A.K., Murty, M.N. and Flynn, P.J. Data clustering: a review. *ACM Computing Surveys*, 31:264–323, 1999.

[150] Jakt, L.M., Cao, L., Cheah, K.S.E., Smith, D.K. Assessing clusters and motifs from gene expression data. *Genome research*, 11:112–123, 2001.

[151] Jazaeri, A.A., Yee, C.J., Sotiriou, C., Brantley, K.R., Boyd, J. and Liu, E.T. Gene expression profiles of BRCA1-linked, BRCA2-linked, and sporadic ovarian cancers. *J. Natl. Cancer Inst.*, 94(13):990–1000, Jul. 2002.

[152] Jiang, D., Pei, J. and Zhang, A. An Interactive Approach to Mining Gene Expression Data. *IEEE Transactions on Knowledge and Data Engineering (TKDE)*, 17(10):1363–1378.

[153] Jiang, D., Pei, J. and Zhang, A. Interactive Exploration of Coherent Patterns in Time-Series Gene Expression Data. In *In Proceedings of the Ninth ACM SIGKDD International Conference on Knowledge Discovery and Data Mining (KDD'03)*, pages 565–570, Washington, DC, USA, Au-

gust 24-27 2003.

[154] Jiang, D., Pei, J. and Zhang A. Towards Interactive Exploration of Gene Expression Patterns. *ACM SIGKDD Explorations (Special Issue on Microarray Data Analysis)*, 5(2):79–90, 2003.

[155] Jiang, D., Pei, J. and Zhang A. Mining Coherent Gene Clusters from Gene-Sample-Time Microarray Data. In *Proceedings of ACM SIGKDD International Conference on Knowledge Discovery and Data Mining (KDD'04)*, pages 430–439, Seattle, WA, August 22-25 2004.

[156] Jiang, D., Pei, J., Ramanathan, M., Lin, C., Tang, C. and Zhang, A. Mining Gene-Sample-Time Microarray Data: A Coherent Gene Cluster Discovery Approach. *Knowledge and Information Systems (KAIS), Springer.*

[157] Jolliffe, I.T. *Principal Component Analysis*. Springer, New York, 1986.

[158] Kandogan, E. Star Coordinates: a Multi-dimensional Visualization Technique with Uniform Treatment of Dimensions. In *Proceedings of the seventh ACM SIGKDD international conference on Knowledge discovery and data mining*, pages 107–116, 2001.

[159] Karp, G. *Cell and Molecular Biology: Concepts and Experiments*. John Wiley & Sons Inc, 2002.

[160] Kasturi, J., Acharya, R. and Ramanathan, M. An information theoretic approach for analyzing temporal patterns of gene expression. *Bioinformatics*, 19(4):449–458, Mar. 2003.

[161] Kaufman, L. and Rousseeuw, P.J. *Finding Groups in Data: an Introduction to Cluster Analysis*. John Wiley and Sons, 1990.

[162] Keim, D. A. Designing Pixel-Oriented Visualization Techniques: Theory and Applications. *IEEE Transactions on Visualization and Computer Graphics*, Vol. 6(1):1–20, January 2000.

[163] Keim, D. A., and Kriegel, H. P. Visualization Techniques for Mining Large Databases: A Comparison. *Transactions on Knowledge and Data Engineering, Special Issue on Data Mining*, Vol. 8(6):923–938, December 1996.

[164] Kemmeren, P., van Berkum, N.L., Vilo, J., Bijma, T., Donders, R., Brazma, A. and Holstege, F.C.P. Protein interaction verification and functional annotation by integrated analysis of genome-scale data. *Molecular Cell*, 9:1133–1143, 2002.

[165] Kerr, K. and Churchill, G. Statistical design and the analysis of gene expression microarrays. *Genetical Research*, 77:123–128, 2001.

[166] Kerr, M.K., Martin, M. and Churchill, G.A. Analysis of variance for gene expression microarray data. *J. Comp. Biol.*, 7(6):819–837, 2000.

[167] Khan, J., Wei, J.S., Ringnr, M., Saal, L.H., Ladanyi, M., Westermann, F., Berthold, F., Schwab, M., Antonescu, C.R., Peterson, C., and Meltzer, P.S. Classification and Diagnostic Prediction of Cancers Using Gene Expression Profiling and Artificial Neural Networks. *Nature Medicine*, Vol.7(6):673–679, 2001.

[168] Khatri, P. and Draghici, S. Ontological Analysis of Gene Expression Data: Current Tools, Limitations, and Open Problems. *Bioinformatics*, 21(18):3587–3595, 2005.

[169] Kirkpatrick, S., Gelatt, C.D. Jr., and Vecchi, M.P. Optimization by simu-

lated annealing. *Science*, 220(4598):671–680, 1983.

[170] Klein, D., Kamvar, S.D. and Manning, C.D. From instance-level constraints to space-level constraints: Making the most of prior knowledge in data clustering. In *Proceedings of 19th International Conference on Machine Learning (ICML'02)*, pages 307–314, 2002.

[171] Kohonen, T. *Self-Organization and Associative Memory*. Spring-Verlag, Berlin, 1994.

[172] Lander, E.S. The new genomics: global views of biology. *Science*, 274(5287):536–539, Oct 1996.

[173] Lander, E.S. Array of hope. *Nat Genet.*, 21(1 Suppl):3–4, Jan 1999.

[174] Landis, J., and Koch, G.G. The Measurement of Observer Agreement for Categorical Data. *Biometrics*, 33:159–174, 1977.

[175] LaTulippe, E., Satagopan, J., Smith, A., Scher, H., Scardino, P., Reuter, V. and Gerald, W.L. Comprehensive gene expression analysis of prostate cancer reveals distinct transcriptional programs associated with metastatic disease. *Cancer Res.*, 62(15):4499–4506, Aug 2002.

[176] Lazzeroni, L. and Owen, A. Plaid models for gene expression data. *Statistica Sinica*, 12(1):61–86, 2002.

[177] Lee, H.K., Hsu, A.K., Sajdak, J., Qin, J. and Pavlidis, P. Coexpression analysis of human genes across many microarray data sets. *Genome Res.*, 14(6):1085–1094, Jun. 2004.

[178] Lee, M.L., Kuo, F.C., Whitmore, G.A., Sklar, J. Importance of replication in microarray gene expression studies: statistical methods and evidence from repetitive cDNA hybridizations. *Proc Natl Acad Sci USA*, 97(18):9834–9839, 2000.

[179] Levine, E., Domany, E. Resampling methods for unsupervised estimation of cluster validity. *Neural computation*, 13:2573–2593, 2001.

[180] Lewin, B. *Genes VIII*. Prentice Hall, 2003.

[181] Li, L., Darden, T.A., Weinberg, C.R., Levine, A.J. and Pedersen, L.G. Gene assessment and sample classification for gene expression data using a genetic algorithm/k-nearest neighbor method. *Combinatorial Chemistry & High Throughput Screening*, 4(8):727–739, 2001.

[182] Li, L., Weinberg, C.R., Darden, T.A., and Pedersen, L.G. Gene selection for sample classification based on gene expression data: study of sensitivity to choice of parameters of the ga/knn method. *Bioinformatics*, 17:1131–1142, 2001.

[183] Lin, Y.M., Furukawa, Y., Tsunoda, T., Yue, C.T., Yang, K.C. and Nakamura, Y. Molecular diagnosis of colorectal tumors by expression profiles of 50 genes expressed differentially in adenomas and carcinomas. *Oncogene*, 21(26):4120–4128, Jun. 2002.

[184] Lipshutz, R.J., Fodor, S.P.A., Gingeras, T.R. and Lockhart, D.J. High density synthetic oligonucleotide arrays. *Nature genetics*, 21(1):20–24, 1999.

[185] Liu, J. and Wang, W. OP-Cluster: Clustering by Tendency in High Dimensional Space. *Proceedings of the Third IEEE International Conference on Data Mining (ICDM'03)*, pages 187–194, November 19-22 2003.

[186] Luan, Y. and Li, H. Clustering of time-course gene expression data using

a mixed-effects model with B-splines. *Bioinformatics*, 19(4):474–482, 2003.

[187] MacDonald, T.J., Brown, K.M., LaFleur, B., Peterson, K., Lawlor, C., Chen, Y., Packer, R.J., Cogen, P. and Stephan, D.A. Expression profiling of medulloblastoma: PDGFRA and the RAS/MAPK pathway as therapeutic targets for metastatic disease. *Nat. Genet.*, 29(2):143–152, 2001.

[188] MacQueen, J.B. Some methods for classification and analysis of multivariate observations. In *Proceedings of the Fifth Berkeley Symposium on Mathematical Statistics and Probability*, volume 1, pages 281–297, Univ.of California, Berkeley, 1967. Univ.of California Press, Berkeley.

[189] Madeira, S.C. and Oliveira, A.L. Biclustering Algorithms for Biological Data Analysis: A Survey. *IEEE/ACM Transactions on Computational Biology and Bioinformatics*, 1(1):24–45, January 2004.

[190] Marcotte, E.M., Pellegrini, M., Thompson, M.J., Yeates, T.O. and Eisenberg, D. A combined algorithm for genome-wide prediction of protein function. *Nature*, 402:83–86, 1999.

[191] Marton, M.J., DeRisi, J.L., Bennett, H.A., Iyer, V.R., Meyer, M.R., Roberts, C.J., Stoughton, R., Burchard, J., Slade, D., Dai, H., Bassett, D.E. Jr., Hartwell, L.H., Brown, P.O. and Friend, S.H. Drug target validation and identification of secondary drug target effects using DNA microarrays. *Nature Med*, 4:1293–1301, 1998.

[192] Mavroudi, S., Papadimitriou, S. and Bezerianos, A. Gene expression data analysis with a dynamically extended self-organized map that exploits class information. *Bioinformatics*, 18:1446–1453, 2002.

[193] Mclachlan, G.J. *Analyzing Microarray Gene Expression Data*. Wiley, 2004.

[194] McLachlan, G.J., Bean, R.W. and Peel, D. A mixture model-based approach to the clustering of microarray expression data. *Bioinformatics*, 18:413–422, 2002.

[195] Meinke, D.W., Cherry, J.M., Dean, C., Rounsley, S.D., Koornneef, M. Arabidopsis thaliana: a model plant for genome analysis. *Nature*, 282(5389):662,679–682, Oct 1998. Review.

[196] Mekalanos, J.J. Environmental signals controlling expression of virulence determinants in bacteria. *J. Bacteriol*, 174:1–7, 1992.

[197] Michell, T.M. *Machine learning*. McGraw Hill, 1997.

[198] Model, F., Adorjan, P., Olek, A. and Piepenbrock, C. Feature selection for DNA methylation based cancer classification. *Bioinformatics*, 17(Suppl 1):S157–164, 2001.

[199] Moon, T. K., and Stirling, W. C. *Mathematical Methods and Algorithms for Signal Processing*. Prentice Hall, Inc., Upper Saddle River, NJ, 2000.

[200] Moreau, Y., Aerts, S., De Moor, B., De Strooper, B. and Dabrowski, M. Comparison and meta-analysis of microarray data: from the bench to the computer desk. *Trends in Genet.*, 19(10):570–577, 2003.

[201] Morrison, N., editor. *Introduction to Fourier Analysis*. John Wiley & Sons, Inc., New York, NY, 1994.

[202] Mullis, K.B. The unusual origin of the polymerase chain rection. *Scientific American*, 256:56–65, 1990.

[203] Murthy, S. K., Kasif, S., and Salzberg, S. A System for Induction of Oblique

Decision Trees. *Journal of Artificial Intelligence Research*, 2:1–33, 1994.

[204] Murthy, S. K., Kasif, S., Salzberg, S., and Beigel, R. OC1: A Randomized Induction of Oblique Decision Trees. In *National Conference on Artificial Intelligence*, pages 322–327, 1993.

[205] Ng, R.T. and Han, J. CLARANS: A Method for Clustering Objects for Spatial Data Mining. *IEEE Transactions on Knowledge and Data Engineering*, 14(5):1003–1016, October 2002.

[206] Nguyen, L.T., Ramanathan, M., Munschauer, F., Brownscheidle, C., Krantz, S., Umhauer, M., et al. Flow cytometric analysis of in vitro proinflammatory cytokine secretion in peripheral blood from multiple sclerosis patients. *J Clin Immunol*, 19(3):179–185, 1999.

[207] Nielsen, T. O., West, R. B., Linn, S. C., Alter, O., Knowling, M. A., O'Connell, J. X., Zhu, S., Fero, M., Sherlock, G., Pollack, J. R., Brown, P. O., Botstein, D., and van de Rijn, M. Molecular Characterisation of Soft Tissue Tumours: a Gene Expression Study. *Lancet*, 359(9314):1301–1307, April 2002.

[208] NIST/SEMATECH. *e-Handbook of Statistical Methods*, chapter Exploratory Data Analysis. http://www.itl.nist.gov/div898/handbook/.

[209] Notterman, D. A., Alon, U., Sierk, A. J., and Levine, A. J. Transcriptional Gene Expression Profiles of Colorectal Adenoma, Adenocarcinoma, and Normal Tissue Examined by Oligonucleotide Arrays. *Cancer Research*, 61:3124–3130, April 2001.

[210] Oppenheim, A. V., Willsky, A. S., Young, I. T. *Signals and Systems*. Prentice-Hall, Inc., Englewood Cliffs, NJ, 1983.

[211] Page, D. and Craven, M. Biological Applications of Multi-relational Data Mining. *SIGKDD Explorations*, 5(1):69–79, July 2003.

[212] Pan, F., Cong, G., Tung, A.K., Yang, Y. and Zaki, M.J. Carpenter: finding closed patterns in long biological datasets. In *Proceedings of the ninth ACM SIGKDD international conference on Knowledge discovery and data mining (KDD'03)*, pages 637–642, Washington, D.C., 2003.

[213] Park, P.J., Pagano, M., and Bonetti, M. A Nonparametric Scoring Algorithm for Identifying Informative Genes from Microarray Data. In *Pacific Symposium on Biocomputing*, pages 52–63, 2001.

[214] Pavlidis, P., Weston, J., Cai, J. and Grundy, W.N. Gene Functional Classification from Heterogeneous Data. In *RECOMB 2001: Proceedings of the Fifth Annual International Conference on Computational Biology*, pages 249–255. ACM Press, 2001.

[215] Pavlidis, P., Weston, J., Cai, J. and Noble, W.S. Learning gene functional classifications from multiple data types. *Journal of Computational Biology*, 9(2):401–411, 2002.

[216] Pei, J., Han, J. and Mao, R. CLOSET: An efficient algorithm for mining frequent closed itemsets. In *Proc. 2000 ACM-SIGMOD Int. Workshop Data Mining and Knowledge Discovery (DMKD'00)*, pages 11–20, Dallas, TX, May 2000.

[217] Pei, J., Jiang, D. and Zhang, A. On Mining Cross-Graph Quasi-Cliques. In *Proceedings of ACM SIGKDD International Conference on Knowledge*

Discovery and Data Mining (KDD'05), pages 228–238, Chicago, IL, August 21-24 2005.

[218] Pei, J., Zhang, X., Cho, M., Wang, H. and Yu, P.S. MaPle: A Fast Algorithm for Maximal Pattern-based Clusterin. *Proceedings of the Third IEEE International Conference on Data Mining (ICDM'03)*, pages 259–266, November 19-22 2003.

[219] Perou, C.M., Jeffrey, S.S., Rijn, M.V.D., Rees, C.A., Eisen, M.B., Ross, D.T., Pergamenschikov, A., Williams, C.F., Zhu, S.X., Lee, J.C.F., Lashkari, D., Shalon, D., Brown, P.O., and Bostein, D. Distinctive gene expression patterns in human mammary epithelial cells and breast cancers. *Proc. Natl. Acad. Sci. USA*, Vol. 96(16):9212–9217, August 1999.

[220] Perou, C.M., Sorlie, T., Eisen, M.B., van de Rijn, M. Jeffrey, S.S., Rees, C.A., Pollack, J.R., Ross, D.T., Johnsen, H.,, Akslen, L.A., Fluge, O., Pergamenschikov, A., Williams, C., Zhu, S.X., Lonning, P.E., Borresen-Dale, A.L., Brown, P.O. and Botstein, D. Molecular portraits of human breast tumours. *Nature*, 406(6797):747–752, 2000.

[221] Peterson, L.E. Factor analysis of cluster-specific gene expression levels from cdna microarrays. *Computer Methods and Programs in Biomedicine*, 69(3):179–188, 2002.

[222] Pevsner, J. *Bioinformatics and Functional Genomics*. Wiley-Liss, Oct. 2003.

[223] Pomeroy, S.L., Tamayo, P., Gaasenbeek, M., Sturla, L.M., Angelo, M., McLaughlin, M.E., Kim, J.Y, Goumnerova, L.C., Black, P.M., Lau, C., Allen, J.C., Zagzag, D., Olson, J.M., Curran, T., Wetmore, C., Biegel, J.A., Poggio, T., Mukherjee, S., Rifkin, R., Califano, A., Stolovitzky, G., Louis, D.N., Mesirov, J.P., Lander, E.S. and Golub, T.R. Prediction of central nervous system embryonal tumour outcome based on gene expression. *Nature*, 415(6870):436–442, Jan 2002.

[224] Purves, W.K., Sadava, D., Orians G.H. and Heller, H.C. *Life: the Science of Biology*. Sinauer Associates Inc. and W.H.FREEMAN and Company, 7 edition, 2003.

[225] Quackenbush, J. Microarray data normalization and transformation. *Nat Genet.*, 32(Suppl):496–501, 2002.

[226] Quinlan, J.R. *C4.5: Programs for Machine Learning*. Morgan Kauffman, San Mateo, CA, 1993.

[227] Ralf-Herwig, P.A., Muller, C., Bull, C., Lehrach, H. and O'Brien, J. Large-Scale Clustering of cDNA-Fingerprinting Data. *Genome Research*, 9:1093–1105, 1999.

[228] Ramaswamy, S., Ross, K.N., Lander, E.S., Golub, T.R. A molecular signature of metastasis in primary solid tumors. *Nat Genet.*, 33(1):49–54, Jan. 2003.

[229] Ramaswamy, S., Tamayo, P., Rifkin, R., Mukherjee, S., Yeang, C.H., Angelo, M., Ladd, C., Reich, M., Latulippe, E., Mesirov, J.P., Poggio, T., Gerald, W., Loda, M., Lander, E.S., Golub, T.R. Multiclass cancer diagnosis using tumor gene expression signatures. *Proc Natl Acad Sci U S A*, 98(26):15149–15154, Dec. 2001.

[230] Ramoni, M.F., Sebastiani, P. and Kohane, I.S. Cluster analysis of gene expression dynamics. *PNAS*, 99(14):9121–9126, July 2002.

[231] Rhodes, D.R., Barrette, T.R., Rubin, M.A., Ghosh, D. and Chinnaiyan, A.M. Meta-analysis of microarrays: interstudy validation of gene expression profiles reveals pathway dysregulation in prostate cancer. *Cancer Res.*, 62(15):4427–4433, Aug. 2002.

[232] Rhodes, D.R., Miller, J.C., Haab, B.B., Furge, K.A. CIT: Identification of Differentially Expressed Clusters of Genes from Microarray Data. *Bioinformatics*, 18:205–206, 2001.

[233] Rose, K. Deterministic annealing for clustering, compression, classification, regression, and related optimization problems. *Proc. IEEE*, 96:2210–2239, 1998.

[234] Rose, K., Gurewitz, E. and Fox, G. A deterministic annealing approach to clustering. *Pattern Recognition Lett.*, 11:589–594, 1990.

[235] Rosenwald, A., Wright, G., Chan, W.C., Connors, J.M., Campo, E., Fisher, R.I., Gascoyne, R.D., Muller-Hermelink, H.K., Smeland, E.B., Giltnane, J.M., Hurt, E.M., Zhao, H., Averett, L., Yang, L., Wilson, W.H., Jaffe, E.S., Simon, R., Klausner, R.D., Powell, J., Duffey, P.L., Longo, D.L., Greiner, T.C., Weisenburger, D.D., Sanger, W.G., Dave, B.J., Lynch, J.C., Vose, J., Armitage, J.O., Montserrat, E., Lopez-Guillermo, A., Grogan, T.M., Miller, T.P., LeBlanc, M., Ott, G., Kvaloy, S., Delabie, J., Holte, H., Krajci, P., Stokke, T., Staudt, L.M.; Lymphoma/Leukemia Molecular Profiling Project. The use of molecular profiling to predict survival after chemotherapy for diffuse large-B-cell lymphoma. *E Engl J Med.*, 346(25):1937–1947, Jun 2002.

[236] Rymon, R. Search through systematic set enumeration. In *Proc. 1992 Int. Conf. Principle of Knowledge Representation and Reasoning (KR'92)*, pages 539–550, Cambridge, MA, 1992.

[237] Sammon, J. W. A nonlinear mapping for data structure analysis. *IEEE Transactions on Computers*, C-18(5):401–409, 1969.

[238] Šášik, R., Hwa, T., Iranfar, N.,Loomis, W.F. Percolation Clustering: A Novel Algorithm Applied to the Clustering of Gene Expression Patterns in Dictyostelium Development. In *Pacific Symposium on Biocomputing*, pages 335–347, 2001.

[239] Schaner, M. E., Ross, D. T., Ciaravino, G., Sorlie, T., Troyanskaya, O., Diehn, M., Wang, Y. C., Duran, G. E., Sikic, T. L., Caldeira, S., Skomedal, H., Tu, I. P., Hernandez-Boussard, T., Johnson, S. W., O'Dwyer, P. J., Fero, M. J., Kristensen, G. B., Borresen-Dale, A. L., Hastie, T., Tibshirani, R., van de Rijn, M., Teng, N. N., Longacre, T. A., Botstein, D., Brown, P. O., and Sikic, B. I. Gene Expression Patterns in Ovarian Carcinomas. *Mol Biol Cell.*, 14(11):4376–4386, November 2003.

[240] Schapire, R.E., Freund, Y., Bartlett, P. and Lee, W.S. Boosting the margine: a new explaination for the effectieness of voting methods. *Annls Stat.*, 26:1651–1686, 1998.

[241] Schena, M. *Micoarray analysis*. Wiley, 2002.

[242] Schena, M., Shalon, D., Davis, R.W. and Brown, P.O. Quantitative moni-

toring of gene expression patterns with a complementary DNA microarray. *Science*, 270:467–470, 1995.

[243] Schena, M., Shalon, D., Heller, R., Chai, A., Brown, P.O., and Davis, R.W. Parallel human genome analysis: Microarray-based expression monitoring of 1000 genes. *Proc. Natl. Acad. Sci. USA*, Vol. 93(20):10614–10619, October 1996.

[244] Schliep, A., Schönhuth, A. and Steinhoff, C. Using hidden Markov models to analyze gene expression time course data. *Bioinformatics*, 19(Suppl. 1):i255–i263, 2003.

[245] Schuler, G.D., Boguski, M.S., Stewart, E.A., Stein, L.D., Gyapay, G., Rice, K., White, R.E., Rodriguez-Tom, P., Aggarwal, A. Bajorek, E. A Gene Map of the Human Genome. *Science*, 274(5278):540–546, Oct. 1996.

[246] Segal, E., Wang, H. and Koller, D. Discovering molecular pathways from protein interaction and gene expression data. *Bioinformatics*, 19:i264–i272, 2003.

[247] Segal, E., Yelensky, R. and Koller, D. Genome-wide discovery of transcriptional modules from DNA sequence and gene expression. *Bioinformatics*, 19:i273–i282, 2003.

[248] Seo, J. and Shneiderman, B. Interactively exploring hierarchical clustering results. *IEEE Computer*, 35(7):80–86, July 2002.

[249] Setubal, J.C. and Meidanis, J. *Introduction to Computational Molecular Biology*. PWS Publishing, 1997.

[250] Shalon, D., Smith, S.J. and Brown, P.O. A DNA microarray system for analyzing complex DNA samples using two-color fluorescent probe hybridization. *Genome Research*, 6:639–645, 1996.

[251] Shamir, R. and Sharan, R. Click: A clustering algorithm for gene expression analysis. In *Proceedings of ISMB '00*, pages 307–316, 2000.

[252] Sherlock, G. Analysis of large-scale gene expression data. *Curr Opin Immunol*, 12(2):201–205, 2000.

[253] Shi, J. and Malik J. Normalized cuts and image segmentation. *IEEE Transactions on Pattern Analysis and Machine Intelligence*, 22(8):888–905, 2000.

[254] Shipp, M.A., Ross, K.N., Tamayo, P., Weng, A.P., Kutok, J.L., Aguiar, R.C., Gaasenbeek, M., Angelo, M., Reich, M., Pinkus, G.S., Ray, T.S., Koval, M.A., Last, K.W., Norton, A., Lister, T.A., Mesirov, J., Neuberg, D.S., Lander, E.S., Aster, J.C. and Golub, T.R. Diffuse large B-cell lymphoma outcome prediction by gene-expression profiling and supervised machine learning. *Nat Med.*, 8(1):68–74, Jan 2002.

[255] Siedow, J.N. Meeting report: Making sense of microarrays. *Genome Biology*, 2(2):reports 4003.1–4003.2, 2001.

[256] Singh, D., Febbo, P.G., Ross, K., Jackson, D.G., Manola, J., Ladd, C., Tamayo, P., Renshaw, A.A., D'Amico, A.V., Richie, J.P., Lander, E.S., Loda, M., Kantoff, P.W., Golub, T.R. and Sellers, W.R. Gene expression correlates of clinical prostate cancer behavior. *Cancer Cell*, 1(2):203–209, Mar 2002.

[257] Slonim, D.K. From patterns to pathways: gene expression data analysis comes of age. *Nat Genet.*, 32(Suppl):502–508, Dec 2002.

[258] Smet, F.D., Mathys, J., Marchal, K., et al. Adaptive quality-based clustering of gene expression profiles. *Bioinformatics*, 18:735–746, 2002.

[259] Sokal, R.R. *Clustering and classification: Background and current directions. In Classifincation and clustering Edited by J. Van Ryzin.* Academic Press, 1977.

[260] Sorlie, T., Perou, C.M., Tibshirani, R., Aas, T., Geisler, S., Johnsen, H., Hastie, T., Eisen, M.B., van de Rijn, M., Jeffrey, S.S., Thorsen, T., Quist, H., Matese, J.C., Brown, P.O., Botstein, D., Eystein Lonning, P. and Borresen-Dale, A.L. Gene expression patterns of breast carcinomas distinguish tumor subclasses with clinical implications. *Proc Natl Acad Sci U S A*, 98(19):10869–10874, Sep. 2001.

[261] Sorlie, T., Tibshirani, R., Parker, J., Hastie, T., Marron, J.S., Nobel, A., Deng, S., Johnsen, H., Pesich, R., Geisler, S., Demeter, J., Perou, C.M., Lonning, P.E., Brown, P.O., Borresen-Dale, A.L. and Botstein, D. Repeated observation of breast tumor subtypes in independent gene expression data sets. *Proc Natl Acad Sci U S A.*, 100(14):8418–8423, Jul. 2003.

[262] Speed, T., editor. *Statistical Analysis of Gene Expression Microarray Data.* Chapman & Hall/CRC, 2003.

[263] Spellman, P.T., Sherlock, G., Zhang, M.Q., Iyer, V.R., Anders, K., Eisen, M.B., Brown, P.O., Botstein, D. and Futcher, B. Comprehensive identification of cell cycle-regulated genes of the yeast saccharomyces cerevisiae by microarray hybridization. *Mol. Biol. Cell*, (12):3273–3297, 1998.

[264] Standafer, E. and Wahlgren, W. *Modern Biology.* Holt, Rinehart and Winston, 2002.

[265] Stark, G.R., Kerr, I.M., Williams, B.R., Silverman, R.H. and Schreiber, R.D. How cells respond to interferons. *Annual Review of Biochemistry*, 67:227–264, 1998.

[266] Stekel, D. *Microarray bioinformatics.* Cambridge University Press, 2003.

[267] Storey, J.D. and Tibshirani, R. Estimating false discovery rates under dependence, with applications to DNA microarrays. Technical Report 2001-28, Department of Statistics, Stanford University, 2001.

[268] Stuart, J.M., Segal, E., Koller, D. and Kim, S.K. A gene-coexpression network for global discovery of conserved genetic modules. *Science*, 302(5643):249–255, Oct. 2003.

[269] Stuart, R. O., Bush, K. T., and Nigam, S. K. Changes in Global Gene Expression Patterns During Development and Maturation of the Rat Kidney. *Proc. Natl. Acad. Sci. USA*, Vol. 98(10):5649–5654, May 2001.

[270] Takahashi, M., Rhodes, D.R., Furge, K.A., Kanayama, H., Kagawa, S., Haab, B.B. and Teh, B.T. Gene expression profiling of clear cell renal cell carcinoma: gene identification and prognostic classification. *Proc Natl Acad Sci U S A*, 98(17):9754–9759, Aug 2001.

[271] Tamayo, P., Solni, D., Mesirov, J., Zhu, Q., Kitareewan, S., Dmitrovsky, E., Lander, E.S. and Golub, T.R. Interpreting patterns of gene expression with self-organizing maps: Methods and application to hematopoietic differentiation. *Proc. Natl. Acad. Sci. USA*, Vol. 96(6):2907–2912, March 1999.

[272] Tanay, A., Sharan, R., Kupiec, M. and Shamir, R. Revealing modularity and organization in the yeast molecular network by integrated analysis of highly heterogeneous genomewide data. *Proc. Natl. Acad. Sci. USA*, 101:2981–2986, 2004.

[273] Tang, C. and Zhang, A. Mining Multiple Phenotype Structures Underlying Gene Expression Profiles. In *In Proceedings of 12th International Conference on Information and Knowledge Management (CIKM'03)*, pages 418–425, New Orleans, LA, USA, November 3-8 2003.

[274] Tang, C., Zhang, A., and Pei, J. Mining phenotypes and informative genes from gene expression data. In *In Proceedings of the Ninth ACM SIGKDD International Conference on Knowledge Discovery and Data Mining (SIGKDD'03)*, pages 655–660, Washington, DC, USA, August 24-27 2003.

[275] Tang, C., Zhang, A., and Ramanathan, M. ESPD: A Pattern Detection Model Underlying Gene Expression Profiles. *Bioinformatics*, 20(6):829–838, 2004.

[276] Tao, H., Bausch, C., Richmond, C., Blattner, F.R., Conway, T. Functional genomics: expression analysis of *Escherichia coli* growing on minimal and rich media. *J. Bacteriol.*, 181(20):6425–6440, 1999.

[277] Tavazoie, S., Hughes, D., Campbell, M.J., Cho, R.J. and Church, G.M. Systematic determination of genetic network architecture. *Nature Genet*, (3):281–285, 1999.

[278] Tefferi, A., Bolander, E., Ansell, M., Wieben, D. and Spelsberg C. Primer on Medical Genomics Part III: Microarray Experiments and Data Analysis. *Mayo Clin Proc.*, 77:927–940, 2002.

[279] The C. elegans Sequencing Consortium. Genome sequence of the nematode C. elegans: a platform for investigating biology. *Science*, 282(5396):2012–8, Dec 1998. Review.

[280] Thomas, J.G., Olson, J.M., Tapscott, S.J. and Zhao, L.P. An Efficient and Robust Statistical Modeling Approach to Discover Differentially Expressed Genes Using Genomic Expression Profiles. *Genome Research*, 11(7):1227–1236, 2001.

[281] Tibshirani, R., Walther, G. and Hastie, T. Estimating the number of clusters in a dataset via the Gap statistic. Technical report, Dept of Statistics, Stanford Univ., 2000.

[282] Tomida, S., Hanai, T., Honda, H. and Kobayashi, T. Analysis of expression profile using fuzzy adaptive resonance theory. *Bioinformatics*, 18:1073–1083, 2002.

[283] Tornow, S., Mewes, H.W. Functional modules by relating protein interaction networks and gene expression. *Nucl. Acids Res.*, 31(21):6283–9, Nov. 2003.

[284] Troyanskaya, O.,Cantor M., Sherlock, G., Brown, P., Hastie, T., Tibshirani, R., Botstein, D. and Altman R. Missing value estimation methods for dna microarrays. *Bioinformatics*, 17(6):520–525, 2001.

[285] Troyanskaya, O.G., Dolinski, K., Owen, A.B., Altman, R.B. and Botstein, D. A Bayesian framework for combining heterogeneous data sources for

gene function prediction. *Proc. Natl. Acad. Sci. USA*, 100:8348–8353, 2003.

[286] Tukey, J.W. The future of data analysis. *An. Math. Stat.*, 33(1-67), 1962.

[287] Tukey, J.W. *Exploratory Data Analysis*. Reading, MA: Addison-Wesley, 1977.

[288] Tukey, J.W. *The Collected Works of John W. Tukey*, volume 3, chapter Philosophy and Principles of Data Analysis, pages 1949–1964. Pacific Grove, CA: Wadsworth and Brooks/Cole, 1986.

[289] Tusher, V.G., Tibshirani, R. and Chu, G. Significance Analysis of Microarrays Applied to the Ionizing Radiation Response. *Proc. Natl. Acad. Sci. USA*, Vol. 98(9):5116–5121, April 2001.

[290] Tuzhilin, A. and Adomavicius, G. Handling very large numbers of association rules in the analysis of microarray data. In *Proceedings of the eighth ACM SIGKDD international conference on Knowledge discovery and data mining*, pages 396–404, Edmonton, Alberta, Canada, 2002.

[291] van 't Veer, L.J., Dai, H., van de Vijver, M.J., He, Y.D., Hart, A.A., Mao, M., Peterse, H.L., van der Kooy, K., Marton, M.J., Witteveen, A.T., Schreiber, G.J., Kerkhoven, R.M., Roberts, C., Linsley, P.S., Bernards, R. and Friend, S.H. Gene expression profiling predicts clinical outcome of breast cancer. *Nature*, 415(6871):530–536, Jan. 2002.

[292] Vapnik, L. *Statistics Learning Theory*. Wiely, New York, 1998.

[293] Virtaneva, K., Wright, F.A., Tanner, S.M., Yuan, B., Lemon, W.J., Caligiuri, M.A., Bloomfield, C.D., Chapelle, A. de la and Krahe, R. Expression Profiling Reveals Fundamental Biological Differences in Acute Myeloid Leukemia with Isolated Trisomy 8 and Normal Cytogenetic. *Proc. Natl. Acad. Sci. USA*, Vol. 98(3):1124–1129, January 2001.

[294] Wagstaff, K. and Cardie, C. Clustering with instance-level constraints. In *Proceedings of 17th International Conference on Machine Learning (ICML'00)*, pages 1103–1110, 2000.

[295] Wang, D.G., Fan, J.B., Siao, C.J., Berno, A., Young, P., Sapolsky, R., Ghandour, G., Perkins, N., Winchester, E., Spencer, J., Kruglyak, L., Stein, L., Hsie, L., Topaloglou, T., Hubbell, E., Robinson, E., Mittmann, M., Morris, M.S., Shen, N., Kilburn, D., Rioux, J., Nusbaum, C., Rozen, S., Hudson, T.J., Lipshutz, R., Chee, M. and Lander, E.S. Large-scale identification, mapping, and genotyping of single-nucleotide polymorphisms in the human genome. *Science*, 280:1077–1082, 1998.

[296] Wang, H., Wang, W., Yang, J. and Yu, P.S. Clustering by Pattern Similarity in Large Data Sets. In *SIGMOD 2002, Proceedings of ACM SIGMOD International Conference on Management of Data*, pages 394–405, 2002.

[297] Wang, K., Gan, L., Jeffery, E., Gayle, M., Gown, A.M., Skelly, M., Nelson, P.S., Ng, W.V., Schummer, M., Hood, L. and Mulligan, J. Monitoring gene expression profile changes in ovarian carcinomas using cDNA microarray. *Gene*, 229(1-2):101–108, Mar 1999.

[298] Ward, M. O. XmdvTool: Integrating Multiple Methods for Visualizing Multivariate Data. In *IEEE Visualization 1994*, pages 326–336, 1994.

[299] Weaver, R.F. *Molecular Biology*. McGraw-Hill, 2001.

[300] Weinstock-Guttman, B., Badgett, D., Patrick, K., Hartrich, L., Hall, D.,

Baier, M., Feichter, J. and Ramanathan, M. Genomic effects of interferon-β in multiple sclerosis patients. *Journal of Immunology*, 171:2694–2702, 2003.

[301] Welford, S.M., Gregg, J., Chen, E., Garrison, D., Sorensen, P.H., Denny, C.T. and Nelson, S.F. Detection of differentially expressed genes in primary tumor tissues using representational differences analysis coupled to microarray hybridization. *Nucl. Acids Res.*, 26:3059–3065, 1998.

[302] Welsh, J.B., Zarrinkar, P.P., Sapinoso, L.M., Kern, S.G., Behling, C.A., Monk, B.J., Lockhart, D.J., Burger, R.A. and Hampton, G.M. Analysis of Gene Expression Profiles in Normal and Neoplastic Ovarian Tissue Samples Identifies Candidate Molecular Markers of Epithelial Ovarian Cancer. *Proc. Natl. Acad. Sci. USA*, Vol. 98(3):1176–1181, January 2001.

[303] West, M., Blanchette, C., Dressman, H., Huang, E., Ishida, S., Spang, R., Zuzan, H., Olson J.A., Marks, J.R. and Nevins J.R. Predicting the clinical status of human breast cancer by using gene expression profiles. *Proc Natl Acad Sci*, 98(20):11462–11467, 2001.

[304] Westfall, P.H. and Young, S.S. *Resampling-based multiple testing: examples and methods for p-value adjustment.* Wiley, New York, 1993.

[305] Wilcoxon, F. Individual comparisons by ranking methods. *Biometrics*, 1:80–83, 1945.

[306] Wodicka, L., Dong, H., Mittmann, M., Ho, M.H., Lockhart, D.J. Genome-wide expression monitoring in Saccharomyces cerevisiae. *Nat. Biotechnol.*, 15(13):1359–1367, 1997.

[307] Xin, W., Rhodes, D.R., Ingold, C., Chinnaiyan, A.M. and Rubin, M.A. Dysregulation of the annexin family protein family is associated with prostate cancer progression. *Am J Pathol.*, 162(1):255–261, 2003.

[308] Xing, E.P. and Karp, R.M. Cliff: Clustering of high-dimensional microarray data via iterative feature filtering using normalized cuts. *Bioinformatics*, Vol. 17(1):306–315, 2001.

[309] Xing, E.P., Ng, A.Y., Jordan, M.I. and Russell, S. *Advances in Neural Information Processing Systems 15*, chapter Distance metric learning with application to clustering with sideinformation, pages 505–512. MIT Press, Cambridge, MA, 2003.

[310] Xiong, M., Jin, L., Li, W. and Boerwinkle, E. Computational methods for gene expression-based tumor classification. *Biotechniques*, 29(6):1264–1268, 2000.

[311] Xu, X. and Zhang, A. Virtual Gene: A Gene Selection Algorithm for Sample Classification on Microarray Datasets. In *Computational Science ICCS 2005: 5th International Conference, 2005 International Workshop on Bioinformatics Research and Applications*, pages 1038–1045. Springer-Verlag GmbH, 2005.

[312] Xu, X. and Zhang, A. Virtual Gene: Using Correlations Between Genes to Select Informative Genes on Microarray Datasets. *Transactions on Computational Systems Biology II, LNBI 3680, Springer-Verlag Berlin Heidelberg*, pages 138–152, 2005.

[313] Xu, Y., Olman, V. and Xu, D. Clustering gene expression data using a graph-theoretic approach: an application of minimum spanning trees.

Bioinformatics, 18:536–545, 2002.

[314] Yang, G. The Complexity of Mining Maximal Frequent Itemsets and Maximal Frequent Patterns. In *Tenth ACM SIGKDD International Conference on Knowledge Discovery and Data Mining*, pages 344–353, Seattle, Washington, Aug. 2004.

[315] Yang, I.V., Chen, E., Hasseman, J.P., Liang, W., Frank, B.C., Wang, S., Sharov, V., Saeed, A.I., White, J., Li, J., Lee, N.H., Yeatman, T.J. and Quackenbush, J. Within the fold: assessing differential expression measures and reproducibility in microarray assays. *Genome Biol.*, 3(11):research 0062, 2002.

[316] Yang, J., Wang, W., Wang, H. and Yu, P.S. δ-cluster: Capturing Subspace Correlation in a Large Data Set. In *Proceedings of 18th International Conference on Data Engineering (ICDE 2002)*, pages 517–528, 2002.

[317] Yang, Y.H., Dudoit, S., Luu, P., Lin, D.M., Peng, V., Ngai, J. and Speed, T.P. Normalization for cDNA microarray data: a robust composite method addressing single and multiple slide systematic variation. *Nucleic Acids Res.*, 30:e15, 2002.

[318] Yeoh, E.J., Ross, M.E., Shurtleff, S.A., Williams, W.K., Patel, D., Mahfouz, R., Behm, F.G., Raimondi, S.C., Relling, M.V., Patel, A., Cheng, C., Campana, D., Wilkins, D., Zhou, X., Li, J., Liu, H., Pui, C.H., Evans, W.E., Naeve, C., Wong, L. and Downing, J.R. Classification, subtype discovery, and prediction of outcome in pediatric acute lymphoblastic leukemia by gene expression profiling. *Cancer Cell*, 1(2):133–143, Mar 2002.

[319] Yeung, K.Y. and Ruzzo, W.L. An empirical study on principal component analysis for clustering gene expression data. Technical Report UW-CSE-2000-11-03, Department of Computer Science & Engineering, University of Washington, 2000.

[320] Yeung, K.Y., Fraley, C., Murua, A., Raftery, A.E., Ruzzo, W.L. Model-based clustering and data transformations for gene expression data. *Bioinformatics*, 17:977–987, 2001.

[321] Yeung, K.Y., Haynor, D.R. and Ruzzo, W.L. Validating Clustering for Gene Expression Data. *Bioinformatics*, Vol.17(4):309–318, 2001.

[322] Yuen, T., Wurmbach, E., Pfeffer, R.L., Ebersole, B.J. and Sealfon, S.C. Accuracy and calibration of commercial oligonucleotide and custom cDNA microarrays. *Nucleic Acids Res.*, 30(10):e48, May 2002.

[323] Zaki, M. J and Hsiao, C.J. Charm: an efficient algorithm for closed itemset mining. In *Proceedings of the Second SIAM International Conference on Data Mining*, pages 457–473, Arlington, VA, 2002. SIAM.

[324] Zhan, F., Hardin, J., Kordsmeier, B., Bumm, K., Zheng, M., Tian, E., Sanderson, R., Yang, Y., Wilson, C., Zangari, M., Anaissie, E., Morris, C., Muwalla, F., van Rhee, F., Fassas, A., Crowley, J., Tricot, G., Barlogie, B., Shaughnessy, J. Jr. Global gene expression profiling of multiple myeloma, monoclonal gammopathy of undetermined significance, and normal bone marrow plasma cells. *Blood*, 99(5):1745–1757, 2002.

[325] Zhang, L. *VizStruct: Visual Exploration for Gene Expression Profiling.* PhD thesis, State University of New York at Buffalo, 2004.

[326] Zhang, L., Tang, C., Shi, Y., Song, Y., Zhang, A. and Ramanathan M. Viz-
       Cluster: An Interactive Visualization Approach to Cluster Analysis and Its
       Application on Microarray Data. In *Second SIAM International Conference
       on Data Mining (SDM'2002)*, pages 19–40, Arlington, Virginia, April 11-13
       2002.
[327] Zhang, L., Zhang, A., and Ramanathan, M. VizStruct: Exploratory Visual-
       ization for Gene Expression Profiling. *Bioinformatics*, 20(1):85–92, January
       2004.
[328] Zhao, L. and Zaki, M. TriCluster: An Effective Algorithm for Mining Co-
       herent Clusters in 3D Microarray Data. In *Proceedings of ACM SIGMOD
       International Conference on Management of Data (SIGMOD'05)*, pages
       694–705, Baltimore, MD, June 13-16 2005.
[329] Zou, T.T., Selaru, F.M., Xu, Y., Shustova, V., Yin, J., Mori, Y., Shibata,
       D., Sato, F., Wang, S., Olaru, A., Deacu, E., Liu, T.C., Abraham, J.M.
       and Meltzer, S.J. Application of cDNA microarrays to generate a molecular
       taxonomy capable of distinguishing between colon cancer and normal colon.
       Oncogene, 21(31):4855–4862, 2002.

Index

About the Author

Dr. Aidong Zhang received her Ph.D degree in computer science from Purdue University, West Lafayette, Indiana, in 1994. She was an assistant professor from 1994 to 1999, an associate professor from 1999 to 2002, and has been a full professor since 2002 in the Department of Computer Science and Engineering at State University of New York at Buffalo. She is the director of the Buffalo Center for Biomedical Computing (BCBC), which was initiated by NIH (National Institute of Health) National Programs of Excellence in Biomedical Computing (NPEBC). Her research interests include bioinformatics, data mining, multimedia systems, and content-based image retrieval. She is an author of over 150 research publications in these areas. Dr. Zhang's research has been funded by National Science Foundation (NSF), National Institute of Health (NIH), National Imaging and Mapping Agency (NIMA), and Xerox.

Dr. Zhang have served on the editorial boards of International Journal of Bioinformatics Research and Applications (IJBRA), ACM Multimedia Systems, the International Journal of Multimedia Tools and Applications, and International Journal of Distributed and Parallel Databases. She was the editor for ACM SIGMOD DiSC (Digital Symposium Collection) from 2001-2003. She was co-chair of the technical program committee for ACM Multimedia 2001. She has also served on various conference program committees. Dr. Zhang is a recipient of the National Science Foundation CAREER award and SUNY (State University of New York) Chancellor's Research Recognition award.